Crop Irrigation Management: Water and Soil

Crop Irrigation Management: Water and Soil

Edited by **Corey Aiken**

New York

Published by Callisto Reference,
106 Park Avenue, Suite 200,
New York, NY 10016, USA
www.callistoreference.com

Crop Irrigation Management: Water and Soil
Edited by Corey Aiken

International Standard Book Number: 978-1-63239-130-8 (Hardback)

Printed in the United States of America.

Contents

Preface   VII

Part 1   **Reuse Water Quality, Soils and Pollution**   1

Chapter 1   **Water Quality at the
Cárdenas-Comalcalco Basin, México**   3
Ángel Galmiche-Tejeda, José Jesús Obrador-Olán,
Eustolia García-López and Eugenio Carrillo Ávila

Chapter 2   **Provision of Essential
Minerals Through Foliar Sprays**   23
Rizwana Jabeen and Rafiq Ahmad

Chapter 3   **Effluent Quality Parameters
for Safe use in Agriculture**   59
Hamid Iqbal Tak, Faheem Ahmad,
Yahya Bakhtiyar and Arif Inam

Chapter 4   **Geospatial Relationships Between Morbidity
and Soil Pollution at Cubatão, Brazil**   73
Roberto Wagner Lourenço,
Admilson Irio Ribeiro, Maria Rita Donalisio,
Ricardo Cordeiro, André Juliano Franco
and Paulo Milton Barbosa Landim

Chapter 5   **Assessment of Geochemistry of Soils for
Agriculture at Winder, Balochistan, Pakistan**   89
Shahid Naseem, Salma Hamza and Erum Bashir

Part 2   **Managing Irrigation of Crops**   111

Chapter 6   **Influence of Irrigation, Soil and Weeding
on Performance of Mediterranean
Cypress Seedling in Nursery**   113
Masoud Tabari

Chapter 7   **Developing Crop-Specific Irrigation**
              Management Strategies Considering Effects
              of Drought on Carbon Metabolism in Plants   121
              Silvia Aparecida Martim, Arnoldo Rocha Façanha
              and Ricardo Enrique Bressan-Smith

Chapter 8   **Growth Characteristics of Rainfed/Irrigated**
              Juniperus excelsa **Planted in an Arid Area**
              at North-Eastern Iran   149
              Masoud Tabari and Mohammad Ali Shirzad

Chapter 9   **Surface Infiltration on Tropical**
              Plinthosols in Maranhão, Brazil   157
              Alba Leonor da Silva Martins, Aline Pacobahyba de Oliveira,
              Emanoel Gomes de Moura and Jesús Hernan Camacho-Tamayo

Part 3   **Examples of Irrigation Systems**   169

Chapter 10   **Irrigation of Field Crops in the Boreal Region**   171
              Pirjo Mäkelä, Jouko Kleemola and Paavo Kuisma

Chapter 11   **Land Flooding Irrigation Treatment System**
              for Water Purification in Taiwan   193
              Yu-Kang Yuan

Chapter 12   **A Review of Subsurface Drip Irrigation**
              and Its Management   219
              Leonor Rodríguez Sinobas and María Gil Rodríguez

              Permissions

              List of Contributors

Preface

Every book is initially just a concept; it takes months of research and hard work to give it the final shape in which the readers receive it. In its early stages, this book also went through rigorous reviewing. The notable contributions made by experts from across the globe were first molded into patterned chapters and then arranged in a sensibly sequential manner to bring out the best results.

A descriptive account based on the field of crop irrigation management, with emphasis on water and soil, has been provided in this up-to-date book. It provides in-depth elucidation regarding the following topics: Reuse Water Quality, Soil and Pollution; Managing Irrigation of Crops; and Examples of Irrigation Systems. All this information has been contributed by reputed authors in their respective fields of expertise. The book should appeal to individuals interested in the safe reuse of water for irrigation purposes in terms of quality of urban drainage basins and effluent quality, as well as those engaged in research work regarding the complications of soils related to health and pollution, infiltration and impacts of irrigation, and managing irrigation systems comprising of subsurface technique of irrigation as well as basin type irrigation. The several examples are indeed a reflection of real world irrigation practices of interest to practitioners, especially when the places of the projects demonstrated herein are spread over a wide range of climatic environments.

It has been my immense pleasure to be a part of this project and to contribute my years of learning in such a meaningful form. I would like to take this opportunity to thank all the people who have been associated with the completion of this book at any step.

Editor

Part 1

Reuse Water Quality, Soils and Pollution

Water Quality at the Cárdenas-Comalcalco Basin, México

Ángel Galmiche-Tejeda[1], José Jesús Obrador-Olán[1],
Eustolia García-López[1] and Eugenio Carrillo Ávila[2]
Colegio de Postgraduados,
[1]Campus Tabasco,
[2]Campus Campeche,
México

1. Introduction

Water is a resource with an economic, social and environmental value. For this reason when decisions concerning water management, analysis and planning are made, the relationships between the economy, society and the environment should be considered. The study of water resources has great importance when it is developed in a basin's geographical frame.

A basin is defined as an area in which rainfall flows through a number of different water channels which in the end converge into one major water body. The water flows towards a common point, which forms a hydrographic unit made up of a group of river bed systems, summits and the outlets, whose limits are marked by "watershed lines" (Dourojeanni and Jouraviev, 2001; Nebel and Wrigth, 1999; ECLAC, 1997).

Cárdenas-Comalcalco basin is located in the hydrological region of Tonalá river, and del Carmen-Machona lagoons. It is bordered by the protected wetland areas to the west of the State of Tabasco and also by the mangrove forests located along the edges of the afore mentioned lagoons.

The objectives of this study were to characterize the water bodies and to determine the chemical and biochemical properties of the water from rivers, lakes and artesian wells of the Cardenas-Comalcalco basin. The results are relevant to water ecosystems management policies and planning since it is a prime resource for both the humans and the wild life which inhabits the area.

2. Materials and methods

2.1 Localization of study area

The study was carried out in the Cárdenas-Comalcalco basin which belongs to the hydrological region of the Tonalá river, and del Carmen-Machona lagoons, located in western Tabasco. It occupies an area of 274 255 ha, situated between the coordinates 17°52′ and 18°39′ latitude north and 93°13′ and 94°00′ longitude west. It covers, totally or partially, four municipalities of state of Tabasco: Huimanguillo, Cárdenas, Comalcalco and Paraíso. The basin has an important hydrological system formed by a number of water bodies such as the lagoons El Carmen, La Machona, Redonda, El Cocal, El Paso del Ostión and El Arrastradero. The rivers San Felipe, Naranjeño and Santa Ana are also part of this system.

2.2 Selection of supervision sites

The hydrological characterization of the basin was carried through the analysis of satellite images and observations in the field, both in land and water ecosystems. The study area was divided into nine subareas, which were assigned a number ordered from west to east and north to south. For the water quality analysis, 57 supervision sites were set. The selection was made according to their representativeness, and previous studies carried out in the area. The sites were distributed as follows: four at Naranjeño river, four at Santa Ana river and four at San Felipe river. Four sites were also established in the lagoons, El Carmen, four at Machona, four at Redonda and four at El Arrastradero. Three sites were established in the lagoon El Cocal, three at El Paso del Ostión and three at Las Palmas. Twenty artesian wells were selected. These were located in places with high urban influence (Figure 1).

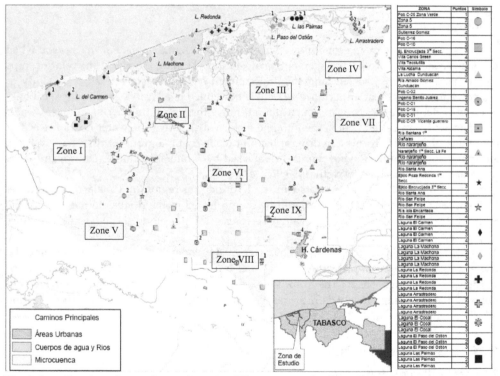

Fig. 1. Localization of samplings places in different water bodies of The Cárdenas-Comalcalco basin.

In order to understand the seasonal variations of water quality, samples were taken during the three distinct season of the year: the dry season, during the 'nortes' (windy) season and rainy season, taking two samples of water in each supervisory place, a superficial one and the other at a depth of 1 meter, with a pvc Alpha Horizontal Van Dhorn bottle, of 4.2 liters

KIT (1140-H42) for water bodies samples, and an acrylic Horizontal Alpha bottle 2.2 liters KIT (W1120-G42) for samples in artesian wells (Arce 2007). Water samples were placed in plastic recipients of 2 liters and 90 milliliters, respectively, which were transported to the laboratory in a cooler with ice to maintain a temperature of 4°C, to be analyzed.

The parameters that were analyzed to determine water quality were: **biochemical parameters**: oxygen biochemistry demand (DBO), dissolved oxygen (OD) and total solids; **chemical parameters**: ammonium, nitrates, total chlorine, phosphates, heavy metals: zinc (Zn), vanadium (V), cadmium (Cd), lead (Pb) and nickel (Ni); potential hydrogen (pH), electric conductivity (CE), dissolved salts: potassium (K), calcium (Ca), magnesium (Mg) and sodium (Na); **biological parameters**: fecal coliforms and the total number of coliforms.

Observations of physical characteristics of the water, such as: depth, turbidity, temperature, color, scent and surrounding vegetation were also carried out. Chemical analyses were carried out with methodology used in the Soil, Waters and Plants Analysis Laboratory, of Colegio de Postgraduados, Campus Tabasco, according to Mexican Official Standards.

3. Results, analysis and discussion

In order to save space and for ease of handling, the information in the following tables shows only the average and median values obtained in analyses.

3.1 Wells

A variation was observed in water quality over the various sampling seasons in several of study sites (Table 1). In most of the sampled wells nitrates levels were above 10 mg/L as defined by the Mexican Official Standard (NOM-127-SSA1-1994) as limit for human consumption and general usage of water. The highest levels were detected in the norte season. Zone 8, that covers the towns C-32, C-21, C-15 and Ingenio Benito Juárez, turned out to be the most polluted in the rainy and norte seasons; while in dry season, in the well of the town Gutiérrez Gómez (Z-5) there were recorded 206.8 mg/L of nitrates. This represents the maximum value obtained in the wells.

Nitrates are the main pollutant of underground water in rural areas, especially in those areas dedicated to intensive cattle raising. High levels of nitrates in the water (greater than 10 mg/L) have been known to cause methemoglobinemia known as blue baby illness. Although nitrates levels that affect small children usually do not affect older children and adults, they are indicative of bacterial contamination and pesticides presence.

Nitrates in drinking water, at levels which range from 100 to 200 mg/L affects the population's health in general, because once consumed, they can be transformed into salpenter that, under proper circumstances, it can combine with amines to form nitrosamines, compounds which are carcinogenic.

There is no a simple way to eliminate nitrates from water since it can not be evaporated with boiling; in fact, boiling water more than 10 minutes can increase their concentration and, if the water is boiled in aluminum recipients, the nitrate can become saltpeter (National Research Council, 1995).

Nitrate levels found indicates a higher risk of contracting cancer by means of water consumption from artesian wells in the mentioned locations, and risk of contracting methemoglobinemia for children of practically the whole basin.

Season	Dry				Rain				Norte			
	Nitrate	Amm.	Phosph	Chlor.	Nitrate	Amm.	Phosph	Chlor.	Nitrate	Amm.	Phosph	Chlor.
Limits	10	0.50	0.0059	15	10	0.50	0.0059	15	10	0.50	0.0059	15
	mg/L											
	Zone 5											
Average	53.35	0	0.45	0.125	29.7	0.225	0.65	0.0625	12.1	0.285	0.59	0.1375
Median	8.8	0	0.4	0	15.4	0.24	0.68	0.1	13.2	0.12	0.56	0.15
	Zone 6											
Average	11	0.225	0.91	0.1875	17.6	0.255	1.02	0.0375	28.05	0.33	0.71	0.125
Median	11	0.18	0.72	0.1	8.8	0.24	0.88	0	30.8	0.24	0.76	0.1
	Zone 7											
Average	19.25	0.105	1.88	0.1625	13.75	0.255	0.72	0	22	0.15	1.08	0.1125
Median	8.8	0.12	2.44	0.2	8.8	0.24	0.52	0	22	0.12	0.68	0.1
	Zone 8											
Average	24.2	0.405	1.17	0.6375	36.3	0.195	0.74	0.025	110.5	0.21	1.8	0.1125
Median	17.6	0.12	1.2	0.2	19.8	0.18	0.44	0	105.6	0.24	1.6	0.1
	Zone 9											
Average	17.325	0.525	0.6	0.05	3.875	0.1625	5.625	0.1375	44	0.09	0.36	0.0875
Median	15.4	0.6	0.56	0.05	3.5	0.15	5	0	30.8	0.12	0.32	0

Table 1. Nitrates, Ammonium, Phosphate and Chlorine Average Values in Wells of the five study areas in the Cárdenas-Comalcalco basin, Tabasco.

Most of sampled wells maintained ammonium levels below the 0.5 mg/L indicated in the Mexican Official Standard (NOM-127-SSA1-1994) and European Standards for Human Consumption Water (1998).

Within the Mexican, nor the World Health Organization or The European Union Standards there are no limits in reference to phosphorus levels in water for human consumption because phosphates are necessary micro nutrients for the operation of the human body and under normal concentrations found in water they pose no risk to health. In The Ecological Criteria of Water Quality (DOF, December 13 1989) for the protection of wild life is set at a maximum of 0.0059 mg/L of phosphates for fresh water, values that were surpassed in all sites and in all three sampling seasons (there were detected up to 3.36 mg/L).

However, these values do not pose threat for human health, since the minimum recommendation in the human diet is of 800 mg/day and a normal diet provides between 1000 and 2000 mg/day; it has been noted that a higher consumption can cause renal illnesses and osteoporosis (http://www.lenntech.com/Periodic-chart-elements/P-en.htm; consulted on 29/04/08).

In the case of Cl, concentrations were below 15 mg/L that OMS recommends as good for human consumption. However, in accordance with Mexican Standard for Wildlife Protection, in some stations concentrations greater than the permissible maximum level of Cl were detected (0.2-1.5 mg/L).

Although there were no significant differences related with turbidity between the different towns and season of year, Table 2 is shown, that in general, values were lower in the dry season, while in the rainy and norte seasons several wells surpassed permissible limits of 5 NTU. As outlined in the Mexican Official Standards for Environmental Health (NOM-127-SSA1-1994) and for Water Quality for Normal Use and Human Consumption.

Season	Dry				Rain				Norte			
	Turbid	OD	DBO	TotSol	Turbid	OD	DBO	TotSol	Turbid	OD	DBO	TotSol
Limits	5			1000	5			1000	5			1000
	NTU	mg/L	mg/L	mg/L	NTU	mg/L	mg/L	mg/L	NTU	mg/L	mg/L	mg/L
Zone 5												
Average	1.45	1.81	2.4	302.56	14.36	1.68	0.93	499	27.31	2.01	0.98	318.63
Median	1.24	2	0	198	3.93	1.51	1.09	462	21.37	2.03	0.765	293
Zone 6												
Average	8.29	2.94	9.02	448.12	12.11	2.37	1.45	574.88	5.66	3.05	0.82	464
Median	1.95	2.38	12	500	6.74	1.21	1.59	627	5.79	3.49	0.51	463.5
Zone 7												
Average	1.44	3.26	7.76	277.05	10.04	3.57	1.24	333.38	21.56	4.083	0.51	233.25
Median	1.21	3.65	8.9	329	6.92	3.43	0.96	346.5	7.5	4.22	0.43	252
Zone 8												
Average	1.08	1.04	7.19	340.17	13.56	2.70	0.95	478.5	8.92	2.62	0.83	408.75
Median	0.60	0.91	4.41	348.5	5.86	2.86	0.96	437	6.78	2.61	0.7	412.5
Zone 9												
Average	5.77	1.29	3.52	493.5	8.84	1.16	0.70	494.38	6.48	2.05	0.75	364.63
Median	3.3	1.24	3.9	490	3.67	0.715	0.735	559	6.12	1.465	0.89	383.5

Table 2. Turbidity, dissolved oxygen, oxygen biochemistry demand and total solids values in artesian wells of the Cárdenas-Comalcalco basin, Tabasco

In the rainy season only four wells were within the guidelines established in the standard at the two sampling depths. While in the norte season, that increased to 5, the number of wells with acceptable turbidity, in general, values increased in this season. Only the well Z5-03 located in Zona Verde was within the quality standard in all three seasons.

In the dry season, the median was of 1.55 NTU which is very below the limit of permissible turbidity. However in the wells located in the village Carlos Green (Z-6) and Cañales (Z-9) were found to have extremely high values, 27.8 and 14.6 NTU, respectively.

In rainy season, turbidity was greater, with an average value of 11.78 NTU. The median was set in 5.55, both values surpassed the permissible limit; in this season only water of wells Zona Verde 3 (Z-5), Poblado C-16 (Z-6), Villa Tecolutilla (Z-8) and C-29 (Z-9) had the levels of turbidity as required by the established norms.

During the norte season turbidity even increased, detecting values of up to 60.6 and 52 NTU in Gutiérrez Gómez (Z-5) and Poblado C-25 (Z-5), respectively. In this case only wells in Zona Verde 3 (Z-5), Poblado C-10 (Z-6), Poblado C-32 and C-21 Ingenio Benito Juárez (Z-8) and Cañales (Z-9) fulfilled the standards of quality over the two sampling depths.

The Average and median values for turbidity, considering all the wells and depths, was 13.99 and 6.44 NTU, respectively, both above the rates found in the established norms. The high turbidity values found are of great concern, since consumption of very muddy waters constitutes a risk factor for human health, since the mud can protect to pathogen organisms of disinfectants effects, besides stimulate bacterial growth and increase chlorine demand. Due to the absorption capacity of some particles, they can have present harmful organic compounds which should not be present in water designated for human consumption (Safe Drinking Water, 1980).

There were not found reference values for OD in the consulted standards and references, because concentrations up to saturation point in water have no effect on human health.

However, in Poblado C-10 (Z-6) and Villa Aldama (Z-7) in the dry and rainy seasons there were detected high oxygen values, which suggests that there exists primary productivity in the wells and, consequently, considerable N and P quantities. Although presence of green algae in water does not necessarily affect human health, primary productivity is generally accompanied by an increment in bacterial populations, which can create a series of infections for consumer.

In general pH was between 6.5 and 8, within limit allowed by Mexican Official Standard NOM-127-SSA1-1994, for Water for Human Consumption (Table 3), although in some places the value surpassed 8 in some of the three seasons studied. In one case a location had high pH values in the three samplings.

Season	Dry					Rain					Norte				
	pH	CE	T°	C-T	C-F	pH	CE	T°	C-T	C-F	pH	CE	T°	C-T	C-F
	6.5-8			0	0	6.5-8			0	0	6.5-8			0	0
Zone 5															
Average	7.39	0.77	26.15	274.6	266.1	7.84	0.91	27	180.8	173.5	7.66	0.58	27.68	179.5	176.6
Median	7.47	0.38	26.15	297.5	287	7.83	0.84	27	161.5	153.5	7.71	0.53	27.7	142.5	142
Zone 6															
Average	7.76	0.79	25.11	209.4	201.8	7.69	1.11	27.4	158	150.8	7.87	0.85	27.9	183.6	178.9
Median	7.84	0.89	25	219.5	210.5	7.73	1.23	27.5	74	72	7.98	0.84	27.8	182	176
Zone 7															
Average	7.60	0.54	25.38	187.5	174	7.68	0.61	28	201.9	195.6	7.74	0.43	28.16	126.6	124.5
Median	7.66	0.60	25.45	187	162	7.84	0.63	27.8	166	162.5	7.77	0.46	28.25	120	117
Zone 8															
Average	7.72	0.72	26.09	143.5	135.5	6.96	1.71	27.5	165.8	160	7.68	0.74	28.68	130.4	129.9
Median	7.83	0.73	25.95	116	108	7.80	0.81	27.5	97	94	7.61	0.75	29.1	128	128
Zone 9															
Average	7.61	0.90	26.04	131.6	125.1	7.51	0.9	27.4	291.3	282.4	7.60	0.66	28.52	99.75	97.8
Median	7.73	0.89	26.25	134.5	131	7.56	1.02	27.4	253.5	248.5	7.61	0.70	28.39	89.5	85
Units. CE: dS m-1; T°: °C; C-T y C-F: NMP/100 mL															

Table 3. pH, electric conductivity (CE), temperature (T°), total (C-T) and fecal (C-F) coliforms values in artesian wells of the Cárdenas-Comalcalco basin, Tabasco.

The values for total and fecal coliforms in the wells sampled were very high in regards to the limits set by Mexican Official Standard (NOM-127-SSA1-1994) and Water Quality Standards of World Health Organization. They outline safe limits between 0 and up to 2 NMP/100 ml of fecal and total coliforms, in water for human consumption. There were found concentrations of up to 505 NMP/100 ml, and values above of 200 NMP/100 ml were the most common. Only in well of Poblado C-31 (Z-9) were detected values smaller than 100 NMP/100 ml in the three samplings.

Although high coliforms concentrations were maintained over the three sampling seasons, a considerable decrease was detected in the norte season, when most of sites had concentrations bellow 200 NMP/100 ml. This decrease is attributed to the increasing phreatic level during that time. The bacteria are diluted in greater quantity of water.

Implications of these results for health of basin's population are considerable. Most of coliforms bacteria cause no illnesses; however, its presence in water is important for public health because they can indicate presence of such pathogenic microorganisms as bacteria, virus and protozoa. Illnesses associated to water polluted with these organisms involve symptoms similar to flu, vomiting, diarrhea and fever (Pontius, 2002).

Salts concentrations (Na 8.7 meq L^{-1}), (Mg 8.33 meq L^{-1}) and (Ca 5 meq L^{-1}) in all wells studied were below established limits; no reference, in the existent legal framework, has been found relating to the maximum K levels in water for human consumption; the other parameters studied were inside the limits allowed (Table 4).

Season	Dry				Rain				Norte			
	K	Ca	Mg	Na	K	Ca	Mg	Na	K	Ca	Mg	Na
Limits	5	12.15	460		5	12.15	460		5	12.15	460	
	Meq L^{-1}											
	Zone 5											
Average	0.20	3.16	2.95	5.45	29.7	0.23	0.65	0.06	12.1	0.29	0.59	0.14
Median	0.18	3.02	1.86	5.49	15.4	0.24	0.68	0.1	13.2	0.12	0.56	0.15
	Zone 6											
Average	0.10	3.47	3.60	5.45	17.6	0.26	1.02	0.04	28.05	0.33	0.71	0.13
Median	0.07	3.36	2.06	5.31	8.8	0.24	0.88	0	30.8	0.24	0.76	0.1
	Zone 7											
Average	0.11	2.65	1.27	2.53	13.75	0.26	0.72	0	22	0.15	1.08	0.11
Median	0.1	2.69	1.24	1.55	8.8	0.24	0.52	0	22	0.12	0.68	0.1
	Zone 8											
Average	0.20	3.91	1.88	3.46	36.3	0.20	0.74	0.03	110.5	0.21	1.8	0.11
Median	0.20	4.00	1.77	3	19.8	0.18	0.44	0	105.6	0.24	1.6	0.1
	Zone 9											
Average	0.035	4.17	3.37	3.57	3.88	0.16	5.63	0.14	44	0.09	0.36	0.09
Median	0.04	4.07	3.50	3.02	3.5	0.15	5	0	30.8	0.12	0.32	0

Table 4. Dissolved salts values: potassium, calcium, magnesium and sodium in wells of the five study areas in the Cárdenas-Comalcalco basin, Tabasco

3.2 Rivers

The obtained results do not show a specific pattern which indicates that any of the rivers have suitable water quality conditions, there were no significant differences in any of the parameters studied neither between rivers and sampling seasons.

Taking as a reference Ecological Criteria of Water Quality (DOF, December 13, 1989) that establish 0.04 mg/L of nitrates, as the maximum permissible for the protection of marine aquatic life, it can be observed that in general, all sampling points were polluted in all three seasons of year. Except for stations 3 and 4 of The Naranjeño river, station 4 of The Santa Ana river and station 4 of The San Felipe river which were below the standards only during the dry season.

In fact, water generally was less polluted by nitrates in dry season (average was in 5.3 and median in 4.4) than in the rainy and norte seasons, when nitrates concentration increased almost twice (Table 5). Evidently, the increase in nitrates concentration coincides with precipitation increase. Also, the study basin is characterized to contain a vast surface of sugar cane cultivation, in which fertilizers and other chemical products are intensively used, some of which are dragged into the rivers by the effects of the rains, causing a marked increase in their concentration in seasons of more pluvial precipitation.

Season	Dry				Rain				Norte			
	Nitrate	Amm.	Phosph	Chlor.	Nitrate	Amm.	Phosph	Chlor.	Nitrate	Amm.	Phosph	Chlor.
Limits	0.04	0,02	0.1	0.011	0.04	0,02	0.1	0.011	0.04	0,02	0.1	0.011
	mg/L											
	Naranjeño river											
Average	7.15	0.2	0.96	0	7.15	0.24	0.69	0.025	6.05	0.12	0.92	0.1
Median	4.4	0.2	0.28	0	8.8	0.24	0.72	0	6.6	0.12	0.92	0.1
	Santa Ana river											
Average	4.95	0.1625	0.69	0.175	7.7	0.345	1.07	0.375	8.25	0.105	1.24	0.1125
Median	4.4	0.15	0.28	0.05	8.8	0.36	0.96	0	8.8	0.06	1.28	0.1
	San Felipe river											
Average	3.85	0.075	1.31	0.1375	7.15	0.72	1.63	0.1	7.625	0.105	1.02	0.0875
Median	4.4	0	0.6	0	8.8	0.48	1.8	0.1	8.5	0.12	0.84	0.1

Table 5. Nitrates, ammonium, phosphates and chlorine values in three rivers studied in The Cárdenas-Comalcalco basin, Tabasco.

Similar behavior was detected in ammonium concentration, although in Mexico there is no official standard that makes reference to tolerance limits. The Ecological Criteria of Water Quality (DOF December 13 1989) only considers ammoniacal nitrogen within the listed pollutants, establishing 0.06 mg/L ammonium as maximum permissible for protection of aquatic life (in fresh water). Although some authors (López et. al, no date) have used 0.50 mg/L as maximum reference value, according to standard (NOM-127-SSA1-1994) in this work it was not used because it refers to water for human consumption.

Aquatic animals are more sensitive to ammonium than mammals. If we apply the previously referred limit of 0.50 mg/L, almost all obtained values would be below it, which would disguise the results obtained in relation to effect on aquatic organisms. Therefore, it was decided to take as reference the tolerance limit for fish, of 0.02 mg L^{-1} of NH_3.

Excepting station 1 of The Naranjeño river, and stations 1, 2 and 3 of The San Felipe river which quantities were non-detectable in dry season, the same as for station 3 of The Naranjeño river and station 3 of The Santa Ana river in that of the norte season, in general, in all seasons, very high levels of ammonium were observed. These levels are very superior to the maximum tolerance for fish (Table 5).

Ammonium averages and median were around 0.1 in the dry and norte seasons, while in rainy season they increased dramatically up to an average of 0.435, showing values of up to 2.16 mg/L. Being a toxic compound, its presence at these concentrations will have noxious effects for aquatic life.

The nitrates and ammonium high concentrations effects on aquatic life have not been studied in this basin. However, it is known that their excess causes water acidification, eutrophication and the occurrence of toxic algae, as well as having a toxic action on aquatic animals in the form of gills epithelium damage, which can cause asphyxia, glycolysis stimulation and Krebs cycle suppression that causes acidosis and reduction of blood's oxygen transporting capacity; unpairing oxidative phosphorylation that causes inhibition of ATP production in brain's basilar region; interference of immune system that increases susceptibility to bacterial and parasitic illnesses (Camargo et. al., 2007).

Practically in all sampling sites and in the three seasons of year phosphates were also superior to the limits allowed in Ecological Criteria of Water Quality (CE-CCA-001/89 12.13.89) for Aquatic Life Protection that is of 0.1 mg/L of total phosphates for rivers and streams and 0.002 mg/L for marine water.

Only the station 1 of The Santa Ana river showed slightly lower levels (0.08) in dry season. As in the case of nitrates and ammonium, values were lower in dry season.

Phosphate excess is one of main causes of eutrophication of natural water bodies (Ramírez et. al., 2005), but few studies exist on effects that high phosphate concentrations in water can have on aquatic animals, although is known that it can negatively affect mollusks reproduction and fish eggs appearance (Russo, 1985; Reynolds and Guillaume, 1998).

A progressive increase was detected in residual chlorine levels from the dry season to that of the rainy season. In the first case, only sample station 3 The Santa Ana river had levels above standard in two depths.

In the rainy season there were 5 stations that surpassed allowed limits (3 The Naranjeño river, 1 The Santa Ana river and 1 and The San Felipe river). During the norte season all sampling stations practically surpassed limits allowed by Ecological Criteria of Water Quality (CE-CCA-001/89 12.13.89) for Aquatic Life Protection that, for the case of fresh water it allows concentrations of up to 0.011 mg/L of residual chlorine.

During in the dry season as well as in that of the rainy season most of sites were inside acceptable turbidity limits for the development of aquatic communities (40 NTU) according to water contamination control (UNESCO, 1995), except for stations 1 The Naranjeño river, station 4 The Santa Ana river and station 4 The San Felipe river that, during the dry season surpassed these limits.

However, in the norte season, sampling stations of the San Felipe river had highest values, of up to 86 NTU, as well as station 3 of the Naranjeño river, with more than 89 NTU, which surpasses maximum permissible limits. In the rainy season there were the lowest values. Turbidity is caused by the leaching of inorganic fertilizers excessively used in agriculture and by erosion provoked by excesses of human activity in higher lands, such as immoderate pruning of forests and mining.

The turbidity can cause a decrease in water O_2 concentration, with a rising negative effect on fish, and reduction of light penetration in water column, which diminishes rate of photosynthesis activity and, consequently, it reduces the primary productivity of the phytoplankton that is the aquatic animals basic food.

A primary productivity reduction can collapse at superior trophic levels. On the other hand, turbidity could also negatively affect populations of invertebrates and to interfere with behavior, feeding and growth of many fish species and cause damages, due gills abrasion and obstruction; it has also been found that increases the susceptibility of fish to illnesses, since mucous secreted by them in answer to the high concentrations of suspended solids attracts bacteria and mushrooms (Scottish Natural Heritage, 1996).

Regulation for waters contamination prevention and control (DOF, March 29 1973) states that dissolved oxygen (OD) level should never be lower than 5.0 mg/L in waters for recreational use, without primary contact and for the exploitation of fish. This concentration is necessary to guarantee aquatic fauna good development.

In general, there were found levels much below this limit in almost all sampling stations and seasons (Table 6), in the three rivers the highest were detected in the dry season, although the San Felipe river always kept below acceptable levels; in the rainy season the concentration, in general, diminished reaching its lowest level in the norte season. This is contrary to expectation, since as water volume increases, nutrients concentrations trend to diminish which works against algae proliferation. In general, low OD concentrations detected seem to be result of accumulation of nutrients leached from the land by rains, those that promote algae excessive development and system eutrophication. Low OD concentration in water interferes with fish growth and reproduction, increases propensity to illnesses and, in extreme cases, it can cause their death (Shoji, et. al 2005). Studies carried out by Breitburg et al. (1997) demonstrated that low OD concentrations alter trophic nets in natural waters bodies when reducing efficiency of larvae and juvenile fish of escaping from their predators.

Season	Dry				Rain				Norte			
	Turbid	OD	DBO	SolTot	Turbid	OD	DBO	SolTot	Turbid	OD	DBO	SolTot
Limits	40	5.0	60	40	40	5.0	60	40	40	5.0	60	40
	NTU	mg/L	mg/L	mg/L	NTU	mg/L	mg/L	mg/L	NTU	mg/L	mg/L	mg/L
Naranjeño river												
Average	41.98	5.16	22.79	154.25	14.5	5.24	1.89	2562.9	32.96	2.57	1.13	978.34
Median	17.15	6.245	20.71	59	13.4	5.12	1.86	363	19.30	2.44	1.13	285.5
Santa Ana river												
Average	25.85	6.16	13.41	45.1	11.01	4.94	1.58	7035.3	20.44	2.77	1.44	5780.6
Median	10.42	6.24	12.20	45	9.37	4.81	1.95	3970	15.75	2.81	1.47	1199
San Felipe river												
Average	18.06	2.39	122.05	71.5	15.63	1.40	0.89	3233.9	55.04	1.43	1.44	1144.3
Median	15.45	1.275	55.99	54	12.70	0.38	0.53	1263	62.75	1.585	1.56	906.5

Table 6. Turbidity, dissolved oxygen, Oxygen Biochemistry Demand and Total Solids values in three rivers studied in The Cárdenas-Comalcalco basin, Tabasco.

Base on The Mexican Official Standard (NOM-001-ECOL-1996), water of the sampled sites in all seasons of year is located in permissible maximum limit for biochemical demand of oxygen (DBO), set at 60 mg/L as the daily average.

However, in the dry season values of three stations of the San Felipe river greatly surpasses that level reaching up to 527 mg/L in site 1 at 1 meter deep. This can be due to the fact that the river is located in the margins of some industry or big earth extensions with intensive agricultural activity that discharges a great quantity of organic waste in the system. A factor could also be a decrease in water flowrates limits, its flow being stagnated. When nutrients do not flow the biological activity concentrates. The high levels of ammonium and phosphates are related with a high content of organic matter, these levels usually cause decrease in values of OD and increase in those of DBO, which is what seems to be the case. The above-mentioned can be linked in particular with high indexes of organic decomposition that cause diverse adverse effects on aquatic organisms, particularly on the fish.

Total suspended solids were very high in the three sampling seasons, although in dry season this diminished in 5 of 12 stations, especially in those of the Santa Ana river (Table 6). This is attributed to the fact that in the rainy season, the rivers of flood plains of Tabasco drag a great quantity of material from soil coming from the mountains of Chiapas and that ends in Gulf of Mexico. The concentration of total solids in water were dramatically far from what the Mexican Official Standard specifies (NOM-001-ECOL-1996) on Permissible Maximum Limits for Basic Pollutants in Residual Water Discharges, that establishes a maximum of 40 mg/L as a daily average for rivers, because in this case an average of 4277 was detected in the rainy season and 2634 in that of the norte.

Most of pH values are within allowed range in the standard for discharges of residual waters (NOM-001-ECOL-1996) that goes from 5 to 10. All sampling sites of the Naranjeño river showed superior values at 8.5 in dry season, probably because of discharges of Ingenio Benito Juárez; in the same season station 1 of The San Felipe river showed acid pH, smaller than 6.5 (Table 7).

Season	Dry					Rain					Norte				
	pH	CE	T°	C-T	C-F	pH	CE	T°	C-T	C-F	pH	CE	T°	C-T	C-F
	6.5-8			<200	<200	6.5-8			<200	<200	6.5-8			<200	<200
Río Naranjeño															
Average	8.65	0.28	27.83	2400.1	2400	7.89	4.71	30.10	95.75	93.25	7.25	1.78	29.73	105.9	104.6
Median	8.89	0.24	27.5	2400	2400	7.92	0.66	29.9	91	89	7.28	0.52	29.65	111	109.5
Río Santa Ana															
Average	7.49	2.35	28.26	2400.	2400	7.57	12.79	30.5	324.9	299.8	7.39	14.26	27.36	68.25	68
Median	7.43	0.51	28.4	2400	2400	7.60	7.20	30.6	203.5	190.5	7.43	4.67	28.75	66	62.5
Río San Felipe															
Average	6.64	1.52	25.40	739.84	457.55	7.67	5.87	29.7	223.6	216.6	7.05	2.08	29.15	172	165.9
Median	7.07	0.84	28.85	299.32	132.46	7.59	2.3	29.6	255.5	246.5	7.02	1.65	29.15	184	176
Units. CE: dS m⁻¹; T°: °C; C-T y C-F: NMP/100 mL															

Table 7. pH, electric conductivity, temperature, fecal and total coliforms values in the three rivers studied.

As for coliforms, the maximum limit of 200 NMP/100 ml set in The Ecological Criteria of Water Quality (CE-CCA-001/89, 12.13.89) for Protection of Aquatic Life of fresh water was surpassed in most of stations of three rivers studied in dry season.

The maximum value reached was 2400 NMP/100 ml. In rainy and norte seasons concentration of total and fecal coliforms decreased dramatically, in fact, in The Naranjeño river permissible levels were reached in these two seasons, while in The Santa Ana river this was only obtained in norte season. The San Felipe river was still classed as polluted, during the norte season. Coliforms reduction observed in rain and norte seasons was due to increases of water volume, these organisms are then dispersed and head out to sea.

The coliforms presence in the basin is of supreme importance for environmental and human health, since many aquatic organisms for human consumption, such as fish and crustaceans, are contaminated causing, in turn, gastrointestinal illnesses to consumers. The high concentrations of these organisms in the rivers are indicative of a continuous flow of black waters toward these water bodies, coming from towns and homes located on the shore. To solve this problem it is necessary to fulfill the normative on the handling of residual waters.

The Ca determined in all sites was below 18 meq/L^{-1} and it allows placing water of three rivers as soft water. The concentrations of Mg had big fluctuations between rivers and sampling seasons, being smaller those of The Naranjeño river and those of the rainy season. The behavior for Na levels was similar, the smallest corresponded to The Naranjeño river, but the season of lowest concentration was the dry season.

In the season of rains there were very high concentrations in the rivers Santa Ana and San Felipe (Table 8). The above-mentioned could be due to that there are carried out agricultural activities with more intensity in the margins of these rivers, since leaching of inorganic fertilizers contributes to salinization of natural water bodies (Wong and Rowell, 1994).

Season	Dry				Rain				Norte			
	K	Ca	Mg	Na	K	Ca	Mg	Na	K	Ca	Mg	Na
Limits		18				18				18		
	Meq L^{-1}											
	Naranjeño river											
Average	0.08	0.98	2.85	1.65	0.88	4.53	11.17	38.73	0.42	1.95	3.36	6.59
Median	0.08	0.43	0.70	1.51	0.13	3.22	1.92	4.30	0.15	1.82	1.09	1.13
	Santa Ana river											
Average	0.44	2.64	5.69	17.05	2.94	7.94	29.41	141.8	3.11	6.41	21.76	81.25
Median	0.09	2.25	1.93	1.46	1.31	5.22	11.48	53.95	0.92	2.85	6.58	20.07
	San Felipe river											
Average	0.21	2.71	2.03	4.95	1.02	4.14	12.48	47.88	0.45	1.19	3.04	9.61
Median	0.21	2.24	2.07	4.96	0.44	2.83	5.02	18.25	0.37	1.14	1.97	5.72

Table 8. Dissolved salts values: potassium, calcium, magnesium and sodium in the three rivers studied.

3.3 Lagoons

The Ecological Criteria of Water Quality (DOF, December 13 1989) set 0.04 mg/L of nitrates as the maximum permissible for protection of marine aquatic life, in this sense, all the lagoons surpassed this value during the rainy and norte seasons. In the dry season all lagoons, except for sampling site 3 of El Cocal lagoon, were below the reference value.

In most of the sampling sites nitrates were found in non-detectable quantities, however, these increased until surpassing reference value in rain season (Table 9), with concentrations between 4.4 and 13.2 mg/L.

This trend continued during the norte season, being detected up to 15.4 mg/L in sampling site 1 of Las Palmas lagoon, although on average concentrations were greater in the rainy season, with 8.2 mg/L than in that of the norte (6.6 mg/L). Although there were extreme values, there were also found sites with very low nitrate concentrations such as sites 2 and 3 of Arrastradero Lagoon.

Increase of nitrates quantities during rainy and norte seasons seems to be result of the influence of a greater water volume, rich in nutritious material coming from human activities that the rivers catch; since in dry season entrance of seawater with its tides allows to take out a great amount of pollutants toward Gulf of Mexico.

Season	Dry				Rain				Norte			
	Nitrato	Amonio	Fosfato	Cloro	Nitrato	Amonio	Fosfato	Cloro	Nitrato	Amonio	Fosfato	Cloro
Límites	0.04	0.02	0.02	0.0075	0.04	0.02	0.02	0.0075	0.04	0.02	0.02	0.0075
	mg/L											
	El Carmen lagoon											
Average	-	0.41	0.22	-	9.35	0.17	0.78	-	5.50	0.24	0.30	0.09
Median	-	0.36	0.16	-	8.80	0.12	0.60	-	4.40	0.24	0.32	0.10
	Machona lagonn											
Average	0.55	0.23	0.25	0.03	7.70	0.15	0.37	0.01	8.35	0.27	0.29	0.05
Median	-	0.24	0.24	-	8.80	0.12	0.40	-	9.90	0.24	0.32	0.05
	Redonda lagoon											
Average	-	0.29	0.32	0.04	7.15	0.32	0.35	0.03	7.15	0.22	0.16	0.11
Median	-	0.24	0.32	-	8.80	0.36	0.32	-	6.60	0.24	0.16	0.10
	Arrastradero lagoon											
Average	0.55	0.27	0.29	0.01	7.15	0.29	0.80	0.01	2.20	0.24	0.86	0.08
Median	-	0.24	0.24	-	6.60	0.24	0.48	-	2.20	0.24	0.84	0.10
	Cocal Lagoon											
Average	2.20	0.22	0.29	0.05	8.07	0.30	0.40	0.02	7.33	0.27	0.19	0.10
Median	2.20	0.24	0.24	0.05	8.80	0.24	0.36	-	8.80	0.24	0.16	0.10
	El Paso del Ostión Lagoon											
Average	-	0.28	0.29	0.07	8.07	0.18	1.17	-	6.60	0.28	0.40	0.12
Median	-	0.24	0.32	0.05	8.80	0.18	0.68	-	6.60	0.30	0.40	0.10
	Las Palmas lagoon											
Average	-	0.46	0.17	0.05	10.27	0.34	0.48	-	9.90	0.28	0.32	0.08
Median	-	0.42	0.20	-	8.80	0.36	0.56	-	11.00	0.30	0.32	0.10

Table 9. Values of nitrates, ammonium, phosphates and chlorine in the seven lagoons studied.

In water, ammonium can be shown in two chemical species whose proportion depends on hydrogen potential (pH) and temperature. One of them is ionized ammonium or NH^{4+} and the other is ammonia or not ionized ammonium (NH_3) which is much more toxic and, by its property of being liposoluble, it can cross biological protection barriers and to cause toxicological damage such as generating methemoglobinemia in fish, which impedes blood oxygenation, as well as degrading cellular membranes, and destroying gills with which fish breathe (Mangas, 2000).

Ammonium detected in the systems studied kept an almost constant level in three sampling seasons (Table 9). Although average values slightly varied, median in each one of sampling seasons was of 0.24 mg/L. All detected values thoroughly surpassed maximum value of ammonia nitrogen set in The Ecological Criteria of Water Quality (DOF, December 13 1989) for Protection of Aquatic Life (seawater) of 0.01 mg/L and tolerance limit for fish of 0.02 mg/L of NH3 (EIFAC, 1973).

Practically in all sampling sites and in the three seasons of year phosphates were also above the limits allowed by the Ecological Criteria of Water Quality (CE-CCA-001/89 12.13.89) for Protection Aquatic Life, which in the case of coastal systems allows concentrations of up to 0.02 mg/L.

In all sampling sites of the seven lagoons and in the three considered seasons, phosphates values greatly surpassed the limits, most of the values detected being around 0.24 mg/L in dry season, 0.4 mg/L in that of rainy season and 0.32 in that of norte season.

Results obtained in this study coincide with that found by INE (2000) in lagoon system Carmen-Machona, where it was reported that orthophosphates (PO_4) and nitrates (NO_3) values exceeded maximum limits for protection of seawater life in The Ecological Criteria.

Residual chlorine levels are relatively low in dry and rain seasons, excepting stations 3 of El Cocal lagoon and 2 of Paso de Ostión lagoon that, in the dry season surpassed limits set in the Ecological Criteria of Water Quality (CE-CCA-001/89 12.13.89) for Protection of Seawater Life, it is of up to 0.0075 mg/L. In the norte season only stations 1 of Machona lagoon and 4 of Redonda lagoon had residual chlorine of non-detectable concentrations, between 0.1 and 0.36 mg/L.

Except for some few cases, mainly in dry season, turbidity generally stayed below the 40 NTU that is maximum acceptable for development of aquatic communities, according to water contamination control (UNESCO, 1995) (Table 10). This was expected due continuous exchange of fresh water and salt water with the sea 'allows diluting' organic matter coming from rivers.

In general, in the lagoons studied there were found to be appropriate levels of O_2 in dry and rain seasons, according to The Ecological Criteria of Water Quality (DOF, December 13 1989) for Protection Seawater Life (marine) that set OD levels superior to 5.0 mg/L.

Contrarily, for the norte season it was found a considerable number of sites with concentrations smaller than those allowed for protection of seawater fauna, specifically in Redonda, Cocal and Las Palmas lagoons. The above-mentioned could be due to a considerable haulage of organic matter toward the lagoon through rivers that promoted proliferation of heterotroph organisms, which consume large quantities of oxygen. That contrasts with DBO data, since in all lagoons, in the three sampling seasons the detected levels were very below the 75 mg/L that the Mexican Official Standard (NOM-001-ECOL-1996) sets as permissible maximum limit as daily average.

Season	Dry				Rain				Norte			
	Turbid	OD	DBO	SolTot	Turbid	OD	DBO	SolTot	Turbid	OD	DBO	SolTot
Limits	40	5.0	75	40	40	5.0	75	40	40	5.0	75	40
	NTU	mg/L	mg/L	mg/L	NTU	mg/L	mg/L	mg/L	NTU	mg/L	mg/L	mg/L
El Carmen lagoon												
Average	8.47	8.21	19.88	12760	10.72	7.16	1.18	19595	10.33	6.56	0.69	19999
Median	8.09	7.92	20.01	14333	8.25	6.83	1.20	19999	9.81	6.63	0.67	19999
Machona lagoon												
Average	7.55	8.12	15.27	13134	12.36	7.36	1.64	19366	9.24	6.75	1.43	19364
Median	7.35	8.10	15.30	16877	12.70	7.46	1.64	19999	8.49	6.97	1.29	19606
Redonda lagoon												
Average	23.63	6.86	18.15	1691.2	17.35	6.13	20.99	19999	15.83	4.77	0.54	16714
Median	23.40	6.76	22.12	1708.5	10.47	6.42	1.75	19999	11.85	4.76	0.50	16766
Arrastradero lagoon												
Average	28.64	6.17	16.17	572.04	14.53	6.51	1.12	4519.2	14.95	6.47	1.19	423.25
Median	26.55	6.56	10.66	715.50	13	6.98	1.09	4652	10.20	6.63	1.14	420
Cocal lagoon												
Average	66.90	7.24	16.62	1642	6.11	1.85	0.40	447.25	6.03	0.81	0.84	16242
Median	60.75	7.23	16.53	1780.5	3.67	1.93	0.26	448	6.12	0.80	0.92	17116
El Paso del Ostión lagoon												
Average	32.87	4.90	11.79	831.21	13.34	5.54	1.33	12743	11.47	5.45	0.93	16242
Median	31.65	4.82	10.78	1008	11.65	5.84	1.39	12485	10.85	5.01	0.83	17116
Las Palmas lagoon												
Average	24.24	6.17	18.59	5951.8	16.50	5.95	1.93	14710	20.45	4.11	2.24	19999
Median	16.60	6.47	16.83	7504	16.50	6.05	1.92	17681	19.40	4.02	2.33	19999

Table 10. Turbidity, dissolved oxygen, biochemistry oxygen demand of and total solids values in the seven lagoons studied

Values of total suspended solids were very above the 125 mg/L (daily average) set as the permissible maximum limit in the same standard (NOM-001-ECOL-1996) for estuaries in the three sampling seasons, especially in those of rain and norte. Increment of total suspended solids in the water coincides with the values obtained in the rivers studied that drag a great quantity of materials from the soil that are carried to coastal lagoons in high precipitation season to be finally deposited in Gulf of Mexico.

Almost all pH values reported are within the allowed range in the standard (NOM-001-ECOL-1996) for residual water discharges in waters and national goods that goes from 5 to 10, and of Canadian Guide of Environmental Water Quality (December, 2003) that allows values between 6.5 and 8.5. The average values of obtained pH were 7.4, 7.5 and 7.3 for the dry, rainy and norte seasons, respectively (Table 11).

The same as for rivers, concentrations of total and fecal coliforms were extraordinarily high in dry season, decreasing up to concentrations bellow the limit in the rainy and norte seasons; the mean values were 931, 195 and 10 NMP/100 ml for dry, rain and norte, respectively. Values of dry season were greater than minimum of 200 NMP/100 ml that are set by The Ecological Criteria of Water Quality (CE-CCA-001/89 12.13.89) for Protection of Seawater Life (marine water).

Época	Secas					Lluvias					Nortes				
	pH	CE	T°	C-T	C-F	pH	CE	T°	C-T	C-F	pH	CE	T°	C-T	C-F
	6.5-8			<200	<200	6.5-8			<200	<200	6.5-8			<200	<200
Laguna El Carmen															
Average	7.35	31.59	27.78	1086	712.9	7.66	43.36	28.26	126.13	119.75	10.33	6.56	27.7	19999	7369.9
Median	7.39	28.63	27.65	716	590.0	7.66	45.21	28.35	129.00	120.50	9.81	6.63	27.5	19999	7321.5
Laguna Machona															
Average	7.61	41.62	28.06	827	450.3	7.60	46.21	28.83	157.38	149.13	9.24	6.75	27.8	19364	7284.9
Median	7.57	32.74	28.75	345	290.5	7.58	48.82	28.85	144.00	137.00	8.49	6.97	27.9	19999	7302.5
Laguna Redonda															
Average	7.64	3.49	28.38	2237	574.4	7.26	44.36	29.59	115.00	108.13	15.83	4.77	28.2	19442	7236.6
Median	7.66	3.23	28.40	2400	500.0	7.22	44.50	29.45	100.50	97.00	11.85	4.76	28.3	19606	7217.5
Laguna Arrastradero															
Average	7.97	1.23	28.85	1992	792.5	7.73	9.27	30.55	125.88	119.88	11.95	6.56	28.7	16714	7614.5
Median	7.81	1.31	28.80	2400	675.0	7.68	8.52	30.20	93.00	91.00	11.10	6.63	29.1	16826	7713.0
Laguna Cocal															
Average	7.74	2.83	28.68	2400	1098.3	7.24	27.19	29.87	295.17	283.00	10.44	4.09	28.5	15213	7089.0
Median	7.72	2.59	28.50	2400	1075.0	7.24	27.04	29.70	295.50	280.00	11.00	3.94	28.5	15636	7042.5
Laguna El Paso del Ostión															
Average	6.72	1.79	28.83	2325	1220.8	7.37	23.15	30.72	494.33	456.67	11.47	5.45	28.1	16242	7305.3
Median	7.50	1.84	28.75	2400	1040.0	7.31	22.69	30.65	602.50	530.00	10.85	5.01	27.9	17116	7381.5
Laguna Las Palmas															
Average	7.28	14.82	28.47	2250	2070.0	7.40	33.99	28.17	243.67	230.17	10.45	4.11	27.6	19999	6989.3
Median	7.23	14.82	28.55	2400	2400.0	7.39	35.68	28.20	215.00	211.00	9.40	4.02	27.9	19999	6960.5
Unidades de medida. CE: dS m-1; T°: °C; C-T y C-F: NMP/100 mL															

Table 11. pH, electric conductivity, temperature, and total and fecal coliforms values in the seven lagoons studied.

The most polluted lagoons, whose higher values surpassed the standard in dry and rainy seasons were El Paso del Ostión, Las Palmas and Cocal, because they are water bodies which are relatively closed and that only have communication with other lagoons in growing season, between October and November, which means that in dry season they do not receive replacement of sea water, and this is the reason why nutrients and microorganisms remain concentrated.

Presence of high concentrations of total and fecal coliforms in lagoons studied represents a serious problem for regions economic activities, in principle because fishing is an important economic activity in surrounding communities.

This has serious effects on fish and captured crustacean quality and it influences negatively in sale price. The most dramatic case is that of oyster production (*Crassostrea viginica*) that, being of high commercial value, it is seen that in the region, it reaches only marginal prices due to accumulation of coliforms consequence of their screening food habits.

Oysters generally are edible in fresh or smoked, in which coliforms imply serious risks to human health. On the other hand, the study area has potential for tourism development, due to its natural beauty; however, coliforms contamination excludes it as non-appropriate area for recreational use, according to the Ecological Criteria of Water Quality (DOF, December 13 1989) that set a maximum of 200 NMP/100 ml.

In the current normative there were not maximum values of concentration for K, Ca, Mg and Na; however, the detected values are normal for lagoons and coastal estuaries. They were considered appropriate for aquatic life (Table 12).

Season	Dry				Rain				Norte			
	K	Ca	Mg	Na	K	Ca	Mg	Na	K	Ca	Mg	Na
	Meq L⁻¹											
	El Carmen lagoon											
Average	6.04	12.50	71.24	261.74	10.00	18.39	105.18	530.98	9.87	18.87	93.75	275.54
Median	5.49	12.38	67.00	251.10	10.00	18.25	107.35	528.25	10.09	19.16	92.52	250.00
	Machona lagoon											
Average	6.41	10.62	58.20	295.16	9.94	18.35	107.93	507.08	9.39	17.41	103.52	229.35
Median	6.08	12.01	62.50	283.50	9.90	18.60	109.35	495.65	10.17	19.81	113.90	258.70
	Redonda lagoon											
Average	0.69	1.99	6.78	28.08	8.98	17.26	101.98	440.21	7.51	12.12	68.77	176.41
Median	0.65	1.96	6.15	25.05	9.05	17.20	102.80	450.00	7.56	12.07	69.08	167.39
	Arrastradero lagoon											
Average	0.23	1.35	2.23	7.80	1.53	5.46	21.10	71.64	5.57	11.22	108.24	242.55
Median	0.25	1.49	2.20	8.18	1.55	4.89	20.15	68.50	5.55	11.17	57.15	238.48
	Cocal lagoon											
Average	0.04	4.17	3.37	3.57	3.88	0.16	5.63	0.14	44.00	0.09	0.36	0.09
Median	0.04	4.07	3.50	3.02	3.50	0.15	5.00	-	30.80	0.12	0.32	-
	El Paso del Ostión lagoon											
Average	0.33	1.57	3.30	12.62	3.75	11.33	51.93	188.78	5.26	9.85	60.03	204.64
Median	0.34	1.54	3.25	12.45	3.70	11.00	50.55	187.80	5.33	10.30	60.03	216.74
	Las Palmas lagoon											
Average	2.60	5.42	25.48	116.52	7.40	15.45	85.67	383.00	9.38	18.32	101.29	276.09
Median	2.52	5.09	28.35	107.60	9.20	17.55	99.95	473.95	9.58	18.69	101.97	267.39

Table 12. Potassium, calcium, magnesium and sodium values in the seven studied lagoons

Finally, it is important to mention that analyses of heavy metals of all wells, rivers and lagoons, were in non-detectable quantities, therefore they are considered clean of these dangerous compounds for wild life and human health. According to The Ecological Criteria of Water Quality (DOF, December 13 1989) maximum permissible values in coastal lagoons for

Pb, Ni and Zn is 0.006, 0.008, 0.09 mg/L. In the actual norm there were no references to safe Va concentrations in water, however, the standards of European Union for Marine Water establish a maximum value of 100 µgV/L (http://www.ukmarinesac.org.uk/activities/water-quality/wq4_1_2.htm).

4. Conclusions and recommendations

4.1 Conclusions

The greatest source of contamination in The Cárdenas-Comalcalco basin are high concentrations of total and fecal coliforms that exceed to a major degree the maximum permissible for the protection of human health and wild fauna. Up to this moment this is a problem of public and environmental health that seems not to be studied by the involved sectors.

Sources of water chemical contamination in the basin are nitrates and the phosphates. The presence of nitrates in wells for human consumption can represent a continuous risk for population, since impact can occur in long term and to go unseen by the population.

Nitrates, ammonium and phosphate concentrations in natural water bodies were very high. Although scope of this study did not contemplate to measure effects on the biological community in the locations studied, there is solid evidence that levels similar to the ones found in the area have an adverse effect on fauna and water productivity. The negative outcome to environmental and to economic activities is unknown, but, according to similar studies, it could be considerable.

Concentrations of organic and inorganic pollutants tend to increase in the rainy and norte seasons due to run off of urban and agricultural organic residuals due the rain flows. Concentrations of these pollutants were smaller in lagoons that are connected to sea due to water exchange between both systems.

4.2 Recomendations

Implementation of a program to improve underground and superficial water quality of basin is necessary. The establishment of a legal base that allows the application of effective standards of environmental quality is required. It is necessary for greater control on agrochemicals used in agricultural activities and the design and construction of septic graves and drainage systems and appropriate treatment of residual waters in rural areas. In the same way, it is necessary to regulate existent industrial activities and that environmental quality standards are met in new industries.

It is recommended the implementation of well handling training programs at homes, and also educational programs focused to prevention of gastrointestinal illnesses caused by water polluted with coliforms and nitrogenated compounds.

It is necessary the realization of studies that allow to measure with more detail conditions of this basin, in which are carried out finer measurements of heavy metals to determine very small concentrations.

In the same way, it is necessary to carry out works that contemplate a greater number of sampling stations and a greater number of measurements throughout the year. Also, it is important to contemplate smaller water bodies that remain closed most part of the year, since results of this work indicate that in fact it is in these type of water bodies, where there are concentrated the greatest quantities of pollutants. The above-mentioned would allow obtaining enough information to propose reparation programs.

5. References

Arce, O. (2006). *Indicadores biológicos del agua*. Universidad Mayor de San Simón, Facultad de Ciencias y Tecnología. Cochabamba, Bolivia.

Breitburg, D.L.; T. Loher, A. Pacey & A. Gerstein. (1997). Varying Effects of Low Dissolved Oxygen on Trophic Interactions in an Estuarine Food Web. *Ecological Monographs*, Vol. 67, No. 4: 489–507

Camargo, J.A. & A. Alonso (Lead Authors); Raphael D. Sagarin (Topic Editor). (2007) "Inorganic nitrogen pollution in aquatic ecosystems: causes and consequences." In: *Encyclopedia of Earth*. Eds. Cutler J. Cleveland (Washington, D.C. Environmental Information Coalition, National Council for Science and the Environment). [First published in the Encyclopedia of Earth April 2, 2007; Last revised April 11, 2007; Retrieved April 29, 2008].

CE-CCA-001/89. (8) 12-13-89 ACUERDO por el que se establecen los *Criterios Ecológicos de Calidad del Aguas*. Instituto Nacional de Ecología. México, D.F.

Dourojeanni, A. y A. Jouraviev. 2001. Crisis de gobernabilidad en la gestión del agua: Desafíos que enfrenta la implementación de las recomendaciones contenidas en el capítulo 18 del Programa 21. CEPAL. Serie Recursos Naturales e Infraestructura. Chile. Pp. 84.

EIFAC, 1973. Water quality criteria for European freshwater fish. Report on ammonia and inland fisheries. *Water Ressources* No. 7: 1010–1022.

INE. Instituto Nacional de Ecología (2000). *La calidad del agua en los ecosistemas costeros de México*. México, D.F. (disponible en http://www.ine.gob.mx)

López T.D., M.O. Pérez, G. Mazari, H. Marisa & M Maas. (No date). *Reserva de la biósfera Chamela-Cuixmala: un estudio de calidad del agua, bajo un enfoque de manejo integrado de cuencas.*

Mangas, R.E. (2000). *Evaluación de los Efectos de la Remoción del Lirio Acuático (Eichhornia crassipes) en la Biota y la Calidad del Agua en el Embalse Manuel Ávila Camacho, en el Edo. De Puebla*. Tesis de Maestría. Instituto de Ciencias. BUAP.

National Research Council. (1995). *Nitrate and Nitrite in Drinking Water*. Subcommittee on Nitrate and Nitrite in Drinking Water, Committee on Toxicology, National Academy Press Washington D.C.

Nebel, B. & R. Wrigth. (1999). *Ciencias ambientales: Ecología y Desarrollo Sostenible*. Sexta Edición. Prentice-Hall. México. Pp. 720.

NOM-001/SEMARNAT-1996. *Norma Oficial Mexicana para Límites máximos permisibles de contaminantes en las descargas de aguas residuales en aguas y bienes nacionales*. DOF.

NOM-127-SSA1-1994. Norma *Oficial Mexicana de Salud ambiental, agua para uso y consumo humano, limites permisibles de calidad y tratamientos a que debe someterse el agua*. DOF.

Pontius, N.L. (2002). *Coliform Bacteria, Regulatory Briefing*. National Rural Water Association. Duncan, UK.

Ramirez, J., F.L. Gutiérrez, & A. Vargas. (2005). Phytoplankton community response to artificial eutrophication experiments conducted in the La Fe reservoir, El Retiro, Antioquia, Colombia. *Caldasia*, 27(1):103-115.

Reynolds, J.D.& H.P. Guillaume. 1998. Effects of phosphate on the reproductive symbiosis between bitterling and freshwater mussels: implications for conservation. *The Journal of Applied Ecology*. Vol. No. 35(4): 575-581

Russo, R.C. (1985) *Ammonia, Nitrite, and Nitrate Fundamentals of Aquatic Toxicology: Methods and Applications.* Hemisphere Publishing Corporation Washington DC. 1985. p 455-471.

Safe Drinking Water Comm. (1980). *Drinking water and health.* Vol 3. Nat. Acad. Press, Washington, 415 pp.

Scotish Natural Heritage. (1996). *Rivers and their catchments: causes and effects of turbid water.* Information & Advisory Note 22

Shoji, J., R. Masuda, Y. Yamashita, & M. Tanaka. (2005) Effect of low dissolved oxygen concentrations on behavior and predation rates on red sea bream *Pagrus major* larvae by the jellyfish *Aurelia aurita* and by juvenile Spanish mackerel *Scomberomorus niphonius*. *Marine Biology*, Vol. 147, No. 4, pp. 863-868.

UNESCO, 1995. *Agua para Todos, agua para la vida.* Informe de las Naciones Unidas sobre el Desarrollo de los Recursos Hídricos en el Mundo (versión española del UN WWDR). Ediciones Mundi-Prensa (www.mundiprensa.com). Gestión de la Calidad del Agua y Control de la Contaminación en América Latina y el Caribe. www.unesco.org/publishing

Wong M.T.F. & D.L. Rowell. (1994). Leaching of nutrients from undisturbed lysimeters of a cleared Ultisol, an Oxisol collected under rubber plantation and an Inceptisol. *Interciencia* 19(6): 352-355.

2

Provision of Essential Minerals Through Foliar Sprays

Rizwana Jabeen[1] and Rafiq Ahmad[2]
[1]Government College for Women, Shahrah-e-Liaquat, Karachi
[1,2]Department of Botany, University of Karachi
Pakistan

1. Introduction

Approximately half of the world's land surface is 'perennial desert or dry lands'. These areas can only be made more productive by irrigation. Indiscriminate use of irrigation water without any management has created salinity problem at many places. Consequently, salinity has become a threat to food supply. Although, currently there is enough food for chronically undernourished (Conway, 1997). Growth of human population will increase by 50%, from 6.1 billion in mid – 2001 to 9.3 billion by 2050, it means that crop production must be increased if food security be ensured, especially for those who live on about $ 1 per day (UN Millennium Declaration, 2000). Therefore, in view an estimates, there is a requirement for raising yield by 20% in Developed Countries and by 60% in Developing Countries (Owen, 2001). Unfortunately, a strong link with salinization throws an immediate question over the sustainability of using irrigation to increase food production and it has been argued elsewhere (Shannon & Noble, 1990; Flowers & Yeo, 1995) that the primary value of increasing the salt tolerance of crops will be to the sustainability of irrigation. In order to achieve this challenge different ways and means must be find out without major increase in the amount of new land under cultivation, which would further threaten forests and biodiversity. In the light of these demographic, agricultural and ecological issues, the threat and effects of salinity become even more alarming.

Most horticultural crops are glycophytes (Greenway & Munns, 1980) and have evolved under conditions of low soil salinity. The glycophytes cannot absorb, transport and utilize mineral nutrients as efficiently or as effectively under saline as non-saline conditions. Therefore, high concentrations of Na^+ and Cl^- in the soil solution may depress nutrients – ion activities and produce extreme ratios of Na^+/Cl^-, Na^+/K^+ Ca^+/Mg^+ and Cl^-, NO^{-3}. As a result plant becomes susceptible to osmotic specific ion injury as well as to nutritional disorders that may results in reduced yield or quality. Therefore, an alternative strategy for coping with salinity could, therefore, is to attempt to supplementary foliar irrigation of sodium antagonistic minerals where the growth medium is known to be or may become saline at some time during the plant growth cycle. Antagonistic behavior of excessive monovalent cations (especially sodium present in rhizosphere under saline condition) with monovalent and divalent cations of essential mineral creates physiological disorders for plant growth.

2. Brief history of mineral nutrition

A brief resume of essential minerals for plant growth is given below in interest to show that their availability of uptake and utilization is adversely affected under excessive salinity (with special reference of sodium) of rhizosphere.

Glimpses of early history for starting research on essential minerals for plant growth appear in literature, which shows that Theophrastus --- a Greek Philosopher performed experiments in crop nutrition during 287-372 B.C. After him along chain of experiment were carried out from different Scientist to know the importance of mineral nutrients for the plant growth.

The question of the nature of the mineral nutrients remained unanswered since the composition of a plant's ashes does not show whether a certain element found is actually necessary for the survival or whether it is merely roughage. The problem was solved when the plant physiologist J.V. Sachs (1832-1897) rediscovered the hydro-culture technique (hydroponics). J.V. Sachs produced the first useable synthetic nutrient solution together with the chemist J. A. Stockhardt. These experiments let Sachs understand the importance of the root hairs for the uptake of solute nutrients. At about the same time, J. A. L. W. Knop (1861) developed the nutrient solution still used very often. The experiments showed that the cations K^+, Ca^{2+}, Mg^{2+} and small amounts of Fe^{2+} or Fe^{3+}, as well as the anions SO_4^{2-}, $H_2PO_4^-$ (or H_3PO_4) and NO_3^- are essential for the growth and survival of the plants. From the 1860's to the 1940's several other scientists studied plant mineral nutrition.

When in the 20th century the demands for the purity of chemicals grew, it did become apparent that plants need a number of additional elements, identified other minerals needed by plants in much smaller amounts, so-called trace elements like boron, copper, manganese, zinc and molybdenum are necessary for the plant normal nutrition. During this time several plant nutrition scientists also developed nutrient recipes for optimum plant growth, together with (Hoagland, 1919; Arnon & Hoagland 1940).

A most comprehensive review elaborating the methodology for determining essential mineral elements appears in the book of Sand and Water Culture Methods Used in the Study of Plant Nutrition by (Hewitt, 1966). Another book Mineral Nutrition of Plants: Principles and Perspectives by (Epstein, 1971) described fundamental concepts about plant nutrition. Horst Marschner in 1986 published an inclusive book "Mineral Nutrition of Higher plants" and its next edition in 1996 narrating full knowledge about plant essential minerals.

3. Problems faced for uptake of essential minerals by plants at saline soil

3.1 Salinity

Plant performance, usually expressed as crop yield, plant biomass, or crop quality, is affected adversely by salinity. Salinity is a major environmental stress and one of the most severe abiotic factors limiting agricultural production, since it alters the availability of water and nutrients. This effect is mostly reported in semi-arid to arid regions due to accumulation of salts at the soil surface where it inhibits the growth and yields of crop plants especially where irrigation is practiced (Greenway & Munns 1980; Tanji, 1990). The physiology of plant responses to salinity and their relation to salinity resistance have been much researched and frequently reviewed in recent years e.g. (Lauchli, 1990; Munns, 1993; & Neumann, 1997.

Salinity has affected, and continues to affect, the land on which crops are, or might be, grown. Although the amount of salt affected land (about 900×10^6 ha) is imprecisely known its extent is sufficient to pose a threat to agriculture (Flower & Yeo, 1995; Munns, 2002) since most plants, and certainly most crop plants, will not grow in high concentrations of salt: only halophytes (by definition) grow in concentrations of sodium chloride higher than about 400 mM.

Crop species in general require a substantial quantity of water with lower salt contents. Moderately and highly salt tolerant crop species and non-conventional wild plants (including halophytes) can survive and grow on water with relatively higher salt contents.

Salts accumulate in the soil will depend upon the irrigation water quality, irrigation management and the adequacy of drainage. If salts become excessive, will result in yield reduction. As water salinity increases, greater care must be taken to leach salts out of the root zone before their accumulation reaches a concentration, which might affect yields. The frequency of leaching depends on water quality and the crop sensitivity to salinity. Salts are present in irrigation water in relatively small but significant amounts. They originate from dissolution or weathering of the rocks and soil, including dissolution of lime, gypsum and other slowly dissolved soil minerals. The suitability of water for irrigation is determined not only by the total amount of salt present but also by the kind of salt.

The problems that result vary both in kind and degree, and are modified by soil, climate and crop, as well as by the skill and knowledge of the water user. As a result, there is no set limit on water quality; rather, its suitability for use is determined by the conditions of use which affect the accumulation of the water constituents and which may restrict crop yield. The more complex the problem, the more difficult it is to formulate an economical management programme for solution.

3.2 Osmotic imbalances

The most common effect of salinity on plant growth is of water stress. Some plants will tolerate high levels of salinity while others can tolerate little or no salinity. This is because some are better able to make the needed osmotic adjustments enabling them to extract more water from a saline soil. The osmotic effects of salinity are result of increased sodium ion concentrations at the root – soil water interface that creates lower water potential. It is well documented that salt stress causes removal of water from the cytoplasm into the extra cellular spaces resulting in a reduction of cytosolic and vacuolar volume (Ashraf, 2004; Munns, 2002). Munns & Termaat, (1986) have shown that the earliest response of a non-halophyte to salinity is that leaves grow more slowly. Although the plants may experience water stress for a short period until they adjust osmotically, water deficit is not the only factor for limited growth, even at relatively high salinities. Growth is reduced as a function of total electrolyte concentration, soil water content and soil matrix effect and is evidenced by reduction in cell division, cell enlargement, cell expansion, cell wall plasticity in the growing region of roots and leaves (Neumann, 1997).

3.3 Ionic toxicity

Ionic toxicity considered responsible for growth inhibitions under excessive saline environment. Salinity reduces plant growth through ionic influences. It has been reported that salinity affects ion activities in solution by changing the ionic strength, by ion-pair formation, and by precipitation (Cramer et al., 1987), resulting in excessive uptake and transport of the salt ions (Na^+, Cl^- and $SO_4^{-2)}$) and/or an inadequate uptake and transport of essential elements, to produce changes in mineral nutrient uptake that affect plant growth and reduce yields and cause crop failure (Bayuelo-Jiménez et al., 2003). Also, significant entry of Na^+ or Cl^- results in severe growth reduction or death in salt-sensitive or glycophytic species, at the same time as producing mild toxicity symptoms in salt-tolerant species (Maathuis & Amtmann, 1999). Ion uptake is the cheapest form of osmotic adjustment under soil saline conditions, but it could also lead to problems of decline in leaf function and ionic imbalance and toxicity (Yildirim et al., 2009). In particular, salinity alters uptake and absorption rates of all mineral nutrients resulting in deficiency symptoms. Bonilla et al. (2004) found that most toxic effects of NaCl can be attributed to Na^+ toxicity. Excessive accumulation of Na^+ can cause a range of ionic and metabolic problems for plants (Hoai et al., 2003). It can be concluded that excessive amounts of any ion (cat or anion) in growth medium can cause toxicity which is species or cultivar specific. However, sodium ion toxicity is more prevalent and more toxic to plants.

3.4 Nutritional imbalance

Salinity acts like drought on plants, preventing roots from performing their osmotic activity where water and nutrients move from an area of low concentration into an area of high concentration. Therefore, because of the salt levels in the soil, water and nutrients cannot move into the plant roots. Salt tolerance of a plant is affected under low nutrient availability. Accumulation of Na^+ and Cl^- in leaves through the transpiration flow is a general and long-term process taking place in salt-stress plants (Munns & Termaat, 1986). Nutrient uptake and accumulation by plants is often reduced under saline conditions as a result of competitive process between the nutrient and a major salt species. However, this depends on the type of nutrients and composition of soil solution (Grattan & Grieve 1999; Homaee et al., 2002). Although plants selectively absorb potassium over sodium, Na^+- induced K^+ deficiency can develop on crops under salinity stress by Na^+ salt (Maas & Grattan, 1999). In some cases the uptake and translocation of ions such as K^+ and Ca^{+2} are affected by salt stress. In few examples under saline conditions uptake of some mineral elements is known to increase, which can sustain the growth of plants when tissue elements is higher, like Nitrogen (Sweby et al., 1994). Phosphorus (Awad et al., 1990), Potassium (Hawkins and Lewis, 1993), Calcium (Rengel, 1992) and Manganese (Cramer & Nowak, 1992). There might still be cases, when the imbalance of minerals might not be detectable depending upon the localization of the element rather than the total tissue concentration (Cramer, 1997).

Most studies related to plant nutrition and salinity interactions have been conducted in sand or solution cultures. A major difficulty in understanding plant nutrition status as affected by soil salinity is reconciling results obtained in experiments conducted in the field and in solution cultures (Grattan & Grieve 1999). While application of fertilizers could improve plant nutritional status, it may also increase the salinity of soil solution. Being antagonistic to other cations, sodium inhibits their entry in root system; hence plants suffer deficiency of other mineral elements, which are essential for growth. An immediate response of salinity induced water potential imbalance is closure of stomates, which on one hand effects on the carbon fixation in leaves and on the other causes deficiency of same essential minerals with specific reference to monovalent potassium cation required for enzyme activation and membrane transport. Antagonistic effect of excessive sodium could be avoided in root zone if these essential mono and divalent cations are provided through foliar irrigation to plants. In view of Foliar application of soluble salts is being undertaken in present work containing cations of essential mineral elements (which are antagonized by sodium) along with some anions, which are essential for plant growth; its main objective was to investigate the interactive effects of salinity and foliar spray of different nutrients compositions on growth of *Gossypium hirsutum*.

4. Control measures for salinity

An integrated, holistic approach is needed to conserve water and prevent soil salinization and water logging while protecting the environment and ecology. Firstly, source control through the implementation of more efficient irrigation systems and practices should be undertaken to minimize water application and reduce deep percolation. Unavoidable drainage waters should be intercepted, isolated and reused to irrigate a succession of crops of increasing salt tolerance, possibly including eucalyptus and halophyte species, so as to reduce drainage water volumes further and to conserve water and minimize pollution, while producing useful biomass. Conjunctive use of saline groundwater and surface water should also be undertaken to aid in lowering water table elevations, hence to reduce the need for drainage and its disposal, and to conserve water.

To achieve these goals, new technologies and management practices must be developed and implemented. Efficiency of irrigation must be increased by the adoption of appropriate management strategies, systems and practices and through education and training. Some practices can be used to control salinity within the crop root zone, while other practices can be used to control salinity within larger units of management, such as irrigation projects and river basins. Additional practices can be used to protect offsite environment and ecological systems - including the associated surface and groundwater resources.

There is usually no single way to achieve salinity control in irrigated lands and associated waters. Many different approaches and practices can be combined into satisfactory control systems; the appropriate combination depends upon economic, climatic, social, as well as edaphic and hydrogeologic situations.

The objective of salinity control is to maintain an acceptable crop yield. Management need not necessarily attempt to control salinity at the lowest possible level, but rather to keep it within limits commensurate with sustained productivity crop; soil and irrigation practices can be modified to help achieve these limits. The problem can be managed through Engineering, Reclamation and Saline Agricultural approach.

4.1 Use of salt tolerant plants

Salt tolerance of a plant may be considered as the ability to germinate, maintain growth and reproduce under persistent or interrupted salt stress. The salt tolerance of plants is a very acute and complex phenomenon, not only because different plants respond to saline conditions in fundamentally different ways, but also because of the great variation in the stress itself. The relative growth of plants in the presence of salinity is termed as their salt tolerance. The ability of plants to tolerate salt is determined by multiple biochemical pathways that facilitate retention and/or acquisition of water, protect chloroplast functions, and maintain ion homeostasis. Essential pathways in this connection are referred those which lead to synthesis of osmotically active metabolites, specific proteins, and certain free radical scavenging enzymes (Parida & Das, 2005). All salts can affect plants growth, but not all inhibit growth within permissible concentration. In addition, salts do not act alone in the soil, but interact in their effects on plants; some of these interactions are simple (e.g. interactions between Na^+ and Ca^+), whereas some are complex (e.g. Carbonates and their effects via increased soil pH). Among the most common effects of soil salinity is growth inhibition by Na^+ and Cl^-. For some plants, especially woody perennials (such as citrus and grapevines), Na^+ retained in the woody roots and stems and it is the Cl^- that accumulates in the shoot and is most damaging to plant often by inhibiting photosynthesis (Flowers, 1988).

However, for many plants (such as Germinaceous crops), Na^+ is the primary case of ion-specific damage (Tester & Davenport, 2003). It attracted the attention of many investigators and practical agricultural workers because of the need to increase yields on saline soil and to develop and utilize new saline areas. Plant species vary in how well they tolerate salt-affected soils. Many lists of salt tolerant cultivars of important crops and those grown for forage, fodder, wood or others economical purposes are available in literature , (Ahmad, R. & Ismail, S. 1993; Francois, L.E., 1994; Maas, E.V., 1996; Marcar, N.E. et al., 1999; Ahmad, R., & Chang, M.H. 2000). Depending upon prevailing range of saline soil or saline irrigation water(being used in irrigation) edaphic and environmental factors one can select a plant for providing economically feasible yield under saline prevailing adverse condition.

4.2 Mineral nutrition through foliar spray

The following information has been cited out of some important reviews related to foliar application of mineral nutrients, which have appeared in literature towards the end of the last century, beginning of the present century since they provide precise knowledge on the subject. Foliar fertilization is an effective method of providing a steady flow of nutrients, in combination with some traditional types of root-uptake fertilizers, to achieve better control of nutrients. Foliar irrigation is widely used to supply specific nutrients to many crops growing under saline environment. Foliar application of nutrients is partially overcoming the negative effect of stress condition influencing root growth and absorption capacity (Salama et al., 1996; El-Flouly & Abou El- Nour, 1998). In this respect, (El-Flouly & El-Sayad, 1997) stated that foliar fertilization of both macro and micronutrient is practiced whenever, nutrients uptake through the root system is restricted due to salt stress. The advantages of foliar spray compared to soil fertilization include: immediate response, convenience of combination spray and comparatively low cost. On the other hand foliar spray have some disadvantages, the main disadvantage is these must be repeatedly applied because of the constant loss of leaf blades to mowing. Other includes, the response is only temporary, only very low doses can be applied and there are limitations due to foliar toxicity. When nutrients are applied directly to the foliage, they must penetrate three barriers: i) The waxy cuticle covering on the epidermal cell, ii) The cell wall of the epidermal cell and iii) The plasma membrane of the epidermal cells. Morphology and organization of leaf tissue is such that it accommodates the uptake of gaseous plant nutrients, whilst that of roots the uptake of water-soluble solutes. These water-soluble plant nutrients are mainly supplied with fertilizers. Only in exceptional cases where nutrients are strongly fixed by soils or where aerial nutrient requirement of a crop is higher than the root uptake rates, foliar application can be adopted as a routine fertilization measure.

Similarly, for maximum stomatal entry, nutrient sprays must be applied when the stomata are open, early morning applications are the best. Also, there is less evaporation during the early morning thus giving a better chance for maximum uptake by leaves. Timing is keeping this point in mind spray so did both in regard to time of the season and time of the day a critical factor in foliar spray. High relative humidity during the time of application will also enhance uptake by minimizing evaporation. Foliar sprays may be effective only during "critical stages" of plants growth cycle and must be applied during or shortly before the critical period to be effective. Since immature foliage does not have well-developed cuticular layer, application of nutrient sprays when there is a significant amount of young foliage present will enhance cuticular entry. Factors that affect foliar absorption include relative humidity, temperature, pH of the nutrient solution, age of leaf, concentration of the nutrient solution, difference in the nutrient compounds (formulations), use of surfactants and addition of non-nutrient facilitating or carriers-mediated agents (Gary & Grigg, 1999).

4.2.1 Foliar spray of macronutrients
The efficiency with which foliar applied macronutrients are utilized depends on the mobility of the specific nutrient throughout the entire plant, mobility comprising long distance transport especially phloem transport as well as the symplastic transport. Potassium and nitrogen are examples of nutrients showing high mobility and when taken up by leaves they can be rapidly distributed throughout the entire plant. Calcium and sulfur show a low mobility and Ca^{2+} taken up by leaves cannot be transported to younger tissues or fruits where it may be required. Most nutrients will move freely in the water stream but the movement of many is restricted in the phloem, hence leaf applications do not meet the requirements of deficient trees.

Foliar application of potassium is an efficient method of potassium supply to plants to avoid interaction both antagonistic and synergistic with essential major secondary and micronutrients (Dibb & Thompson, 1985). Foliar K^+ may be a supplemental nutrient management practice when conditions reduce plant K^+ uptake from soil, therefore, foliar application of potassium may be possible management tool to alleviate reduced yields caused by K^+ deficiency under saline irrigation. Foliar spray did not only increase the crop yields but also reduce the quantities of fertilizer applied through soil. Islam et al., (2003) used 0.1% KNO_3 as foliar spray on jute plant leaves and obtained good results whereas, 250 ppm of KNO_3 produce promising results in *Lagenaria siceraria* , (Ahmad & Jabeen, 2005).

Similarly, foliar spray of nitrogen also provides best platform to enhance the plant growth when growing in saline strata. Foliar application of nitrogen results in increased grain protein content and bread making quality of wheat when applied at or after anthesis (Gooding & Davies, 1992). Rajput et al., (1995) concluded that foliar application increased heading, maturity, grain and biological yield and gain highest return. Due to high importance of foliar fertilizer and the initial role of nitrogen, in the present study KNO_3 was selected which can provide both K and N sources to the plant.

4.2.2 Foliar spray of micronutrients
Role of essential mineral element required in traces for growth and development of plants is known since long in literature (Hewitt, 1966). Small amount of Cu, Zn, B, Fe, Mo, and Mn are essential for growth and quality of the crop because they control most of the physiological activities of the crop by interrupting the level of chlorophyll content in leaves, which ultimately influence the photosynthetic activity of the plant (Kanwar & Randhawa, 1967).

Jamro et al., (2002), showed that the effect of foliar application of micronutrients significantly increased the cane length at lowest rates of zinc and copper (1.5 kg and 2.5 kg /ha), produced highest cane length of 145.40 cm and 144.93 cm, respectively, whereas, the lowest cane length of 113.07 cm was recorded in untreated plants.

Gregoriou et al., (1983) found that the quickest and most successful treatment of trees suffering from iron chlorosis on calcareous soils was obtained by incorporating Sequestrene 138 Fe - EDDHA in the soil. Kassab (2005) indicated that foliar spray of zinc, manganese and iron significantly increased growth parameter yield and its components of mung bean plants. In addition, spraying salinity stressed plants with micronutrients can reduce the undesirable effect of salinity through improving growth and nutrient status of plants as well. Micronutrient requirements can generally be better met by foliar application than requirements of macronutrients because in absolute terms higher quantities of macronutrients are needed. Abou El-Nour (2002) reported that plants irrigated with 5.6 dS/m irrigation water and sprayed with supplementary micronutrients foliar spray with an EDTA micronutrient compound contained 2.8% Fe + 2.8% Mn + 2.8% Zn + 14% N applied to maize showed significant increment in root dry weight as compared to control, where the increment reached to 19%.

4.2.3 Preparation of spray medium

Recipe of spray medium is also important in which surfactant /adjuvant are mixed with desired minerals to spread liquid on the surface of the leaf and let it stay there for some time for stomatal absorption (Mengel & Kirkby, 1987). Surfactants (surface active agents) are a type of substances designed to improve the dispersing/emulsifying, absorbing spreading, sticking and / penetrating properties of the spray mixture. Pure water will stand as a droplet with a small area of contact with the waxy leaf surface. Water droplet containing a surfactant will spread in a thin layer over a waxy leaf surface. Surfactant lowers the contact angle of spray droplets on the leaves thus enhancing absorption.

It is commonly believed that the optimal pH values of spray solutions for the maximum uptake of most mineral nutrients are within the range of 3.0-5.5 (Kannan, 1980). Acidic foliar sprays can penetrate leaf surfaces more effectively, but it is possible for a negative effect to occur when too much acidity is present. Each type of organic acid has its own pH disassociation range with the mineral as the pH drops (increased acidity). Blanpied (1979) reported that maximum Ca^{+2} absorption by apple leaves are at pH 3.3 - 5.2. Reed & Tukey (1978) found that maximum phosphorus absorption by Chrysanthemum (*Dendranthema gradiflora*) leaves was at pH 3-6 for Na-Phosphate and pH 7-10 for K-phosphate. Howard et. al., (2000) sprayed *Cotton* with buffer solutions of pH 4 and 6 containing boric acid potassium nitrate separately or in mixture. The highest yield was found when buffer solution of pH 4 was sprayed containing both the above-mentioned chemicals.

5. Materials and methods

Keeping in view that broad leave plants will have better chance of retaining minerals given through foliar spray medium, commercially important plant *Gossypium hirsutum* belonging to family Malvaceae and grown for lint and oil was selected for present work. Plants were grown with saline water irrigation at sand and foliar application of some sodium antagonistic essential minerals was practiced at different stages of growth. The cotton seeds were obtained from Central Cotton Institute of Multan, Pakistan.

Experiments on growth of *Gossypium hirsutum* was conducted at Biosaline Nursery, Department of Botany (University of Karachi) in large size plastic pots using various combinations of nutrients in foliar spray medium. Some essential trace elements were included which is being di and trivalent show antagonism with monovalent Na^+. They were given with K^+, which is used as growth promoter for plants raised under saline condition. The pots were filled with 18 kg of costal sand each, having basal outlet for drainage and capable of retaining 3 liter of water at saturation. Any additional amount of water easily leaches out from the drainage outlet. The practice of over irrigation avoids salt accumulation in the rhizosphere.

Experiment was divided into 12 sets, viz., 1. Non-spray 2. Foliar spray with water 3. Foliar spray with Fe-EDTA (5-ppm) 4. Foliar spray with $MnCl_2$ (5-ppm) 5. Foliar spray with MoO_3 (5-ppm) 6. Foliar spray with KNO_3 (500-ppm) 7. Foliar spray with KCl 500-ppm) 8. Foliar spray with Urea (1000-ppm) 9. Foliar spray with KNO_3 (500-ppm) + Fe-EDTA (5-ppm) 10. Foliar spray with KNO_3 (500-ppm) + $MnCl_2$ (5-ppm) 11. Foliar spray with KNO_3 (500-ppm) + MoO_3 (5-ppm) 12. Foliar spray with KNO_3 (500ppm) + Fe-EDTA (5-ppm) + $MnCl_2$ (5-ppm) +MoO_3 (5-ppm). Out of a total 180 pots used in present experiment 15 were used in each set, exposed to three different irrigation regimes given to 5 pots under each treatment viz., i) Nonsaline water (E.C$_{iw}$: 0.6 dS/m), ii) 0.4% sea-salt solution (E.C$_{iw}$:6.2 dS/m) and iii) 0.8%: sea-salt solution (E.C$_{iw}$: 10.8 dS/m).

The seeds of *Cotton* variety CIM 496 were used for the current investigations. The seeds were delinted with concentrated H_2SO_4 for one minute to remove the fiber and immediately washed with running water. The seeds were then surface sterilized with 0.1% $HgCl_2$ for 5 minutes. Five seeds were sown in each plastic pot irrigated with non- saline water. Irrigation with gradually increasing concentrations of sea- salt(S.S) in irrigation water was started in plants at five leaf stages (including cotyledonary leaves) and continued till it reached to the salinity levels of 6.2 and 10.8 dS/m. Pot was irrigated with 3-litre tap water/ salt solution twice a week. Three plants were kept in each pot. Cow dung manure was added in the soil at 9:1 ratio to plastic pots. Whereas NPK (1:2:1) was given in three split dozes. Insecticide and fungicide was used whenever needed. Spray medium was containing 10 ppm of liquid soap as a surfactant. Foliar spray was started at five leaf stage, and followed by at just beginning flowering, and intermediate fruiting stage, plants were completely sprayed with 300 ml/plant of respective spray nutrient solution.

Complete data on growth of various vegetative parameters i.e. plant height (cm) ,number of leaves and monopodial and sympodial branches, total leaf area, fresh and dry biomass was taken and reported in PhD Thesis (Jabeen, 2009),but due to limitations in number of printed pages only the data on fresh and dry biomass is presented in this chapter. Whereas reproductive parameters is presented in terms of number of squares, flower and balls/ plant, seed and lint weight; seed number per plant, seed cotton yield and lint/ seed ratio was recorded at termination of experiment. Fiber characteristics are reported in PhD Thesis (Jabeen, 2009).

Samples of leaf and stem, were taken at grand period of growth, and were dried separately overnight in oven at 70⁰C for the analysis of Na^+ and K^+ (A.O.A.C., 1984). Concentration of Na^+ and K^+ cations in samples was measured using a Petracourt PFP.1 Photometer.

Leaves samples were collected at grand period of growth, from 3rd /4th node below the apex for biochemical analysis. i) Chlorophyll was extracted from the leaves in 80%acetone and measured at 663 nm and 645 nm in a Spectrophotometer as outlined by Machlaclam & Zalik, (1963). ii) A total Soluble Carbohydrates content was measured in an aqueous extract of leaf sample according to Ciha & Brun, (1978). Extraction was done in extraction solution (glacial acetic acid: methanol: water, 1:4:5) and optical density was recorded at 490 nm. iii) Protein was estimated by Hartree (1972). Extraction was done in 5% Trichloroacetic acid (TCA) and estimated after color reaction with Folin Ciocalteu's reagent at 650 nm.

Soil samples were collected fortnightly for salinity measurements. They were dried, saturated with de – ionized water, kept overnight followed by water extraction under vacuum (USDA, 1954). This extract was used for pH and electrical conductivity (dS/m) measurements using a Canterbury Conductivity meter (Model AGR 1000).

Statistical analysis of the data was carried out as outlined by Little & Hills (1975) and Gomez & Gomez (1976). All the data were statistically analyzed by computer program Costat 3.03. and *SPSS VERSION 11*.Mean separation of data was carried out using Duncan Multiple Range Test (Duncan, 1955).

6. Results

6.1 Vegetative growth

Interaction of sea salt irrigation and foliar spray of different minerals on vegetative biomass (gm) per plant are presented in Figure 1-2.

6.1.1 Vegetative biomass

Following conclusions are made on cotton vegetative biomass after consulting results of interaction of sea salt irrigation with foliar spray of different compositions (Figure 1-2).

i.   Vegetative biomass of the plants growing with irrigation water of different sea salt dilutions without any foliar spray remained comparatively less than that of sprayed with water.

ii.   Those undergoing with single salt spray of micronutrient (Fe, Mn, Mo) show increase in plant biomass in comparison with non-spray or only water spray, whereas plant biomass among themselves was in the order of Mo< Mn <Fe respectively.

iii.   The plants undergoing spray medium for supply of Nitrogen (N) through potassium nitrate or urea though show increase in biomass in comparison with spray of single micronutrients in mentioned above. Whereas, increase in biomass in KNO_3 spray was significantly more than irrespective of salinity treatments.

iv.   Supply of potassium (K) through potassium nitrate and potassium chloride shows that biomass of plants sprayed with former is increased than the later. This could be attributed as a result of accompanying Nitrogen.

v.   Spray medium of potassium in combination with individual micronutrient (i.e. Fe, Mn and Mo) show significant effect of Fe in increasing plant biomass among these treatments. Their grading would be KNO_3 + Mo < KNO_3 + Mn < KNO_3 + Fe for performance at this parameter.

vi.   Spray medium of K in combination with all the three micronutrient (Fe, Mn and Mo) show significant increase in biomass in comparison with all the above-mentioned composition.

vii.   The increase in biomass shown by different medium in control plants follow similar pattern in plants growing at 6.2 and 10.8 dS/m sea salt irrigation water. The slight fluctuation shown in KCl and KNO_3 +Fe + Mn + Mo micronutrient spray at earlier period of growth is non significant.

viii.   ANOVA for fresh and dry biomass production showed significant difference at level P<0.0001 in respect to salinity and spray, whereas their interaction was not significant.

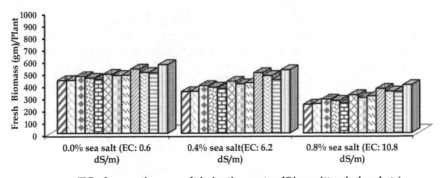

(EC of respective sea salt irrigation water dS/m written in brackets)

Fig. 1. Effect of foliar spray of different mineral nutrients and irrigation water of different salinity levels on Fresh Biomass (gm) in *Gossypium hirsutum*.

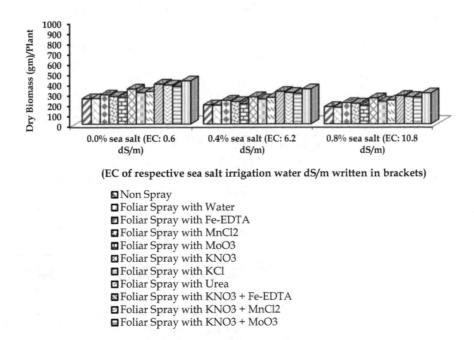

(EC of respective sea salt irrigation water dS/m written in brackets)

Non Spray
Foliar Spray with Water
Foliar Spray with Fe-EDTA
Foliar Spray with MnCl2
Foliar Spray with MoO3
Foliar Spray with KNO3
Foliar Spray with KCl
Foliar Spray with Urea
Foliar Spray with KNO3 + Fe-EDTA
Foliar Spray with KNO3 + MnCl2
Foliar Spray with KNO3 + MoO3

Fig. 2. Effect of foliar spray of different mineral nutrients and irrigation water of different salinity levels on Dry Biomass (gm) in *Gossypium hirsutum*.

6.2 Reproductive growth
6.2.1 Number of flowers, balls and seed per plant, seed and lint weight / per plant, seed cotton yield, seed / lint ratio

Reproductive parameters presented in Table 1 are based on production of number of flowers, bolls and seed per plant, weight of seed and lint / per plant, seed cotton yield and seed / lint Ratio as a result of foliar spray of different compositions at cotton plants raised by sea salt irrigation water.

Promotion /reduction percentage calculated on the basis above mentioned parameters is given in the Table 1.

Salinity stresses seemed to have reduced the yield on all the above-mentioned reproductive parameters with increase in salinity of irrigation water. Growth was promoted up to various degrees by foliar spray of mineral nutrients under nonsaline as well as saline conditions.

Increasing salinity of rooting medium has proportionally decreased the yield at all the above-mentioned parameters, which was offset up to various degree by the spray of different mineral nutrients. This effect is well documented in light of seed cotton yield, which is considered main parameter for determining the growth.

ANOVA for reproductive data exhibited significant difference at level $P<0.001$ in respect to salinity, at level $(P<0.0001)$ spray and while their interaction was not significant.

Treatment	Flowers/plant	Balls/plant	Seeds/plant	Seed weight (g)/plant	Lint weight (gm)/plant
Non Spray					
0.6 dS/m	58.00 a ±4.08	41.00 a ±2.95	902.00 a ±5.32	77.49 a ±0.01	56.98 a ±0.01
	-	-	-	-	-
6.2 dS/m	51.00 a b ±3.8	32.00 a ±3.44	512.00 b ±3.39	54.08 b ±0.01	38.09 b ±0.01
	(-15.00)	(-23.08)	(-55.84)	(-48.90)	(-51.08)
10.8 dS/m	35.00 a ±2.64	21.00 a ±2.68	252.00 c ±2.68	29.19 c ±0.01	18.71 c ±0.01
	(-30.00)	(-46.15)	(-92.21)	(-89.49)	(-90.85)
LSD 0.05	19.97	19.84	19.7	0.25	0.21
Foliar spray with water					
0.6 dS/m	60.00 a ±3.1	45.00 a ±3.19	945.00 a ±4.78	85.95 a ±4.78	63.47 a ±0.02
	(+10.00)	(+15.38)	(+22.73)	(+29.93)	(+30.42)
6.2 dS/m	52.00 a ±3.96	34.00 a ±3.96	518.00 b ±2.18	58.14 b ±0.01	41.29 b ±0.12
	(-5.00)	(-7.69)	(-37.66)	(-22.45)	(-25.39)
10.8 dS/m	2.58 a ±5.77	25.00 a ±3.04	300.00 c ±2.83	37.25 c ±0.01	24.81 c ±0.07
	(-15.00)	(-23.08)	(-68.83)	(-54.95)	(-59.30)
LSD 0.05	19.97	19.97	1.99	0.01	0.27
Foliar spray with Fe-EDTA (5 ppm)					
0.6 dS/m	74.00 a ±4.6	57.00 a ±4.41	1425.00 a ±3.55	114.57 a ±0.01	86.42 a ±0.27
	(+20.00)	(+30.77)	(+57.14)	(+62.13)	(+63.93)
6.2 dS/m	69.00 a ±4.12	52.00 a ±5.30	1040.00 b ±1.18	94.12 b ±0.01	68.23 b ±0.11
	(+5.00)	(+7.69)	(-11.69)	(+5.82)	(+2.77)
10.8 dS/m	55.00 a ±2.89	41.00 a ±4.22	615.00 c ±1.99	61.91 c ±0.01	43.32 c ±0.18
	(-10.00)	(-15.38)	(-61.04)	(-43.69)	(-49.13)
LSD 0.05	19.97	19.97	1.99	1.99	0.67

Treatment	Flowers/plant	Balls/plant	Seeds/plant	Seed weight (g)/plant	Lint weight (gm)/plant
Foliar spray with MnCl$_2$ (5 ppm)					
0.6 dS/m	69.00 [a] ±4.02	51.00 [a] ±3.96	1224.00 [a] ±3.97	101.49 [a] ±0.01	75.72 [a] ±0.26
	(+25.00)	(+38.46)	(+87.01)	(+80.50)	(+83.76)
6.2 dS/m	61.00 [a] ±4.14	47.00 [a] ±4.59	893.00 [b] ±1.52	38.78 [b] ±22.67	60.36 [b] ±0.14
	(+10.00)	(+15.38)	(+11.04)	(+21.77)	(+19.32)
10.8 dS/m	53.00 [a] ±2.66	34.00 [a] ±3.69	476.00 [c] ±1.99	50.66 [c] ±0.00	33.52 [c] ±0.02
	(-5.00)	(-7.69)	(-45.45)	(-32.43)	(-38.95)
LSD $_{0.05}$	19.25	125	14.25	14.81	0.21
Foliar spray with MoO$_3$ (5 ppm)					
0.6 dS/m	66.00 [a] ±3.94	48.00 [a] ±4.34	836.00 [a] ±4.44	24.70 [a] ±0.02	55.21 [a] ±0.012
	(+40.00)	(+61.54)	(+173.51)	(+127.89)	(+132.79)
6.2 dS/m	59.00 [a] ±3.95	41.00 [a] ±4.82	697.00 [b] ±2.36	71.75 [b] ±0.012	51.31 [b] ±0.02
	(+20.00)	(+30.77)	(+42.86)	(+50.49)	(+48.10)
10.8 dS/m	49.00 [a] ±2.59	30.00 [a] ±3.87	360.00 [c] ±2.36	44.70 [c] ±0.16	29.76 [c] ±0.01
	(+10.00)	(+15.38)	(-12.34)	(+2.79)	(-2.88)
LSD $_{0.05}$	19.97	19.24	14.25	0.14	0.21
Foliar spray with KNO$_3$ (500 ppm)					
0.6 dS/m	85.00 [a] ±3.83	67.00 [a] ±3.64	1876.00 [a] ±3.68	156.11 [a] ±0.07	122.63 [a] ±0.01
	(+105.00)	(+161.54)	(+409.09)	(+393.12)	(+426.62)
6.2 dS/m	78.00 [a] ±4.53	68.00 [a] ±3.95	1564.00 [b] ±2.75	144.84 [b] ±0.09	110.86 [b] ±1.25
	(+70.00)	(+107.69)	(+213.64)	(+238.10)	(+251.80)
10.8 dS/m	66.00 [a] ±3.00	51.00 [a] ±3.11	918.00 [c] ±2.71	93.33 [c] ±012	67.71 [c] ±0.15
	(+55.00)	(+84.62)	(+110.39)	(+148.98)	(+146.04)
LSD $_{0.05}$	19.25	17.25	14.25	1.01	0.21
Foliar spray with KCl (500 ppm)					
0.6 dS/m	77.00 [a] ±4.36	60.00 [a] ±4.63	1560.33 [a] ±3.38	127.20 [a] ±0.10	97.22 [a] ±0.07
	(+94.74)	(+150.00)	(+320.78)	(+304.54)	(+326.72)
6.2 dS/m	71.00 [a] ±4.3	55.00 [a] ±4.59	1155.00 [b] ±1.51	105.60 [b] ±0.01	78.24 [b] ±0.09
	(+76.47)	(+130.00)	(+153.90)	(+159.56)	(+165.57)
10.8 dS/m	58.00 [a] ±3.0	44.00 [a] ±3.9	748.00 [c] ±1.95	71.28 [c] ±0.01	49.36 [c] ±0.11
	(-10.71)	(-14.29)	(+40.26)	(+56.01)	(+50.46)
LSD $_{0.05}$	19.87	19.21	19.24	0.05	0.01
Foliar spray with Urea (1000 ppm)					
0.6 dS/m	81.00 [a] ±4.47	64.00 [a] ±4.63	1728.00 [a] ±3.48	142.72 [a] ±0.01	110.58 [a] ±0.17
	(+82.35)	(+140.00)	(+203.90)	(+188.44)	(+199.69)
6.2 dS/m	74.00 [a] ±4.13	61.00 [a] ±4.59	1403.00 [b] ±1.87	123.21 [b] ±0.01	92.82 [b] ±0.18
	(+107.14)	(+214.29)	(+118.18)	(+132.20)	(+133.50)
10.8 dS/m	62.00 [b] ±2.99	47.00 [b] ±3.9	846.00 [c] ±2.05	80.85 [c] ±0.01	57.55 [c] ±0.01
	(+9.09)	(+13.33)	(+21.43)	(+34.69)	(+26.62)
LSD $_{0.05}$	19.97	19.97	1.99	0.01	0.02

Treatment	Flowers/plant	Balls/plant	Seeds/plant	Seed weight (g)/plant	Lint weight (gm)/plant
Foliar spray with KNO$_3$ (500 ppm) + Fe-EDTA (5 ppm)					
0.6 dS/m	95.00[a]±4.66	78.00[a]±4.53	2496.00[a]±6.33	213.72[a]±0.21	174.86[a]±1.02
	(+120.00)	(+184.62)	(+483.77)	(+474.07)	(+521.27)
6.2 dS/m	90.00[a]±4.38	74.00[a]±5.23	1998.00[b]±4.63	187.79[b]±10.2	150.78[b]±1.25
	(+85.00)	(+130.77)	(+274.03)	(+308.16)	(+331.65)
10.8 dS/m	78.00[b]±3.43	61.00[b]±4.41	1342.00[c]±3.28	142.74[c]±1.36	112.68[c]±1.47
	(+3.33)	(+23.08)	(+122.08)	(+165.31)	(+168.24)
LSD $_{0.05}$	19.78	19.97	4.25	4.25	0.25
Foliar spray with KNO$_3$ (500 ppm) + MnCl$_2$ (5 ppm)					
0.6 dS/m	91.00[a]±5.18	74.00[a]±5.38	2220.00[a]±4.97	195.37[a]±10.4	195.38[a]±14.25
	(+113.64)	(+166.67)	(+562.34)	(+578.46)	(+822.51)
6.2 dS/m	84.00[a]±4.94	71.00[a]±5.56	1775.00[b]±3.25	173.25[b]±1.25	137.68[b]±1.25
	(+121.05)	(+191.67)	(+370.78)	(+434.85)	(+478.21)
10.8 dS/m	76.00[a]±3.56	58.00[b]±5.01	1160.00[c]±2.69	124.14[c]±3.2	95.25[c]±3.4
	(+58.33)	(+82.35)	(+224.68)	(+304.38)	(+321.38)
LSD $_{0.05}$	19.97	19.97	1.99	0.01	0.02
Foliar spray with KNO$_3$ (500 ppm) + MoO$_3$ (5 ppm)					
0.6 dS/m	89.00[a]±5.21	70.00[a]±5.2	2030.00[a]±3.9	171.52[a]±3.10	136.81[a]±2.36
	(+129.17)	(+182.35)	(+772.73)	(+769.84)	(+866.91)
6.2 dS/m	81.00[a]±4.52	68.00[a]±4.25	1632.00[b]±2.76	153.07[b]±3.11	119.47[b]±3.42
	(+113.64)	(+166.67)	(+496.10)	(+552.76)	(+612.85)
10.8 dS/m	70.00[a]±4.23	54.00[a]±3.25	1026.00[c]±2.59	105.38[c]±3.4	78.36[c]±.014
	(+64.00)	(+88.89)	(+300.0)	(+395.24)	(+429.50)
LSD $_{0.05}$	1.28	5.24	1.99	0.12	0.04
Foliar spray with KNO$_3$ (500 ppm) + Fe-EDTA (5 ppm)+ MnCl$_2$ (5 ppm)+ MoO$_3$ (5 ppm)					
0.6 dS/m	105.00[a]±2.58	85.00[a]±5.21	2975.00[a]±9.19	255.01[a]±0.12	212.47[a]±10.2
	(+225.00)	(+346.15)	(+1081.82)	(+1079.17)	(+1236.07)
6.2 dS/m	95.00[a]±4.56	78.00[a]±4.25	2340.00[b]±7.24	143.60[b]±1.25	179.49[b]±10.6
	(+155.00)	(+238.46)	(+640.26)	(+704.23)	(+798.25)
10.8 dS/m	80.00[a]±5.96	65.00[a]±3.25	1625.00[c]±5.39	162.51[c]±1.25	130.59[c]±1.24
	(+140.00)	(+215.38)	(+468.18)	(+561.38)	(+619.42)
LSD $_{0.05}$	1.25	1.97	19.9	0.01	0.27

Values are means of five replicates ± SE. Different letters in the same column are significantly different at P <0.05 level, as determined by Duncan's Multiple Range Test. Figures in parenthesis indicate % promotion (+) and reduction (-) over control.

Table 1. Effect of foliar spray of mineral nutrients and irrigation water of different salinity levels on Promotion and Reduction percentage of reproductive parameters in *Gossypium hirsutum*.

6.3 Analytical analysis
6.3.1 Minerals analysis

Na^+ and K^+ ion concentration (ppm) and Na+/ K+ ratio performed in the stem and leaf samples are presented in Table 2.

i. In leaf and stem Na^+ concentrations increased at both 6.2 and 10.8 dS/m salinity levels of sea salt irrigation water.

ii. Though degree of reduction was reduced by different mineral nutrients spray up to greater extent and by only water spray up to lesser extent.

iii. Foliar spray of potassium with the mixture each single (Fe, Mn, Mo) microelement at control plants and those growing under both the salinities in above mentioned plant parts showed best result as compare to potassium nitrate with the mixture single microelement, alone KNO_3 , KCl, Urea, Fe, Mn and Mo spray which is evident from (Table 2.) as well.

iv. Increase in Na^+, decreased is found in the K^+ accumulation in leaf and stem under low and high salinity levels, irrespective of any foliar treatment.

v. Due to increase in Na^+ concentrations, K+ concentrations decrease in leaves and stems, resulted increase in Na^+ / K^+ ratio under increasing salinities which adversely affect growth but the spray of KNO_3 KCl, Urea, Fe, Mn and Mo alone and potassium with the mixture each single (Fe, Mn, Mo) microelement decreased this ratio suppressing the inhibitory effect of excessive sodium on growth.

Treatment	LEAF			STEM		
	Na⁺ (ppm)	K⁺ (ppm)	Na⁺/ K⁺	Na⁺(ppm)	K⁺ (ppm)	Na⁺/ K⁺
Nonspray						
0.6 dS/m	181.67 ᶜ±0.88	52.00 ᵃ±0.58	3.49 ᶜ±0.02	184.27 ᶜ±0.88	48.40 ᵃ±0.58	3.81 ᶜ±0.03
6.2 dS/m	232.33 ᵇ±0.88	47.00 ᵇ±0.58	4.95 ᵇ±0.08	234.93 ᵇ±0.88	43.40 ᵇ±0.58	5.42 ᵇ±0.09
10.8 dS/m	244.33 ᵃ±0.67	47.00 ᵇ±0.58	5.20 ᵃ±0.08	247.49 ᵃ±0.67	43.40 ᵇ±0.58	5.69 ᵃ±0.09
LSD ₀.₀₅	2.82	1.99	0.22	2.82	1.99	0.26
Foliar spray with water						
0.6 dS/m	184.33 ᶜ±1.53	58.00 ᵃ±1.00	3.18 ᶜ±0.03	186.93 ᶜ±1.53	54.40 ᵃ±1.00	3.44 ᶜ±0.04
6.2 dS/m	193.33 ᵇ±0.88	57.00 ᵇ±0.58	3.39 ᵇ±0.03	195.93 ᵇ±0.88	53.40 ᵃ±0.58	3.67 ᵇ±0.03
10.8 dS/m	224.67 ᵃ±0.67	53.00 ᵇ±0.58	4.24 ᵃ±0.04	227.27 ᵃ±0.67	49.40 ᵇ±0.58	4.60 ᵃ±0.04
LSD ₀.₀₅	2.82	1.99	0.10	2.82	1.99	0.12
Foliar spray with Fe-EDTA (5 ppm)						
0.6 dS/m	130.00 ᵇ±0.58	85.00 ᵃ±0.58	1.53 ᶜ±0.02	132.60 ᵇ±0.58	81.40 ᵃ±0.58	1.63 ᶜ±0.02
6.2 dS/m	131.00 ᵃᵇ±0.58	82.00 ᵇ±0.58	1.60 ᵇ±0.02	133.60 ᵃᵇ±0.58	78.40 ᵇ±0.58	1.70 ᵇ±0.02
10.8 dS/m	132.33 ᵃ±0.33	79.33 ᶜ±0.88	1.67 ᵃ±0.02	134.93 ᵃ±0.33	75.73 ᶜ±0.88	1.78 ᵃ±0.02
LSD ₀.₀₅	1.76	2.40	0.06	1.76	2.40	0.06
Foliar spray with MnCl₂ (5 ppm)						
0.6 dS/m	152.00 ᵃ±0.58	78.00 ᵃ±0.58	1.95 ᶜ±0.01	154.60 ᵃ±0.58	74.40 ᵃ±0.58	2.08 ᶜ±0.01
6.2 dS/m	143.67 ᵇ±0.88	66.00 ᵇ±0.58	2.18 ᵇ±0.01	146.27 ᶜ±0.88	62.40 ᵇ±0.58	2.34 ᵇ±0.01
10.8 dS/m	148.33 ᶜ±0.88	63.33 ᶜ±0.88	2.34 ᵃ±0.04	150.93 ᵇ±0.88	59.73 ᶜ±0.88	2.53 ᵃ±0.05
LSD ₀.₀₅	2.74	2.40	0.088	2.74	2.40	0.09
Foliar spray with MoO₃ (5 ppm)						
0.6 dS/m	153.33 ᵃ±0.88	62.33 ᵃ±0.88	2.46 ᶜ±0.02	155.93 ᵃ±0.88	58.73 ᵃ±0.88	2.66 ᶜ±0.02
6.2 dS/m	157.00 ᵃᵇ±0.58	60.67 ᵇ±0.88	2.59 ᵇ±0.03	159.60 ᵃᵇ±0.58	57.07 ᵇ±0.88	2.80 ᵇ±0.03
10.8 dS/m	162.33 ᵃ±0.88	59.67 ᶜ±0.67	2.72 ᵃ±0.02	164.93 ᵃ±0.88	56.07 ᶜ±0.67	2.94 ᵃ±0.02
LSD ₀.₀₅	1.76	2.40	0.06	1.76	2.40	0.06

	LEAF			STEM		
Treatment	Na$^+$ (ppm)	K$^+$ (ppm)	Na$^+$/ K$^+$	Na$^+$(ppm)	K$^+$ (ppm)	Na$^+$/ K$^+$
Foliar spray with KNO$_3$ (500 ppm)						
0.6 dS/m	90.33 a±0.88	124.00 a±0.58	0.73 c±0.01	92.93 a±0.88	120.40 a±0.58	0.77 c±0.01
6.2 dS/m	110.00 b±5.04	121.67 b±0.88	0.90 b±0.04	112.60 c±5.04	118.07 b±0.88	0.95 b±0.04
10.8 dS/m	122.67 c±0.67	101.33 c±0.67	1.21 a±0.01	125.27 b±0.67	97.73 c±0.67	1.28 a±0.01
LSD 0.05	2.74	2.40	0.08	2.74	2.40	0.09
Foliar spray with KCl (500 ppm)						
0.6 dS/m	85.33 c±3.18	99.00 a±3.79	0.86 c±0.04	87.93 c±3.18	95.40 a±3.79	0.92 c±0.04
6.2 dS/m	126.00 b±0.58	103.00 a±1.53	1.22 b±0.01	128.60 b±0.58	99.40 a±1.53	1.29 b±0.02
10.8 dS/m	134.33 c±1.20	100.00 a±0.58	1.34 a±0.01	136.93 a±1.20	96.40 a±0.58	1.42 a±0.00
LSD 0.05	6.88	8.23	0.07	6.88	8.25	0.08
Foliar spray with Urea (1000 ppm)						
0.6 dS/m	103.67 c±1.86	93.33 a±0.88	1.11 c±0.01	106.27 c±1.86	89.73 a±0.88	1.18 c±0.01
6.2 dS/m	133.00 b±1.53	93.00 a±2.08	1.43 b±0.03	135.60 b±1.53	89.40 a±2.08	1.52 b±0.03
10.8 dS/m	141.33 a±1.86	91.67 a±1.45	1.54 a±0.01	143.93 a±1.86	88.07 a±1.45	1.63 a±0.01
LSD 0.05	6.06	5.36	0.06	6.06	5.36	0.07
Foliar spray with KNO$_3$ (500 ppm) + Fe-EDTA (5 ppm)						
0.6 dS/m	76.33 c±1.20	168.00 b±3.00	0.45 c±0.01	78.93 c±1.20	164.40 b±3.00	0.48 c±0.02
6.2 dS/m	102.00 b±1.53	158.33 a±1.20	0.64 b±0.01	104.60 b±1.53	154.73 a±1.20	0.68 b±0.02
10.8 dS/m	117.00 a±0.58	146.67 a±2.19	0.80 a±0.01	119.60 a±0.58	143.07 a±2.19	0.84 a±0.01
LSD 0.05	4.05	7.79	0.04	4.05	7.79	0.04
Foliar spray with KNO$_3$ (500 ppm) + MnCl$_2$ (5 ppm)						
0.6 dS/m	81.00 c±0.58	131.33 b±0.67	0.62 c±0.00	83.60 c±0.58	127.73 b±0.67	0.65 c±0.00
6.2 dS/m	111.00 b±0.58	136.33 a±0.33	0.81 b±0.00	113.60 b±0.58	132.73 a±0.33	0.86 b±0.00
10.8 dS/m	121.00 a±0.58	134.33 a±0.88	0.90 a±0.00	123.60 a±0.58	130.73 c±0.88	0.95 a±0.00
LSD 0.05	1.99	2.36	0.004	1.99	2.30	0.016
Foliar spray with KNO$_3$ (500 ppm) + MoO$_3$ (5 ppm)						
0.6 dS/m	84.67 b±7.76	126.33 a±0.88	0.67 b±0.06	87.27 b±7.76	122.73 a±0.88	0.71 b±0.07
6.2 dS/m	84.33 b±1.45	126.00 a±8.09	0.67 b±0.04	86.93 b±1.45	122.40 a±8.09	0.72 b±0.04
10.8 dS/m	126.67 a±1.77	135.00 a±0.58	0.94 a±0.01	129.27 a±1.77	131.40 a±0.58	0.98 a±0.02
LSD 0.05	16.1	16.28	0.15	16.12	16.28	0.16
Foliar spray with KNO$_3$ (500 ppm) + Fe-EDTA (5 ppm)+ MnCl$_2$ (5 ppm)+ MoO$_3$ (5 ppm)						
0.6 dS/m	57.33 b±0.88	187.00 a±1.16	0.31 c±0.00	59.93 b±0.88	183.40 a±1.16	0.33 c±0.00
6.2 dS/m	86.67 a±5.70	179.67 ab±4.98	0.48 b±0.02	89.27 a±5.70	176.07 ab±4.98	0.51 b±0.02
10.8 dS/m	89.67 a±1.20	168.33 b±2.73	0.53 a±0.00	92.27 a±1.20	164.73 a±2.73	0.56 a±0.01
LSD 0.05	11.7	11.57	0.04	11.76	11.57	0.03

Values are means of five replicates ± SE. Different letters in the same column are significantly different at *P* <0.05 level, as determined by Duncan's Multiple Range Test.

Table 2. Effect of foliar spray of mineral nutrients and irrigation water of different salinity levels on sodium and potassium compositions in *Gossypium hirsutum*.

6.3.2 Biochemical analysis (chlorophyll, protein and carbohydrate content)

Biochemical estimation (i.e. chlorophyll content, carbohydrates and proteins) performed in the leaf samples collected at grand period of growth are presented in Table 3. Chlorophyll content, carbohydrates and proteins were proportionally reduced with increase of salinity of sea salt irrigation water irrespective of any foliar spray medium. Foliar spray of KNO_3 with the mixture each single (Fe, Mn, Mo) microelement at control plants and those growing under both the salinities chlorophyll content, carbohydrates and proteins showed increase as compare to KNO_3 with the mixture single microelement, alone KNO_3 KCl, Urea, Fe, Mn and Mo spray. The spray of only water shows non-significant increase over control.

Treatment	Chlorophyll "a"	Chlorophyll "b"	Total Chlorophyll	Chlorophyll a/b	Total Sugars	Total Protein
	(mg/gm fresh weight)				(mg/gm dry weight)	
Non spray						
0.6 dS/m	0.637 [a]±0.003	0.903 [a]±0.003	1.490 [a]±0.005	0.738 [a]±0.005	25.450[a]±0.483	28.403 [a]±0.641
6.2 dS/m	0.531 [a]±0.082	0.827 [b]±0.012	1.308 [a]±0.091	0.666 [a]±0.098	23.283[a]±0.606	26.243 [a]±0.572
10.8 dS/m	0.308 [b]±0.006	0.509 [c]±0.000	0.767 [b]±0.006	0.613 [a]±0.014	20.100[b]±1.094	21.047 ±1.041
LSD 0.05	0.16	0.02	0.18	0.19	2.67	2.69
Foliar spray with water						
0.6 dS/m	0.667 [a]±0.003	0.933 [a]±0.003	1.520 [a]±0.005	0.768 [a]±0.005	25.950 [a]±0.483	28.903 [a]±0.641
6.2 dS/m	0.561 [a]±0.082	0.857 [b]±0.012	1.338 [a]±0.091	0.696 [a]±0.098	23.783 [a]±0.606	26.743 [a]±0.572
10.8 dS/m	0.338 [b]±0.006	0.539 [c]±0.000	0.797 [b]±0.006	0.643 [a]±0.014	20.600 [b]±1.094	21.547 [b]±1.041
LSD 0.05	0.16	0.02	0.18	0.19	2.67	2.69
Foliar spray with Fe-EDTA (5 ppm)						
0.6 dS/m	0.707 [a]±0.003	0.973 [a]±0.003	1.560 [a]±0.005	0.808 [a]±0.005	26.750 [b]±0.483	29.703 [a]±0.641
6.2 dS/m	0.601 [a]±0.082	0.897 [b]±0.012	1.378 [a]±0.091	0.736 [a]±0.098	24.583 [a]±0.606	27.543 [b]±0.572
10.8 dS/m	0.378 [b]±0.006	0.579 [c]±0.000	0.837 [b]±0.006	0.683 [a]±0.014	21.400 [a]±1.094	22.347 [a]±1.041
LSD 0.05	0.163	0.024	0.18	0.198	2.67	2.69
Foliar spray with MnCl₂ (5 ppm)						
0.6 dS/m	0.699 [a]±0.003	0.965 [a]±0.003	1.552 [a]±0.005	0.800 [a]±0.005	26.450 [b]±0.483	29.403 [b]±0.641
6.2 dS/m	0.593 [a]±0.082	0.889 [b]±0.012	1.370 [a]±0.091	0.728 [a]±0.098	24.283 [a]±0.606	27.243 [a]±0.572
10.8 dS/m	0.370 [b]±0.006	0.571 [c]±0.000	0.829±0.006	0.675 [a]±0.014	21.100 [a]±1.094	22.047 [a]±1.041
LSD 0.05	0.16	0.024	0.18	0.198	2.67	2.69
Foliar spray with MoO₃ (5 ppm)						
0.6 dS/m	0.687 [a]±0.003	0.953 [a]±0.003	1.540 [a]±0.005	0.788 [a]±0.005	26.340 [a]±0.483	29.293 [a]±0.641
6.2 dS/m	0.581 [a]±0.082	0.877 [b]±0.012	1.358 [a]±0.091	0.716 [a]±0.098	24.173 [a]±0.606	27.133 [a]±0.572
10.8 dS/m	0.358 [b]±0.006	0.559 [c]±0.000	0.817 [b]±0.006	0.663 [a]±0.014	20.990 [b]±1.094	21.937 [cb]±1.041
LSD 0.05	0.16	0.024	0.18	0.19	2.67	2.69
Foliar spray with Urea (1000 ppm)						
0.6 dS/m	0.722 [a]±0.003	0.988 [a]±0.003	1.575 [a]±0.005	0.823 [a]±0.005	27.020 [a]±0.483	29.973 [a]±0.641
6.2 dS/m	0.616 [a]±0.082	0.912 [b]±0.012	1.393 [a]±0.091	0.751 [a]±0.098	24.853 [a]±0.606	27.813 [a]±0.572
10.8 dS/m	0.393 [b]±0.006	0.594 [c]±0.000	0.852 [b]±0.006	0.698 [a]±0.014	21.670 [b]±1.094	22.617 [b]±1.041
LSD 0.05	0.16	0.02	0.18	0.19	2.67	2.69
Foliar spray with KCl (500ppm)						
0.6 dS/m	0.717 [a]±0.003	0.983 [a]±0.003	1.570 [a]±0.005	0.818 [a]±0.005	26.950 [b]±0.483	29.903 [b]±0.641
6.2 dS/m	0.611 [a]±0.082	0.907 [b]±0.012	1.388 [a]±0.091	0.746 [a]±0.098	24.783 [a]±0.606	27.743 [a]±0.572
10.8 dS/m	0.388 [b]±0.006	0.589 [c]±0.000	0.847 [b]±0.006	0.693 [a]±0.014	21.600 [a]±1.094	22.547 [a]±1.041
LSD 0.05	0.16	0.02	0.18	0.198	2.67	2.69

Treatment	Chlorophyll "a"	Chlorophyll "b"	Total Chlorophyll	Chlorophyll a/b	Total Sugars	Total Protein
	(mg/gm fresh weight)				(mg/gm dry weight)	
Foliar spray with KNO$_3$ (500 ppm)						
0.6 dS/m	0.732 [a]±0.003	0.998 [a]±0.003	1.585 [a]±0.005	0.833 [a]±0.005	27.450 [b]±0.483	30.403 [b]±0.641
6.2 dS/m	0.626 [a]±0.641	0.922 [b]±0.012	1.403 [a]±0.091	0.761 [a]±0.098	25.283 [a]±0.606	28.243 [a]±0.572
10.8 dS/m	0.403 [b]±0.006	0.604 [c]±0.000	0.862 [b]±0.006	0.708 [a]±0.014	22.100 [a]±1.094	23.047 [a]±1.041
LSD $_{0.05}$	0.16	0.02	0.18	0.198	2.67	2.69
Foliar spray with KNO$_3$ (500 ppm) +Fe-EDTA (5 ppm)						
0.6 dS/m	0.745 [a]±0.003	1.012 [a]±0.003	1.598 [a]±0.005	0.846 [a]±0.005	28.450 [a]±0.483	31.403 [a]±0.641
6.2 dS/m	0.639 [a]±0.082	0.936 [b]±0.012	1.416 [a]±0.091	0.775 [a]±0.098	26.283 [a]±0.606	29.243 [a]±0.572
10.8 dS/m	0.416 [b]±0.006	0.617 [c]±0.000	0.875 [b]±0.006	0.721 [a]±0.014	23.100 [b]±1.094	24.047 [b]±1.094
LSD $_{0.05}$	0.16	0.02	0.18	0.19	2.67	2.69
Foliar spray with KNO$_3$ (500 ppm) + MnCl$_2$ (5ppm)						
0.6 dS/m	0.742 [a]±0.003	1.009 [a]±0.003	1.596 [a]±0.005	0.843 [a]±0.005	28.340 [a]±0.483	31.293 [a]±0.641
6.2 dS/m	0.636 [a]±0.082	0.933 [b]±0.012	1.414 [a]±0.091	0.772 [a]±0.098	26.173 [a]±0.606	29.133 [a]±0.572
10.8 dS/m	0.414 [b]±0.006	0.614 [c]±0.000	0.872 [b]±0.006	0.718 [a]±0.014	22.990 [b]±1.094	23.937 [b]±1.041
LSD $_{0.05}$	0.16	0.024	0.18	0.198	2.37	2.69
Foliar spray with KNO$_3$ (500 ppm) + MoO$_3$ (5ppm)						
0.6 dS/m	0.737 [a]±0.003	1.003 [a]±0.003	1.590 [a]±0.005	0.838 [a]±0.005	28.020 [b]±0.483	30.973 [b]±0.641
6.2 dS/m	0.631 [a]±0.082	0.927 [b]±0.012	1.408 [a]±0.091	0.766 [a]±0.098	25.853 [a]±0.606	28.813 [a]±0.572
10.8 dS/m	0.408 [b]±0.006	0.609 [c]±0.000	0.867 [b]±0.006	0.713 [a]±0.014	22.670 [a]±1.094	23.617 [a]±1.041
LSD $_{0.05}$	0.16	0.024	0.18	0.19	2.67	2.69
Foliar spray with KNO$_3$ (500 ppm) + Fe- EDTA (5ppm)+ MnCl$_2$ (5ppm) + MoO$_3$ (5ppm)						
0.6 dS/m	0.747 [a]±0.003	1.013 [a]±0.003	1.600 [a]±0.005	0.848 [a]±0.005	29.040 [a]±0.483	31.993 [b]±0.641
6.2 dS/m	0.641 [a]±0.082	0.937 [b]±0.012	1.418 [a]±0.091	0.776 [a]±0.098	26.873 [a]±0.606	29.833 [a]±0.572
10.8 dS/m	0.418 [b]±0.006	0.619 [c]±0.000	0.877 [b]±0.006	0.723 [a]±0.014	23.690 [a]±1.094	24.637 [a]±1.041
LSD $_{0.05}$	0.16	0.02	0.18	0.19	2.67	2.69

Values are means of five replicates ± SE. Different letters in the same column are significantly different at $P <0.05$ level, as determined by Duncan's Multiple Range Test.

Table 3. Effect of foliar spray of mineral nutrients and irrigation water of different salinity levels on chlorophyll, sugar and protein content in *Gossypium hirsutum*.

6.3.3 Electrical conductivity and pH of Soil

Changes in electrical conductivity and pH of irrigation water, leachate and soil was monitored at different stages of growth during the course of experiment and only data of at termination of experiment is presented in Table 4. Increase in EC was seen as the concentration of sea salt irrigation increased but in spite of good amount of salt drained out through leachate with subsequent irrigation. The schedule of irrigation kept the increase in EC of soil about twice that of irrigation water. The resultant EC of the rooting medium was about twice at low and thrice at high levels than threshold values 6.2 and10.8 dS/m. Foliar application seems to have non-significant effect on the above-mentioned parameter irrespective of any salinity.

Treatment	Irrigation Water		Leachate		Soil	
	EC_{iw} (dS/m)	pH	EC (dS/m)	pH	EC_e (dS/m)	pH
Nonspray						
0.6 dS/m	0.6 [c] ±0.06	7.34±0.07	3.20 [c] ±0.12	8.31 [a] ±0.17	1.53 [c] ±0.07	7.00 [a] ±0.81
6.2 dS/m	6.4 [b] ±0.09	8.22 [a] ±0.25	18.95 [b] ±0.09	8.21 [a] ±0.12	14.01 [b] ±0.09	8.00 [a] ±0.12
10.8 dS/m	10.9 [a] ±0.09	8.55 [a] ±0.16	24.51 [a] ±0.09	8.51 [a] ±0.02	21.03 [a] ±0.09	8.31 [a] ±0.02
LSD $_{0.05}$	0.71	0.608	0.90	0.42	1.16	1.63
Foliar spray with water						
0.6 dS/m	0.67 [b] ±0.03	7.31 [b] ±0.04	3.20 [c] ±0.12	8.34 [a] ±0.08	1.60 [c] ±0.06	8.14 [a] ±0.28
6.2 dS/m	6.2 [a] ±0.12	8.35 [a] ±0.20	17.89 [b] ±0.12	8.35 [a] ±0.18	14.33 [b] ±0.12	8.28 [a] ±0.18
10.8 dS/m	10.8 [a] ±0.09	8.40 [a] ±0.19	25.14 [a] ±0.09	8.55 [a] ±0.06	22.97 [a] ±0.09	8.21 [a] ±0.06
LSD $_{0.05}$	0.55	0.55	0.37	0.40	1.63	0.66
Foliar spray with Fe-EDTA (5ppm)						
0.6 dS/m	0.67 [c] ±0.03	7.34 [a] ±0.04	3.20 [c] ±0.12	8.34 [a] ±0.08	1.60 [c] ±0.06	8.23 [a] ±0.28
6.2 dS/m	6.3 [b] ±0.15	8.24 [a] ±0.22	18.54 [b] ±0.15	8.24 [a] ±0.09	14.07 [b] ±0.15	8.12 [a] ±0.09
10.8 dS/m	10.9 [a] ±0.09	8.54±0.16	24.12 [a] ±0.09	8.54 [a] ±0.02	22.98 [a] ±0.09	8.12 [a] ±0.02
LSD $_{0.05}$	1.65	1.85	0.38	1.66	1.63	1.62
Foliar spray with MnCl$_2$ (5 ppm)						
0.6 dS/m	0.77 [c] ±0.12	7.64 [b] ±0.22	3.27 [c] ±0.18	8.52 [a] ±0.23	1.83 [c] ±0.12	8.44 [a b] ±0.09
6.2 dS/m	6.3 [b] ±0.09	8.21 [a] ±0.19	17.89 [b] ±0.09	8.32 [a] ±0.06	14.12 [b] ±0.09	8.12 [a] ±0.06
10.8 dS/m	10.8 [a] ±0.15	8.72 [a] ±0.05	24.65 [a] ±0.15	8.72 [a] ±0.18	21.07 [a] ±0.15	8.52 [a] ±0.18
LSD $_{0.05}$	1.64	0.58	0.48	0.59	1.64	041
Foliar spray with MoO$_3$ (5 ppm)						
0.6 dS/m	0.61 [c] ±0.07	7.47 [b] ±0.01	3.24 [c] ±0.14	8.21 [b] ±0.12	1.57 [c] ±0.09	8.12 [b] ±0.10
6.2 dS/m	6.2 [b] ±0.17	8.25 [a] ±0.25	18.93 [b] ±0.17	8.24 [a] ±0.12	14.10 [b] ±0.17	8.02 [b] ±0.12
10.8 dS/m	10.8 [a] ±0.09	8.66 [a] ±0.12	25.63 [a] ±0.09	8.64 [a] ±0.06	21.12 [a] ±0.09	8.45 [a] ±0.06
LSD $_{0.05}$	1.64	0.44	0.61	0.27	1.64	0.24
Foliar spray with KNO$_3$ (500 ppm)						
0.6 dS/m	0.73 [c] ±0.09	7.45±0.01	3.37 [c] ±0.27	8.24 [a] ±0.12	1.57 [c] ±0.09	8.12 [a] ±0.10
6.2 dS/m	6.2 [b] ±0.10	8.34 [a] ±0.15	18.21 [b] ±0.10	8.34 [a] ±0.03	14.00 [b] ±0.10	8.11 [a] ±0.03
10.8 dS/m	10.78 [a] ±0.10	8.64 [a] ±0.16	24.12 [a] ±0.10	8.64 [a] ±0.06	21.90 [a] ±0.10	8.42 [a] ±0.06
LSD 0.05	1.68	0.36	0.65	0.33	1.65	0.43
Foliar spray with Urea (1000 ppm)						
EC: 0.6 dS/m	0.67 [c] ±0.07	7.52 [b] ±0.09	3.40 [a] ±0.30	8.60 [c] ±0.16	1.63 [a] ±0.15	8.73 [a] ±0.21
EC: 6.2 dS/m	6.27 [b] ±0.58	8.36 [a] ±0.09	14.97 [b] ±0.09	8.39 [a] ±0.05	14.52 [b] ±0.58	8.19 [a] ±0.05
10.8 dS/m	10.25 [a] ±0.58	8.55 [a] ±0.13	22.97 [a] ±0.09	8.59 [a] ±0.02	21.36 [a] ±0.58	8.39 [a] ±0.02
LSD $_{0.05}$	1.64	0.50	0.62	0.43	1.66	0.47
Foliar spray with KCl (500 ppm)						
0.6 dS/	0.72 [c] ±0.06	7.63 [b] ±0.10	3.34 [c] ±0.24	8.57 [a] ±0.18	1.73 [c] ±0.17	8.48 [a] ±0.20
6.2 dS/m	6.50 [b] ±0.58	8.23 [a] ±0.22	15.10 [b] ±0.17	8.27 [a] ±0.09	14.52 [b] ±0.58	8.07 [a] ±0.09
10.8 dS/m	10.20 [a] ±0.58	8.59 [a] ±0.08	23.00 [a] ±0.10	8.62 [a] ±0.07	22.65 [a] ±0.58	10.20 [a] ±0.07
LSD $_{0.05}$	1.64	0.29	0.85	0.49	1.65	0.33

Treatment	Irrigation Water		Leachate		Soil	
	EC_{iw} (dS/m)	pH	EC (dS/m)	pH	EC_e (dS/m)	pH
Foliar spray with KNO_3 (500 ppm) + Fe-EDTA (5 ppm)						
0.6 dS/m	0.61 [c]±0.06	7.62 [c]±0.10	3.34 [c]±0.24	8.51 [a]±0.18	1.73 [c]±0.17	8.41 [a]±0.20
6.2 dS/m	6.32 [b]±0.17	8.23 [b]±0.22	17.58 [b]±0.17	8.23 [a]±0.09	14.10 [b]±0.17	8.01 [a]±0.09
10.8 dS/m	10.7 [a]±0.10	8.54 [a]±0.08	25.63 [a]±0.10	8.63 [a]±0.07	21.00 [a]±0.10	8.41 [a]±0.07
LSD 0.05	1.63	0.29	0.85	0.49	1.65	0.33
Foliar spray with KNO_3 (500 ppm) + $MnCl_2$ (5 ppm)						
0.6 dS/m	0.61 [c]±0.07	7.51 [b]±0.09	3.40 [c]±0.30	8.64 [a]±0.16	1.63 [c]±0.15	8.71 [a]±0.21
6.2 dS/m	6.1 [b]±0.09	8.31 [a]±0.09	16.39 [b]±0.09	8.31 [a]±0.05	14.12 [b]±0.09	8.11 [a]±0.05
10.8 dS/m	10.8 [a]±0.09	8.51 [a]±0.13	23.36 [a]±0.09	8.51 [a]±0.02	21.03 [a]±0.09	8.31 [a]±0.02
LSD 0.05	1.63	0.505	0.62	0.43	1.66	0.46
Foliar spray with KNO_3 (500 ppm) + MoO_3 (5 ppm)						
0.6 dS/m	0.62 [c]±0.07	7.54 [b]±0.08	3.50 [c]±0.40	8.52 [a]±0.22	1.67 [c]±0.15	8.41 [a]±0.12
6.2 dS/m	6.2 [b]±0.10	8.24 [a]±0.09	18.65 [b]±0.10	8.22 [a]±0.09	14.12 [b]±0.10	8.01 [a]±0.09
10.8 dS/m	10.8 [a]±0.10	8.54 [a]±0.08	24.78 [a]±0.10	8.62 [a]±0.06	21.36 [a]±0.10	8.41 [a]±0.06
LSD 0.05	1.63	0.56	0.49	0.49	1.64	0.42
Foliar spray with KNO_3 (500 ppm) + Fe-EDTA (5 ppm) + $MnCl_2$ (5 ppm) + MoO_3 (5 ppm)						
0.6 dS/m	0.62 [c]±0.06	7.57 [a]±0.12	3.27 [c]±0.18	8.33 [a]±0.14	1.53 [c]±0.09	8.53 [a]±0.06
6.2 dS/m	6.2 [b]±0.03	8.17 [a]±0.20	19.36 [b]±0.03	8.12 [a]±0.12	14.17 [b]±0.03	8.97 [a]±0.12
10.8 dS/m	10.8 [a]±0.10	8.44 [a]±0.16	24.56 [a]±0.10	8.42 [a]±0.17	21.3 [a]±0.10	8.94 [a]±0.17
LSD 0.05	1.63	1.99	1.99	1.99	1.99	1.99

Values are means of five replicates ± SE. Different letters in the same column are significantly different at $P < 0.05$ level, as determined by Duncan's Multiple Range Test.

Table 4. Effect of foliar spray of mineral nutrients and irrigation water of different salinity levels on Electrical conductivity (EC) and pH of irrigation water, leachate and soil in *Gossypium hirsutum*.

7. Discussion

7.1 Vegetative growth

Growth is an end result between anabolic and catabolic reactions within a plant. Saline environment has shown reduction in growth depending upon their degree of salt tolerance. The degree of reduction in growth by increase in electric conductivity unit of growth medium (due to presence of salt) has also been worked out by different research work (Maas, 1986; Maas & Hoffmann, 1977).

Vegetative growth vigor as result of sea salt irrigation water and foliar application of minerals determined by measuring fresh and dry biomass in present investigations is described below:

Biomass production is a measure of net photosynthesis and factors limiting plant growth limited net photosynthesis (Reddy et al., 1997). (Kuznetsov et al., 1990) found that rhizosphere salinity of rooting medium caused decrease in the biomass production in cotton plants. Qadir & Shams (1997) reported decrease in biomass production in cultivars of cotton grown at ECe: 10 – 20 dS/m soil salinity.

Above-mentioned problem created by presence of extra sodium ions in root zone could be avoided if essential sodium antagonistic mineral are provide through foliar spray as shown by different workers in following references text. Oosterhuis (1998) reported that foliar feeding of a nutrient might actually promote the root absorption of the same nutrient. Spraying nutrients not only can increase the crop yield but also can reduce the quantities of fertilizer applied through soil Ahmad (1998). Being given through spray medium of single salt composition, there is an advantage of not facing the problems of ion antagonistism, which is encountered in mineral uptake through roots under saline environment.

The method of foliar application is practical only in those plants that are compassionate to aerial spray and are not injured by this treatment. Examples of plants which accept foliar application are orchid, forest trees, cereals crops like wheat, maize, rice and barley; oil seeds crops; potato, tomato sugar beet and many other vegetables (Kochhar & Krishnamorthy, 1988). Saline substrate is found to decline values of potassium in xylem vessels of plants (Wolf et al., 1990). Inhibition of cation uptake in presence of excessive sodium through root system with special reference to monovalent potassium ion is well documented in literature by, Lopez & Satti, 1996; in spinach, Chow et al., 1990 in fennel, Botella et al., 1997 in maize. Favorable growth response of including K^+ in composition of foliar spray has been demonstrated by many research workers and is confirmed by the work reported in present investigation even under saline environment.

Fageria (2001) reported following regarding reasonable supply of essential nutrients is one of the most significant factors in increasing crop yields. In crop plants, the nutrient relations are generally considered in terms of growth response and change in concentration of nutrients. Upon addition of two nutrients, an increase in crop yield that is more than adding only one, the interaction is constructive (synergistic). Similarly, if adding the two nutrients together produced fewer yields as compared to individual ones, the relations are unconstructive (antagonistic). However, most interactions are multipart; a nutrient interacting simultaneously with more than one nutrient this may induce deficiencies, toxicities, modified growth responses, and/or modified nutrient composition. Better understanding of nutrient interactions may be useful in understanding importance of balanced supply of nutrients and consequently improvement in plant growth or yields

Foliar application of essential microelements like iron, manganese and copper may be more practical than application to soil, where they are adsorbed on the soil particles and hence are less obtainable to the root system.

Selection of microelement was done on the basis of their specific role on plant growth. Iron forms two types iron containing protein haem proteins and iron sulphur - proteins in plant metabolism. Cytocrome are haem proteins, which are constituents of the redox system in chloroplast and mitochondria. While in case of iron sulphur proteins ferredoxin is the most prominent iron sulphur protein, which acts as an electron transmitter in number of basic metabolic, processes. In iron deficient leaves the rate of photosynthesis decreases unit leaf per area but not per unit chloroplast (Terry, 1980). Chelates of iron (III) and occasionally of iron (II) are therefore the dominant forms of soluble iron in soil and nutrient solution. As a rule iron (II) is the species taken up. Iron (III) therefore has to be reduced at the root surface before transport into the cytoplasm (Roemheld & Marschner, 1983).

Manganese is absorbed mainly as Mn (II) and translocated predominantly as the free divalent cation in the xylem from the root to the shoot. The specific role of Mn as a mineral nutrient is presumably related to its tightly bound form in metalloprotein, where it acts as a structural constituent, as an active binding site. The most well known and extensively studied function of Mn in green plants is its involvement in photosynthetic oxygen evolution. It is now established that Mn is required in both lower and high plants for the Hill reaction – the water splitting and oxygen evolving system in photosynthesis (Chenaie & Martin, 1968).

Molybdenum is a metal, it occurs in aqueous solution mainly as a molybdate oxyanion, MoO_4^{-2}, in its highest oxidized form [Mo (VI)]. The requirement of plants for Mo is lower than that for any of the other mineral nutrient. Nitrogenase and nitrate reductase are two well-defined enzymes containing Mo. Mo requirement of higher plants therefore depends on the mode of nitrogen supply.

In present investigations potassium in foliar spray medium was given in concentration 500ppm keeping in view the leaf morphology of cotton leaves. In addition, three bi and trivalent essential minerals (i.e. iron, manganese and molybdenum) were included in spray medium, considering possibility of inhibition in their uptake from sodium rich substrate thus not being sufficiently available for growth. Urea was included to see the effect of different sources for supply of nitrogen through foliar application on growth. The foliar spray of KNO_3 or KCl individually is expected to throw some light on the effect of nitrogen and chlorine on growth. The following discussion deals with the effect of foliar spray comprising of potassium and other micronutrients in plants growing at saline substrate with reference to our work.

Foliar nutrient spray had beneficial effect on plant fresh and dry biomass which persists even in salinity. Foliar spray of with KNO_3 with three micronutrient (Fe, Mn, Mo) was of highest order whereas, foliar spray of KNO_3 with Fe, KNO_3 with Mn and KNO_3 with Mo, occupies second, third and fourth position respectively. Foliar spray of alone KNO_3, Urea, KCl, Fe, Mn, Mo and water, occupies fifth, sixth, seven, eight, ninth, tenth and eleventh position respectively.

Kaya et al., (2001) reported that fresh biomass of Spinach significantly reduced at 60 mM salinity, but foliar sprays of 5 mM KH_2PO_4 mitigated the detrimental effect of high salt. Kaya et al., (2001a) reported the same results in Cucumber and Pepper and Leidi & Saiz, 1997 in Cotton. Foliar spray of Ca $(NO_3)_2$, $MnSO_4$ and K_2HPO_4 partially minimized the salt induced nutrient deficiency increased in dry matter grown in different salinity levels Sultana et al., (2002). Whereas, Bernardo Murillo-Amador et al., (2005) reported that salt-stressed plants had less dry matter in the root and shoot when sprayed with the foliar Ca $(NO_3)_2$ sprays.

Foliar spray of KNO_3 alone or in combination with one or three microelement the accumulation of dry matter increased in contrast to control nonspray, water spray or micro nutrient alone, indicating that toxic ions such Na^+, Cl^- in the leaves, may interfere with phloem loading restricting the uptake of nutrients from roots to shoots. The rate of foliar absorption of Cl^- increases in the following order: sorghum < cotton, sunflower < cauliflower < sesame, alfalfa, sugar beet < barley, tomato < potato, safflower (Maas et al., 1982). However, the above order does not apply to foliar injury. Thus, when nutrients are applied to the leaves, and restricts the inhibition due to toxic effect of Na^+, Cl^- or minimizing the salinity induced nutrient deficiency.

Findings in our experiment showed that both the sprays of 500 ppm KNO_3 and 500 ppm KCl resulted in significant growth promotion under non-saline as well as condition. In addition former up to greater extent and later in smaller extant show considerable inhibition in offsetting sodium-induced toxicity of saline rooting medium. Provision of nitrogen attached with potassium in KNO_3 may have contributed to this better performance, as some research worker considers chlorine attached with KCl being non-essential element is considered growth inhibitor in higher concentration. It is evident that salt stress has a significant effect on nitrogen nutrition in plants. Salinity reduces the uptake of NO^{-3} in many plant species mostly due to high Cl^- content of saline soil (Khan & Srivastava, 1998). Recent preliminary studies indicate that adequate levels of chloride in the nutrient solution may reduce the amount of nitrogen required without effecting plant growth or yield. The negatively charged chloride anion also acts as a counter ion to the positively charged cations in the cell. Chloride is involved in regulating turgor pressure and growth of cells and is important in drought resistance. Chloride may also be beneficial in disease prevention, especially of the roots, by promoting healthy growth of the plant while creating a root zone environment (pH and osmotic properties) detrimental to pathogens (disease causing organisms) Mengel & Kirkby (1982).

Supply of Nitrogen foliar application through KNO_3 and urea has shown betterment in growth irrespective at non-saline as well as saline rooting medium, but growth under spray of former salt was better than later. This effect could be probably due to presence of growth promoting essential mineral K^+ attached with it. Since plants do not directly utilize urea nitrogen in comparison with nitrate nitrogen during uptake. This behavior could be probably due to there being inorganic or organic nature. It appears that salt bearing sodium antagonistic potassium along with inorganic nitrogen provides a better spray material for promoting growth.

Irvin, 1995 reported the effect of foliar Nitrogen (N) applications on Blueberries, the N derived from the foliar sprays comprised only a small percentage of the total N in leaves, and leaves contained more foliar derived N than shoots. Plants did absorb more N from urea than KNO_3 applications. Nevin et al., (1990) reviewed urea foliar fertilization of avocado and found better growth with supply of foliar supplied urea.

Salinity stress has been reported stimulatory as well as inhibitory effects on the uptake of some micronutrients by plants. The uptake of Fe, Mn, Zn and Cu generally increases in crop plants under salinity stress (Alam, 1994). The detrimental effects of NaCl stress on the nutrition of bean plants are reflected in higher concentrations of Cl and Mn in roots and Cl, Fe and Mn in leaves and Cl and Fe in fruits Carbonell-Barrachina et al., (1998). Briefly, it is reasonable to believe that numerous salinity-nutrient interactions are occurring at the same time but whether these ultimately affect crop yield or quality depends upon the salinity level and composition of salts, the crop species, the nutrient in question and a number of environmental factors.

Foliar application of micronutrients showed encouraging effects on vegetative growth and nutrient uptake either before or after the salinization treatment El-Fouly et al., (2006). While in our studies spray of individual micronutrient (i.e. Fe, Mn, and Mo) show significant growth-promoting effect specially that of Fe in increasing various vegetative and reproductive growth parameters. Their grading in would be Mo <Mn <Fe respectively under control as well as high salinity level. Whereas spray of potassium in combination with individual micronutrient (i.e. Fe, Mn, and Mo) shows significant growth promoting specially affect specially that of Fe in increasing various vegetative and reproductive growth parameters. Its grading would be (KNO_3 + Mo) < (KNO_3 + Mn) < (KNO_3 + Fe) respectively. Similarly, when spray medium of K was done with all the three micronutrient Fe, Mn and Mo show significant increase at various vegetative and reproductive growth parameters in comparison with their individual spray of above-mentioned elements under control as well as high salinity level. Supply of essential mineral element (Nitrogen, Potassium, Chlorine, Iron, Zinc, Manganese and Molybdenum) contributed for an increase in vegetative parameters irrespective of Nonsaline and saline conditions in all the three plants studied in present investigations. The toxic effect of excessive sodium was of course inhibited due to spray of above-mentioned mineral nutrients.

7.2 Reproductive growth

The following discussion is based on reproductive growth vigor with reference to number of flowers, balls and seed per plant, weight of seed and lint / per plant, seed cotton yield, seed/ lint ratio.

Plants normally take up nutrients from soils sediments through their roots although nutrients can be also supplied to plants as fertilizers by foliar sprays. Dhingra et al., (1995) reported that salinity of rhizosphere has been found accountable for reduction in reproductive yield. Reduction could be cumulative effect of various factors such as decline in number of flowers (Bishnoi et al., 1990; Sharma, 1992) faulty development of pollen grain and ovules resulting improper fertilization and denature embryo, reduction in number of pods per plant and seeds per pod, production of shrived seeds etc. Kumar et al., (1980). Early flower initiation was noticed in present study at 6.2 dS/m in *Gossypium hirsutum* over control. Increased production of flowers alone does not help in achieving high yield both in terms of number of fruits or weight of seeds (Dhingra & Varghese, 1997).

Foliar spray of different nutrient solutions used in present investigations reduced the inhibitory effect of saline water irrigation on various reproductive parameters. No doubt these foliar spray were responsible for increasing reproductive growth in non saline medium as well, but inspite of the growth inhibition caused by salinity their application retained supremacy over the growth retarding toxic effects of excessive sodium in rooting medium. It appears that inhibition in reproductive yield due to salinity presented in terms of number and weight of seed per plant is reduced due to shy bearing of flowers, shedding of flowers and balls, development of pollen grain and ovules, fertilization, filling of seeds/ball etc.

According to Sarkar & Malik (2001) foliar spray of KNO_3 as well as Ca $(NO_3)_2$ exerted growth promoting effects on *Lathyrus sativus* L. (Grasspea). They further showed that foliar spray of at 0.50% KNO_3 during 50% flowering stage resulted in higher rate of pods formation /plant, increase in length of pod, number of seeds/pod and weight of 1000 seed in comparison with spray of 0.25 and 1.00% KNO_3 water spray and nonspray (control). However the spray of 0.406% Ca $(NO_3)_2$ gave result equivalent to 0.50% spray of KNO_3.

Brar & Tiwari (2004) reported increase in yield of cotton by 22%, 27% and 36% due to foliar application KCl, Urea and KNO_3 respectively. In the present investigation it is observed that number of flowers and balls per plant decreased at 6.2 and 10.8 dS/m respectively. The salinity of the rooting medium also reduces seed cotton yield 12.1% and 30.0% at 6.2 and 10.8 dS/m respectively. The foliar spray of KNO_3 along with mixture of three microelement (Fe, Mn, and Mo) occupies 1st position increasing seed cotton yield whereas the spray medium of KNO_3 with individual microelement Fe, Mn, Mo occupied 2nd, 3rd and 4th position respectively. The spray of all the above-mentioned individual nutrient namely KNO_3, Urea, KCl, Fe, Mn, Mo were capable of reducing the effect of sodium toxicity of rooting medium up to smaller extent but their spray was still promoting growth over water spray and nonspray treatments.

Hodgson & MacLeod (2006) reported proportionate increase in the yield of *Cotton* due foliar spray of 2.8, 5.9, 8.4 and 10.5 kg/hectare of Nitrogen. Ali et al., (2007) found increase in seed cotton yield by 6.31% and 12.30% due to extra supply of soil urea 50 and 75 kg/acre Urea through soil respectively as compare 25 kg/acre.

Table 5 A and B was compiled for the purpose of discussion out results presented in various figures of some important reproductive parameters to determine extent of promoting various spray medium. Taking into consideration "Seed *Cotton yield*" which is the main parameter for determining tonnage of production, one can reach to the following profitable salient features.

Concentrations of irrigation water	Seed Cotton yield (gm) /plant	Reduction percent in yield for seed cotton yield
0.6 dS/m	134.48	-
6.2 dS/m	92.16	31.64
10.8 dS/m	47.88	64.39

Table 5.A. Reduction percent Seed Cotton yield in Cotton plants undergoing sea salt irrigation water of different salinity levels.

Spray Treatment	0.6 dS/m		6.2 dS/m		10.8 dS/m	
	Seed Cotton yield (gm) /plant	% Increase	Seed Cotton yield (gm) /plant	%Increase	Seed Cotton yield (gm) /plant	%Increase
Non Spray	134.48		92.16		47.88	
Foliar Spray with water	149.40	9.99	99.28	7.17	62.00	22.77
Foliar Spray with Mo	163.20	17.60	123.00	25.07	74.40	35.65
Foliar Spray with Mn	177.48	24.23	144.76	36.34	84.32	43.22
Foliar Spray with Fe	200.64	32.97	162.24	43.20	104.96	54.38
Foliar Spray with KCl	224.40	40.07	183.70	49.83	120.56	60.29
Foliar Spray with Urea	253.44	46.92	215.94	57.32	138.18	65.35
Foliar Spray with KNO₃	278.72	51.75	255.68	63.95	161.16	70.29
Foliar Spray with KNO₃ + Mo	308.00	56.34	272.00	66.12	183.60	73.92
Foliar Spray with KNO3 + Mn	390.72	65.58	310.98	70.36	219.24	78.16
Foliar Spray with KNO₃ + Fe	419.64	67.95	338.92	72.81	254.98	81.22
Foliar Spray with KNO₃ + Fe +Mn + Mo	467.50	71.23	397.80	76.83	292.50	83.63

Promotion % calculates over the values obtained under nonspray treatment.

Table 5.B. Percent increase Seed Cotton yield due to foliar spray of different mineral nutrients in Cotton plants undergoing sea salt irrigation water of different salinity levels.

i.   Reduction in seed cotton yield found 31.46 % at 6.2 dS/m and 64.39 % at 10.8 dS/m under sea salt water irrigation.
ii.   The seed cotton yield under nonsaline condition in comparison with nonspray control plants after showing various figures of increase under different foliar spray medium shows a maximum of 71.23% when sprayed with the mixture of all the nutrients.
iii. Seed cotton yield in plants irrigated with sea salt solution 6.2 dS/m after showing various figures of increase under different foliar spray medium shows a maximum of 76.83%. Hence in reality total improvement in growth under above mentioned saline condition first by overcoming the toxic effect of salinity being 31.46%(Table 5 A), plus the improvement due to spray of a mixture of all the nutrient medium being 76.83% will be a total of 108.29% under above mentioned treatment.
iv.   Seed cotton yield in plants irrigated with sea salt solution 10.8 dS/m after showing various figures of increase under different foliar spray medium shows a maximum of 83.63%. Hence in reality total improvement in growth under above mentioned saline condition first by overcoming the toxic effect of salinity being 64.39%(Table 5 A), plus the improvement due to spray of a mixture of all the nutrient medium being 83.62% will be a total of 148.01% under above mentioned treatment.

The overall comparative pattern of increase in different reproductive growth parameters in relation to their interaction with irrigation of different sea salt concentration and spray of different various mineral elements studied is given below:

Non-spray< water spray< Mo < Mn< Fe< KCl<Urea< KNO$_3$ < (KNO$_3$ + Mo) < (KNO$_3$ + Mn) < (KNO$_3$+Fe) < (KNO3 +Fe + Mn + Mo)

7.3 Mineral analysis

The effect of sea salt irrigation water on presence of Na$^+$ and K$^+$ in aerial vegetative parts of cotton plants was undertaken to find out their uptake from roots at saline rhizosphere and visualize uptake of K given through leaves along with different mineral composition.

In present investigation Na$^+$ concentration significantly increased in both stem and leaf with increase in salinity of substrate at 6.2 and 10.8 dS/m. Humera (2003) reported increase in Na$^+$ content in different plant parts with increase in salinity levels of substrate in the different species of family Crucifarea. Increase of Na$^+$ in the plant parts could be due to many reasons. i) Roots may be unable to check entry of sodium and their upward translocation due to its excessive presence in the rooting medium. ii) Plants may respond to accumulate high sodium ions to maintain osmotic adjustments against the low water potential in the saline soil.

The concentration of K$^+$ significantly decreased in both stem and leaf of in above-mentioned with increase in salinity levels of rooting medium. The influx of Na$^+$ to the root competes with K$^+$ uptake, since the uptake mechanisms for both ions are similar (Niu et al., 1995) but Na$^+$ ions having lower atomic weight and less electron positivity have better opportunity for uptake. High concentrations of Na$^+$ in the rooting medium of plants have been reported having antagonistic effect on K$^+$ uptake (Greenway & Munns, 1980; Jeschke, 1984). Jafri, 1990; Ahmad et al., 2002 reported an increase in Na$^+$ and decrease in K$^+$ uptake in cotton plants with increase in salinity of rooting medium. Maggio Albino et al., 2007 found leaf Na$^+$ increases whereas potassium and calcium ions decreased in tomato plant at increasing salinity which indicates that possibility of adsorption of K$^+$ and other di and trivalent cations at root are reduce in the presence of higher levels of Na$^+$ in rooting medium. Accumulation of toxic ions such as Na$^+$ and Cl$^-$ was found accompanied by a reduction in K$^+$ content and the Na$^+$/ K$^+$ ratio of leaf blades in salt-sensitive sorghum increased with increase in salinity levels (Lacerda et al., 2003).

The failure to maintain required Na^+/ K^+ ratio reduces the survival potential of the plant under higher salinity regimes.

Some workers used K^+ / Na^+ ratio instead of Na^+/ K^+ ratio in their experiment and have shown that K^+, Na^+ decreases with increases in salinity (Akhavan-Kharazian et al., 1991; Cachorro et al., 1993).

The situation of providing monovalent K^+ and some essential di and trivalent ions through foliar uptake is changed in plants growing under high sodium rhizosphere; the following text throws some light with reference to present investigation.

Na^+ concentrations increased as the salinity level of irrigation water increased whereas K^+ concentrations were reduced. Foliar spray of potassium with the mixture each single (Fe, Mn, Mo) microelement at control plants and those growing under both the salinities at all the above mentioned plant parts occupy 1st position growth performance as compare to potassium with the mixture single microelement (Fe), (Mn) and (Mo), respectively having 2nd, 3rd and 4th position whereas alone potassium, Urea, KCl, Fe, Mn and Mo spray occupied 5th, 6th, 7th 8th , 9th and 10th position.

Na^+ / K^+ ratio have been discussed together due to application of many minerals in foliar medium, Na^+ / K^+ ratio in these plants under increasing salinities which adversely effect growth but the spray of foliar spray medium of different mineral nutrients alone or in combination with KNO^3 decreased this ratio suppressing the inhibitory effect of excessive sodium on growth.

Kaya et al., (2001) reported that a K^+ concentration of spinach was significantly reduced at 60 mM salinity, but foliar sprays of KH_2PO_4 mitigated the detrimental effect of high salt. Foliar spray of Ca (NO_3) $_2$, $MnSO_4$ and K_2HPO_4 in *rice* plant is reported to partially minimized the salt induced nutrient deficiency and increase potassium content grown in different salinity levels (Sultana et al. 2002). Kaya et al., (2007) while working on *Cucumis melo* found that 150 mM NaCl levels significantly increases Na^+ concentrations and decreases K^+ concentrations, but supplementary 5 mM $KNO^3$and 10 mM proline significantly ameliorated the adverse effects of salinity resulting increase in plant growth. Levent et al., (2007) reported in wheat cultivars that increasing levels of NaCl significantly increase in Na^+ concentrations and decrease K^+ concentrations but spray of soluble silicon significantly ameliorated the adverse effects of salinity resulting increase in plant growth.

7.4 Chlorophyll, protein and carbohydrate

The amount of chlorophyll, protein and carbohydrate was proportionally reduced with increase of salinity of irrigation water. Whereas the spray of different sodium antagonistic essential mineral elements recorded an increase in their quantity that suppressing the inhibitory effect of salt on the growth. However the pattern of decrease in the amount of these biochemicals persists proportionate to increase in salinity treatment. Changes in the quantity of above mentioned biochemical's in plants subjected to spray medium of different chemical composition under same as well as nonsaline environment is given below:

Foliar spray of potassium with the mixture each single (Fe, Mn, Mo) microelement at control plants and those growing under both the salinities Chlorophyll content, carbohydrates and proteins showed significant increase as compare to potassium with the mixture single microelement, alone KNO_3 KCl, Urea, Fe, Mn and Mo spray. The spray of only water shows non-significant increase over control. Decrease in chlorophyll content at high salt concentrations is reported by Ahmad & Abdullah (1979) in cotton. The mechanism of salt effect on pigment is not yet clearly understood but decrease in the leaf pigment under higher salinity of rooting medium is attributed to the inhibition of iron containing enzymes, which inactivate the biosynthesis of chlorophyll (Rubin and Chernavina, 1960). On the other hand increase in chlorophyll content is reported by (Reddy et al., 1992) in some salt tolerant plants under saline substrate. Kaya et al., (2007) have shown that 150 mM NaCl levels significantly decreased chlorophyll content, in *Cucumis melo* but supplementing 5 mM KNO_3 and 10 mM proline in spray medium significantly ameliorated the adverse effects of salinity resulting increase in plant growth. According to Sarkar & Malik (2001) reported that improvement in growth due to foliar spray of 0.5% KNO_3 in Grasspea at 0.50%, which resulted in sufficient, supply of Nitrogen, which increase, chlorophyll and protein content in plants. The concentration of some organic solutes such as proline, polyamines, amino acids, soluble sugars, and sugar alcohols increases in leaves under saline conditions, contributing in the osmoregulation of plants Kafi et al., (2003).

Diego et al., (2004) reported that total soluble carbohydrates increased only in roots but proline content was decreased in both roots and leaves of *Prosopis alba* under increasing saline rooting medium. Carbohydrates can be accumulated and used by leaves for osmotic adjustment under salt stress (Cheesman, 1988). Under saline conditions, an increase observed in soluble carbohydrate composition of olive leaves (Tattini et al., 1996).

7.4 Changes in EC and pH values in soil

Changes in EC and pH values in soil due to salt accumulation during saline water irrigation have been presented in tabulated form. An increase in ECe values with the increase of salt of irrigation water is evident. The presence of sodium in irrigation water increases the exchangeable sodium in the colloidal system of the soil, which results in the deterioration of soil physical properties, and affects the plant growth and productivity (El-Saidi, 1997). Increasing amount of EC in leachate shows that salts accumulated in soil due to saline water irrigation are being regularly washed down in subsequent irrigation. Hence the plants are in reality growing under resultant ECe of rhizosphere, which is about twice that of irrigation water which considering in terms of reported EC values at threshold points (Maas & Hoffmann, 1977). The foliar sprays of Sodium antagonistic essential minerals have definitely extended these limits.

8. Conclusion

Considerable improvement was observed by spray of essential minerals used in present investigations on various vegetative and reproductive growth parameters in *Gossypium hirsutum* raised at saline rooting medium created by increasing concentrations of sea salt irrigation. Their overall performance is concluded below:

i.   Irrigation with water of different sea salt (S.S) concentrations without any foliar spray resulted in growth inhibition in the order of increasing sea salt concentrations of rooting medium both on vegetative and reproductive parameter.

ii.   Foliar spray of only water under nonsaline as well as saline irrigation resulted in some growth promotion both on vegetative and reproductive parameter over their respective non-spray treatments, but the enhancement in growth due to foliar spray of various mineral compositions increased considerably.

iii. Control plants (non saline) as well as those undergoing sea salt irrigation provided with single salt spray of micronutrient (Fe/Mn/ Mo) show various degrees of increase at different vegetative and reproductive growth parameters in comparison with non-spray or those sprayed only with water, whereas plant growth among themselves was in the order of Fe> Mn> Mo respectively.

iv. Sprays of KNO_3 and KCl both have shown significant growth improvement both under non-saline and saline condition. Provision of nitrogen attached with potassium in KNO_3 may have contributed to this better performance, as chlorine attached with KCl is considered growth inhibitor. Provision of Potassium was definitely antagonistic to toxic sodium helping in water relation and intermediary metabolism.

v. Supply of Nitrogen in the foliar application through KNO_3 and urea has shown betterment in growth irrespective at non-saline as well as saline rooting medium, but growth under spray of former salt was better than later, probably due to presence of growth promoting K^+ attached with it.

vi. Spray of potassium nitrate in combination with individual micronutrient (i.e. Fe, Mn, and Mo) show significant growth promoting effect specially that of Fe in increasing various vegetative and reproductive growth parameters. Their grading would be $(KNO_3 + Fe)> (KNO_3 + Mn)> (KNO_3 + Mo)$ respectively.

vii. Spray medium of potassium nitrate in combination with all the three micronutrient Fe, Mn and Mo show significant increase at various vegetative and reproductive growth parameters in comparison with individual spray of above-mentioned elements.

viii. The overall comparative pattern of promotion in different vegetative and reproductive growth parameters in relation to interaction with irrigation of different sea salt solution and spray of different various mineral elements is given below:
Non-spray< water spray< Mo < Mn< Fe< KCl< Urea < KNO_3 < (KNO_3 + Mo) < (KNO_3 + Mn) < (KNO_3+Fe) < (KNO_3 +Fe + Mn + Mo)

ix. The amount of chlorophyll, protein and carbohydrate was proportionally reduced with increase of salinity of irrigation water but the spray of above mentioned sodium antagonistic essential mineral elements suppressed the inhibitory effect of salt and increased their quantities following the same order as mentioned for various growth parameters.

x. Ionic distribution indicated greater uptake of Na^+ to the aerial parts of the plants under increasing salinity, which adversely affected the Na^+ /K^+ ratio but the spray of sodium antagonistic different essential mineral elements decreased this ratio suppressing the inhibitory effect of salt thus increasing growth following the same order as mentioned for different growth parameters.

xi. An increase in ECe values of soil with increase in salt of irrigation water is evident. Increasing amount of EC in leachate shows that salt accumulated in soil due to saline water irrigation is being regularly washed down in subsequent irrigation. Hence the plants are in reality growing under resulted ECe of rhizosphere, which is about twice that of irrigation water under prevalent soil texture.

xii. In general, growth of *Gossypium hirsutum* was inhibited under soil salinity beyond their threshold values whereas foliar spray of potassium alone or with other essential microelement released sodium induced toxic effect, increasing growth vigor.

9. References

A.O.A.C. (1984). Official Methods of Analysis 10th Ed. Association Official Analytical Chemists, Washington DC., USA.2004.

Abou El-Nour, E.A.A. (2002). Can Supplemented Potassium Foliar Feeding Reduce the Recommended Soil Potassium ? Pak. Jour. Biol. Sci., 5(3): 259-262.

Ahmad, N. (1998). Foliar fertilization of cotton with potassium in the USA. Proc.Symp. " Foliar Fertilization: A Technique to /improve Production and Decrease Pollution"10-14 Dec. 1995. Cario, Egypt.pp.7-15. (El- Fouly, M.M., Abdalla, F.E. and Abdel- Maguid. eds.), Publ. NRC, Cario.

Ahmad, R. & Abdullah, Z. (1979). Saline agriculture under desert conditions. In: Advances in Desert And Arid Land Technology And Development (A. Bishay and W.G.Mc Ginnies, eds.), Vol. 1, pp. 593-618, Harwood Academic Publishers, N.Y.

Ahmad, R. & Ismail, S. 1993. Studies on selection of salt-tolerant plants for food, fodder and fuel from world flora. In: Leith, H. and AI-Masoum, A, ed., Towards the rational use of high salinity tolerant plants. Kluwer Academic Publishers, Vol. 2.295-304.

Ahmad, R., & M.H. Chang. (2000). Measures to institute, Las Cruces, New Mexico, US.

Ahmad, R. & Jabeen, R. (2005). Foliar spray of mineral elements antagonistic to sodium- A Technique to induce salt tolerance in plant growing under saline conditions. Pak. Jour. Bot., 37(4): 913-920.

Ahmad, S., Ashraf, M. & Khan, M.D. (2002). Intra-Specific variation for salt tolerance in Cotton (Gossypium hirsutum L.). In: Prospects for Saline Agriculture. (Ahmad, R and Malik, K.A., eds.), pp. 199-207, Kluwer Academic Publ., Dordrecht, London.

Akhavan-Kharazian,M.,Campbell, W.F., Jurinak, J.J. & Dudley, L.M. (1991). Calcium amelioration of NaCl effects on plant growth, chlorophyll, and ion concentration in Phaseolus vulgaris. Arid Soil Res. Rehab., 5: 9-19.

Alam, S.M. (1994). Nutrient by plants under stress conditions. In: Pessarakli, M (Ed.), Handbook of Plant and Crop Stress. Marcel Dekker, New York, pp.227-246.

Ali, M. A., Ali, M., Yar, K., Mueen-ud-Din & Yamin, M. (2007). Effect of nitrogen and plant population levels on seed cotton yield of newly introduced cotton variety CIM-497. Jour. Agric. Res., 45(4): 289-298.

Arnon D.I, & Hoagland D.R(1940). Crop production in artificial culture solutions and in soils with special reference to factors influencing yields and absorption of inorganic nutrients. Soil Sci 50:463–483.

Ashraf, M. (2004). Some important physiological selection criteria for salt tolerance in plants. Flora, 199: 361-376.

Awad, S.S., Edwards, D.G. & Campbell, L.C. (1990). Phosphorus enhancement of salt tolerance of tomato. Crop Sci., 30: 123-128.

Bayuelo-Jime nez, J. S., Debouck, D. G. & Lynch, J. P. (2002). Salinity Tolerance in Phaseolus Species during Early Vegetative Growth Crop Sci. 42:2184–2192.

Bernardo Murillo-Amador., Flores-Hernandez. A., Garcia-Hernandez, J. L., Valdez-Cepeda, R. D., Avila-Serrano, N. Y., Troyo-Dieguez, E. & Ruiz-Espinoza, F. H. (2005). Soil Amendment With Organic Products Increases the Production of Prickly Pear Cactus as a Green Vegetable (Nopalitos). Jour. PACD.,pp 97-109.

Bishnoi, U.R., Mays, D.A. & Fabasso, M. T. (1990). Response of no-till and conventionally planted grain sorghum to weed control method and row spacing. Plant Soil., 129(2):117-120.

Bonilla I., El-Hamdaoui A., & Bolaños L. (2004): Boron and calcium increase Pisum sativum seed germination and seedling development under salt stress. Plant and Soil, 267: 97–107.

Blanpied, G.D. (1979). Effect of artificial rain water pH and calcium concentration on the calcium and potassium in apple leaves. Hortscience 14:706708.

Botella, M.A., Martinez, V., Pardines, J. & CerdaA, A. (1997). Salinity induced potassium deficiency in maize plants. Jour. Plant Physiol. 150: 200-205.

Brar, M.S. & Tiwari, K.N. (2004). Boosting Seed Cotton Yields in Punjab with Potassium: A Review. Better Crops., 88(3):28-31.

Cachorro, P., Ortiz, A. & Cerda, A. (1993). Growth, water relations and solute composition of *Phaseolus vulgaris* L. under saline conditions. Plant Sci., 95(1):23-29.

Carbonell-Barrachina, A.A., Burlo, F. & Mataix, I. (1998). Response of bean micronutrient to arsenic and salinity. Jour. Plant Nutr, 21, 1287-1299.

Cheeseman, J.M.(1988). Mechanism of salinity tolerance in plants. Plant Physiol., 87: 547-550. Cheniae, G.M. & Martin, I.F. (1969). Photoreactivation of manganese catalyst in photosynthetic oxygen evolution. Plant Physiol., 44:351-360.

Cheniae, G.M. & Martin, I.F. (1969). Photoreactivation of manganese catalyst in photosynthetic oxygen evolution. Plant Physiol., 44:351-360.

Chow, W.S., Ball, M.C. & Anderson, J.M. (1990). Growth and photosynthetic responses of spinach to salinity: Implications of K. nutrition for salt tolerance. Aust. Jour. Plant Physiol., 17: 563-578.

Ciha, A., J. & Brun, W.A. (1978). Effect of pod removal on nonstructural carbohydrate concentration in soyabean tissue. Crop Sci. 18:773-776.

Conway G. (1997). The Doubly Green Revolution: Food for All in the Twenty-First Century. Ithaca, New York: Comstock Publishing Associates. Penguin Books.

Cramer, G.R. (1997). Uptake and role of ions in salt tolerance. In: Strategies for improving salt tolerance in higher plant. (Jaiwal, P.K., Singh, R.P. and Gulati, A., eds.), pp. 55-86. Oxford And IBH Publishing Co. Pvt. Ltd., New Delhi.

Cramer, G.R. & Novak, R.S. (1992). Supplemental manganese improves the relative growth, net assimilation and photosynthetic rates of salt-stressed barley. Physiol.Plant., 84: 600-605.

Cramer, G. R., Lynch, J., Läuchli, A., & Epstein, E. (1987) Influx of Na+, K+, and Ca2+ into roots of salt stressed cotton seedlings: effects of supplemental Ca2+. Plant Physiology 83, 510–516.

Dhingra, H.R. & Varghese, T.M. (1997). Flowering and sexual reproductive under salt stress. In: Strategies for improving salt tolerance in higher plants. (Jaiwal, P.K., Singh, R.P. and Gulati, A., eds.), pp. 221-245. Oxford And IBH Publishing Co. Pvt. Ltd., New Delhi.

Dhingra, H.R., Chhabra, S., Kajal, S. & Varghese, T.M. (1995). Interactive effect of some growth regulators and salinity applied at two growth stages on yield characteristics of chickpea. In: Proceedings International conferences on Sustainable Agriculture and Environmental, HAU, Hisar.

Dibb, D.W. & Thompson, W.R. (1985). Interaction of potassium with other nutrients. Jour. Agron., p. 515-533.

Diego, A.M., Marta, R.G., Carlos, A.M. & Macro, A.O. (2004). The effects of salt stress on growth, nitrate reduction and proline and glycinebetaine accumulation in *Prosopis alba*. Brazi. Jour. Plant Physiol., 16(1):39-46.

Duncun, D.B. (1955). Multiple range and multiple F-test. Biometrics. 11: 1-42.

El – Saidi, M. T. (1997). Salinity and its effects on growth, yield and some physiological processes of crop plants. In: Strategies for improving salt tolerance in higher plants. (Jaiwal, P.K., Singh, R.P. and Gulati, A., eds.), pp. 111-127. Oxford And IBH Publishing Co. Pvt. Ltd., New Delhi.

El-Fouly,M.M. & A.A. El-Sayad(1997). Potassium status in the oil and crops, recommendations and present use in Eygpt. Proc. of The Regional Workshop of the International potash Institute, 26-30 May, 1997,Bornovoa, Izmir, Turkey,, Ed. Johnston,A.E., 50-65.

El-Flouy, M.M. & Abou El-Nour, E.A.A. (1998). Registration and use of foliar fertilizers in Egypt. Proc. Sym. Foliar Fertilization: A Technique to Improve Production and Decrease Pollution 10-14 Dec., 1995. Cairo.(El-Flouy, M.M., Abdalla, F.E. and Abdel-Maguid, A.A. eds.).NRC. Cario. Egypt, pp 1-5.

El-Fouly, M. M., Moustafa, H. A. & Attia, K. A. (2006). Chemical composition of soyabeans from plants treated with growth retardants. Pest Sci.,1(5):189-190.

Epstein, E. (1977). Genetic potentials for solving problems of soil mineral stress: adaptation of crops to salinity. In: Plant adaptation to mineral stress in problem soils. (Wright, M. J. eds.), Cornell University Agricultural Experiment Station., pp.73-123. Ithaca, New York.

Fageria, V. D. (2001). Nutrient interactions in crop plants. Jour. Plant Nutri., 24(8):1269-1290.

Flowers, T.J. (1988). Chloride as a nutrient and as an osmoticum. In: Tinker B, Lauchli A, eds. Advances in plant nutrition , vol. 3. New York Praeger, 55-78.

Flowers, T.J. & Yeo, A.R. (1995). Breeding for salinity resistance in crop plants: Where next? Aust. Jour. Plant Physiol. , 22: 875-884.

Francois, L. E. (1984). Salinity effects on germination, growth and yield of turnips, Hort science, 19: 82-84.

Gomez, K.A. & Gomez, A.A. (1976). Statistical procedures for agricultural research with emphasis on rice. International Rice Research Institute. Los Banos, Phillipines, 294 pp.

Gooding, M. J. & Devies, W. P. (1992). Foliar urea fertilization of cereals. Nutrient Cycling in Agroeccosystem., 32: 209-222.

Grattan, S.R. & Grieve, C.M. (1999). Salinity - mineral nutrient relations in horticultural crops. Sci. Hort., 78: 127-157.

Greenway, H. & Munns, R. (1980). Mechanism of salt tolerance in non-halophytes, Annu.Rev,Plant Physiol., 31: 149-190.

Gregoriou, C., Papademetriou, M. & Christofides, L. (1983). Use of chelates for correcting iron chlorosis in avocados growing in calcareous soil in Cyprus. California Avocado Society Yearbook 67: 115-122.

Gary, Y. & Grigg, C.G.C.S, M.G. (1999). Foliar nutrition of turfgrasses: Superintendents can save money and reduce leaching by feeding grass through the leaves. GSM, 1-4. Growers' Association Yearbook, 19: 31-32.

Hartree, E. F. (1972). Determination of protein: A modification of the Lowry method that gives a linear photometric response. Analy. Bioche.,48(2):422-427.

Hawkins, H.J. & Lewis, Q.A.M. (1993). Effect of NaCl salinity, nitrogen form, calcium and potassium concentration on nitrogen uptake and kinetics in Triticum aestivum L. cv. gamtoos New Phtol., 124: 171-177.

Hewitt, E. J. (1966). Sand and Water Culture Methods Used in the Study of Plant Nutrition, Tech. Commun. No. 22 (Commonwealth Agric.Bureaux, Farnham Royal, U.K.), 2nd Ed.

Hoai N.T.T., Shim I.S., Kobayashi K., & Kenji U. (2003): Accumulation of some nitrogen compounds in response to salt stress and their relationships with salt tolerance in rice(Oryza sativa L.) seedlings. Plant Growth Regulation, 41: 159–164.

Hodgson, A.S. & MacLeod, D.A. (2006). Effects of foliar applied nitrogen fertilizer on cotton waterlogged in cracking grey clay. Aust. Jour. Agric. Res., 38(4): 681-688.

Homaee.M; Feddes. R.A & Dirksen, C. (2002). Simulation of root water uptake. II. Non-uniform transient water stress using different reduction functions. *Agricultural Water Manag.* 57, 111-126.

Howard, D.D., Essington, M.E., Gwathmey, C.O. & Percell, W.M. (2000). Buffering of Foliar Potassium and Boron Solutions for No-tillage Cotton Production. Jour. Cotton Sci., 4: 237-244.

Humera, G. (2003). Salt tolerance in canola with special reference to its reproductive physiology under saline conditions. Ph.D. Thesis, Department of Botany, University of Karachi.

Irvin, E. W. & Kwantes, M. (1995). Ontogenic changes in seed weight and carbohydrate composition as related to growth of cucumber (Cucumis sativus L.) fruit.Scientia Hort., 63(3-4):155-165.

Islam, A., Sayeed, A., Absar, N., Ibrahim, M., Mondal, H. & Alam, S. (2003). Effect of NPK fertilizers and cowdung in combination with foliar spray of chemicals on growth and quality of Jute plant. Jour. Biol. Sci., 3(11):1016-1025.

Jabeen, R. (2009). Foliar supply of sodium antagonistic essential minerals on broad leaf plants to make them grow under saline condition. Ph.D. Thesis, Department of Botany, University of Karachi. http://eprints.hec.gov.pk/6226/

Jafri, A. Z. (1990). Physiology of salt tolerance in cotton. Ph.D. Thesis, Department of Botany, University of Karachi.

Jamro, G. H., Kazi, B. R., Oad, F. C., Jamali, N. M. & Oad, N. L. (2002). Effect of Foliar Application of Micro Nutrients on the Growth Traits of Sugarcane Variety Cp-65/357 (Ratoon Crop). Asian Jour. Plant Sci.,1(4): 462-463.

Jeschke, W.D. (1984). K+- Na+ exchange at cellular membranes intracellular compartmentation of cations and salt tolerance. In: Salinity tolerance in plants Strategies for crop improvement, (Staple, R. C. and Toenniessen, G. H., eds.), 37 p., John Wiley and Sons. New York.

Kafi, M., Stewart, W. S. & Borland, A. M. (2003). Carbohydrate and Proline Contents in Leaves, Roots, and Apices of Salt-Tolerant and Salt-Sensitive Wheat Cultivars1.Russian Jour. Plant Physiol., 50(2):155-162.

Kanan, S. (1980). Mechanisms of foliar uptake of plant nutrients: accomplishments and prospects. J. Plant Nutr. 2: 717735. Bl a n p i e d G.D. 1979. Effect of artificial rain water pH and calcium concentration on the calcium and potassium in apple leaves. Hortscience 14: 706708.

Kanwar, J.S. & Randhawa, N.S. (1967). Micronutrients research in soil and plants in India. A review Indian Council of Agric. Res. New Dehli.

Kassab, A.O. (2005). Soil moisture stress and micronutrients foliar application effects on the growth and yield of mung bean plants. Jour.Agric.Sci., 30:247-256.

Kaya, C., Higgs, D. & Kirnak, H. (2001). The effects of high salinity (NaCl) and supplementary phosphorus and potassium on physiology and nutrition development of spinach Bulg. Jour. plant physiol., 27:(3–4), 47–59.

Kaya, C., Kirnak, H. & Higgs, D. (2001a). The effects of supplementary potassium and phosphorus on physiological development and mineral nutrition of cucumber and pepper cultivars grown at high salinity (NaCl). J. Plant Nutr., 24(9): 25-27.

Kaya, C., A.L. Tuna, A. Muhammad & H.A. Hunlu (2007). Improved salt tolerance of melon(Cucumis milo L.) by the addition of proline and potassium nitrate. Environmental and Experimental Botany, 60(3): 397-403.

Khan, M.G & Srivastava, H.S. (1998). Changes in Growth and Nitrogen Assimilation in Maize Plants Induced by NaCl and Growth.Biol.Plantr., 41(1):93-99.

Kochhar, P.L. & Krishnamorthy, H. N. (1988. A text Book of Plant Physiology, pp:174, 175, 165-168, 179,180.

Kumar, J., Gowda, C.L.L., Saxena, N.P., Sethi., S.C. & Singh, U. (1980). Effects of salinity on the seed size and germinability of chickpea and protein content. Int. Chickpea Newsletter, 3:10.

Kuznetsov, V. V., Roschupkin, B. V., Khydrove, B.T. & Borisova, N. N. (1990). Interaction between starting and adaptive resistance of plants under conditions of salinity. Doklady. Bot. Sci. 313: 63-69.

Lacerada, C.F., Cambraia, J., Oliva, M.A. & Ruiz, H.A. (2003). Plant growth and solute accumulation and distribution in two sorghum genotypes, under NaCl stress. Rev. Bras. Fisiol. Veg., 13:270-284.

Leidi, E.O. & Saiz, J.F. (1997). Is salinity tolerance related to Na+ accumulation upland cotton (Gossypium hirsutum L.) seedling. Plant Soil, 190:67-75.

Levent, T. A., aya, C., Ashraf, M., Altunlu, H ., Yokas, I. & Yagmur, B. (2007). The effects of calcium sulphate on growth, membrane stability and nutrient uptake of tomato plants grown under salt stress. Environ. Exper. Bot.,59(2):173-178 .

Little, T.M. & Hills, F.J. (1975). Statistical methods in agriculture research (2nd print) Uni.of California, 242 pp.

Lopez, M.V. & Satti, S.M.E. (1996). Calcium and potassium-enhanced growth and yield of tomato under sodium chloride stress. Plant Sci. 114: 19-27.

Maathuis, F. J. M. & Amtmann, A. (1999) K+ nutrition and Na+ toxicity: the basis of cellular K+/Na+ ratios. Annals of Botany 84, 123– 133.

Marcar, N.E., Ismail, S., Hossain, A.K.M.A., & Ahmad, R. (1999). Trees, shrubs and grasses for saltlands:an annotated bibliography. ACIAR Monograph No. 56, 316 pp.

Maas, E.V. (1986). Crop tolerance to saline soil and water. In: Prospects for Biosaline Research, (Ahmad, R. and Pietro, A.S., eds.), pp. 205-219.

Maas, E.V. & Hoffmann, G.J. (1977). Crop salt tolerant- Current assessment. Jour. Irrig. and Drainage Div, ASCE, 103: 115-134.

Maas, E.V., Clark, R.A. and Franciose, L.E. (1982). Sprinkling-induced foliar injury to pepper plants: Effects of irrigation frequency, duration and water composition. Irrig.Sci., 3: 101-109.

Maas, E.V. (1996). Plant Response to Soil Salinity. In 4th National conference and Workshop on the"Productive use and Rehabilitation of Saline Lands" held at Albany, western australia, 25-30 March, 1996. pp 385-391.

Machlachlam, S. & Zalik, S. (1963). Plastids structure, cholorophyll concentration and frr amino acid composition of cholorophyll mutant of Barley. Can. Jour. Bot., 41: 1053-1062.

Maggio, A. , Raimondi, G., Martino A. & De Pascale, S. (2007). Salt stress response in tomato beyond the salinity tolerance threshold. Environ. Experim. Bot., 59(9-3): 276-282.

Mengel, K. & Kirkby, E.V. (1982). Principles of Plant nutrition, 3rd ed. Inter. Potash Inst. Bern.

Mengel, K. & Kirkby, E.A. (1987). Principles of plant nutrition. 4th Edition. Int. Potash Inst. Worblaufen-Bern, Switzerland.

Munns, R. (1993). Physiological processes limiting plant growth in saline soils: Some dogmas and hypothesis. Plant Cell Environment., 16: 15-24

Munns, R. (2002). Comparative physiology of salt and water stress. Plant Cell and Environ.25: 239-250.

Munns, R. & TermaAt, A. (1986). Whole plant responses to salinity. Aust.J. plant physiol., 13; 143-160.

Neumann, P.M. (1997). Salinity resistance and plant growth revisited. Plant cells and Environ., 20:1193-1198.

Nevin, J.M., Lovatt, C.J. & Embleton, T.W. (1990). Problems with urea-N foliar fertilization of avocado. Acta Hort., 275: 535-541.

Niu, X., Bressan, R. A., Hasegawa, P. M. & Pardo, J. M.(1995). Ion Homeostasis in NaCl Stress Environments' Plant Physiol. 109: 735-742.

Oosterhuis, D.M. (1998). Foliar fertilization of cotton with potassium in the USA. Pro. Symp. " Foliar Fertilization: A Technique to /improve Production and Decrease Pollution" 10-14 Dec. (1995). Cario, Egypt.pp.49-64. (El- Fouly, M.M., Abdalla, F.E. and Abdel-Maguid. eds.), Publ. NRC, Cario.

Owen, S. (2001). Salt of the earth. Genetic engineering may help to reclaim agricultural land lost due to salinization. *EMBO Rep.*, 2: 877-879.

Paridaa, A.K. & Dasa, A.B. (2005). Salt tolerance and salinity effects on plants: A Review.Ecotoxic. and Environm., 60(3):324-349.

Qadir, M. & Shams, M. (1997). Some agronomic and physiological aspects of salt tolerance in cotton (Gossypium hirsutum L.), Jour. Agron. Crop Sci., 179: 101-106.

Rajput, A.L., Singh, D.P. & Singh, S.P. (1995). Effect of soil and foliar application of nitrogen and zinc with farmyard manure in late sown wheat. Ind. Jour. Agron., 40:598-600.

Reddy, M. P., Rao, U.S. & Iyengar, E.R.R. (1997). Carbon metabolism under salt stress. In : Strategies for improving salt tolerance in higher plant. (Jaiwal, P.K., Singh, R.P. and Gulati, A., eds.), pp. 159-190. Oxford And IBH Publishing Co. Pvt. Ltd., New Delhi.

Reed, D.W. , Tu k e y J .R. 1978. Effect of pH on foliar absorption of phosphorus compounds by chrysanthemum. J. Amer. Soc. Hort. Sci. 103: 337340.

Rengel, Z. (1992). The role of calcium in salt toxicity. Plant Cell Environ. 15: 625-632.

Romheld, V. & Marschner, H. (1983). Mechanism of iron uptake by peanut plants I. Fe^{+3} reduction, chelate splitting, and release of phenolics. Plant Physiol.,71: 949-954.

Rubin, B.A. & Ladygina, M.E. (1959). Effect of streptomycin on greening of seedlings. Doklady AN SSR, 124 (5): 1163-1166.

Salama, Z., Shaaban, M.M. and Abou El-Nour,E.A. (1996). Effect of iron foliar application on increasing tolerance of maize seedlings to saline irrigation water. Egyptian Jour. Of Appl. Sci., 11:169-175.

Sarkar, R.K. & Malik, G.C. (2001). Effect of foliar spray of KNO$_3$ and Ca(NO$_3$)^2 on Grasspea (Lathyrus sativus L.) grown in rice fallows. Lathyrus Lathyrism Newsletter 2: 47-48.

Shannon, M.C, & Noble, C.L. (1990). Genetic approaches for developing economic salt-tolerant crops. In: Agricultural Salinity Assessment and Management. (Tanji, K.K. eds.), 71: 161-184.

Sharma, P.K. (1992). Study on reproductive behovior of mungbean [Vigna radiata (L.) Wilczek]under saline conditions. M.Sc.Thesis, HAU, Hisar.

Sultana, N., Ikeda, T. & Kashem, M.A. (2001). Effect of foliar spray of nutrient solutions on photosynthesis, dry matter accumulation and yield in seawater – stressed rice. Envirn. And Exper. Bot. 46: 129-140.

Sweby, D.L., Huckett, B.I. & Watt, M.P. (1994). Effect of nitrogen nutrition on salt stressed Nicotiana tobacum var. samsun in vitro plantlets. Jour.Exp. Bot., 45: 995-1008.

Tanji, K.K. (1990). Nature and extent of agricultural salinity. In: Agricultural Salinity Assessment and Management (Tanji , K K eds.), pp. 1-13. ASCE, New York.

Tattini, M., Gucci, R., Romani, A., Baldi, A. & Everard, J.D. (1996). Changes in non-structural carbohydrates in olive (Olea europaeaa L.) leaves during root zone salinity stress. Plant Physiol., 98: 117-124.

Terry, N. (1980). Limiting factors in photosynthesis I. Use of iron stress to control photochemical capacity in vivo. Plant Physiol., 65: 114-120

Tester, N. & Davenport R. (2003). Na$^+$ tolerance and Na$^+$ transport in higher plants. Ann. Bot., 91: 503-527.

U. S. Salinity Laboratory Staff. (1954. Diognosis and improvement of saline and alkali soils. USDA. Handbook 60, Wahington, D.C., U.S.A.

UN, Millennium, (2000). UN, Millennium Declaration, UN. A.Res:55/2

Weinberg, R. (1975. Effects of growth in highly salinized media on the enzymes of the photosynthetic apparatus in pea seedlings. Plant Physiol., 56: 8-12.

Wolf, O., Munns, R., Tonnet, W.D. & Jeschke, W.D. (1990). Concentration and transport of solutes in xylem and phloem along the leaf axis of NaCl- treated Hordeum vulgare (L). Jour. Exp. Bot., 41:1133-1141.

Yildirim E., Karlidag H., &Turan M. (2009): Mitigation of salt stress in strawberry by foliar K, Ca and Mg nutrient supply. Plant, Soil and Environment, 55: 213–221.

Effluent Quality Parameters
for Safe use in Agriculture

Hamid Iqbal Tak[1], Yahya Bakhtiyar[2], Faheem Ahmad[1] and Arif Inam[3]
[1]Department of Biological Sciences, Faculty of Agriculture, Science and Technology,
North-West University, Mafikeng Campus, Mmabatho,
[2]Department of Zoology, Jammu University,
[3]Department of Botany, Aligarh Muslim University,
[1]South Africa
[2,3]India

1. Introduction

"When the well is dry, we know the worth of water."
Benjamin Franklin, (1706-1790), Poor Richard's Almanac, 1746

Fast depletion of groundwater reserves, coupled with severe water pollution, has put governments all over the world in a difficult position to provide sufficient fresh water for our daily use. Ismail Serageldin vice president of World Bank in 1995 predicted that "if the wars of this century were fought over oil, the wars of the next century will be fought for water". Thus it signifies the role water is going to play in the current century we live in. At the same time, the need for sustained food production to feed the hungry mouths of the ever increasing population is apparent. In many arid and semi-arid countries since water is becoming increasingly scarce resource and planners are forced to consider alternate sources of water which might be used economically and effectively. The use of wastewater (WW) for crop irrigation as an alternative for effluent water disposal and for freshwater (FW) usage is common worldwide in countries in which water is scarce. Disposal of wastewater is also a problem of increasing importance throughout the world including India. Both the need to conserve fresh water and to safe and economically dispose of wastewater makes its use in agriculture a very feasible option. Furthermore, wastewater reuse may reduce fertilizer rates in addition to low cost source of irrigation water. In many parts of the world, treated municipal wastewater and raw sewage wastewater and even industrial wastewater has been successfully used for the irrigation of various crops (Asano and Tchobanoglous 1987, Adriel et al., 2007; Tak et al., 2010). It is well known that the enteric diseases, anaemia and gastrointestinal illnesses are high among sewage wastewater farmers. In addition, the consumers of vegetable crops which are eaten uncooked and grown without any treatment are also at risk. This chapter particularly envisages the review on the safe and quality parameters of wastewater for sustainable use in agriculture.

The use of sewage effluents for agricultural irrigation is an old and popular practice in agriculture (Feigin et al., 1984). Irrigation with wastewater has been used for three purposes:

i.   complementary treatment method for wastewater (Bouwer & Chaney, 1974);
ii. use of marginal water as an available water source for agriculture (Al-Jaloud et al., 1995; Tanji, 1997) – a sector demanding ~ 70% of the consumptive water use.

iii.   use of wastewater as nutrient source (Bouwer & Chaney, 1974; Vazquez- Montiel et al., 1996) associated with mineral fertilizer savings and high crop yields (Feigin et al., 1991; Tak et al., 2010).

Irrigated agriculture is dependent on an adequate water supply of usable quality. Water quality concerns have often been neglected and the situation is now changing in many areas. To avoid problems when using these poor quality water supplies, there must be sound planning to ensure that the quality of water available is put to the best use. The objective of this article is to help the reader in better understanding of the effect of water quality upon soil and crops and to assist in selecting suitable alternatives to cope with potential water quality related problems that might reduce production under prevailing conditions of use. Thus knowledge of irrigation water quality is critical in understanding management for its long-term usage and productivity. Conceptually, water quality refers to the characteristics of water that will influence its suitability for a specific use, i.e. how well the quality meets the needs of the user. Quality is defined by certain physical, chemical and biological characteristics. Even a personal preference such as taste is a simple evaluation of acceptability. For example, if two drinking waters of equally good quality are available, people may express a preference for one supply rather than the other; the better tasting water becomes the preferred supply. In irrigation water evaluation, however, the emphasis is placed on the chemical and physical characteristics of water and only rarely is any other factor considered important. There have been a number of different water quality guidelines related to use of wastewater in agriculture. Each has been useful but none has been entirely satisfactory because of the wide variability in field conditions. The guidelines presented in this paper have also relied on previous ones but are modified for evaluating and managing water quality-related problems of irrigated agriculture.

2. Irrigation water quality criteria

Soil scientists use various physico chemical parameters to describe irrigation water effects on crop production and soil quality. These include, Salinity hazard - total soluble salt content, Sodium hazard - relative proportion of sodium to calcium and magnesium ions, pH - acidic or basic, Alkalinity - carbonate and bicarbonate, Specific ions: chloride, sulfate, boron, and nitrate. However, another potential irrigation water quality parameter that may affect its suitability for agricultural system is microbial pathogens, which has often been neglected.

2.1 Salinity hazard /electrical conductivity

The most influential water quality guideline on crop productivity is the water salinity hazard as measured by electrical conductivity (EC_w). The primary effect of high EC_w water on crop productivity is the inability of the plant to compete with ions in the soil solution for water a condition known as Osmotic drought (physiological drought). Higher the EC, lesser is the water available to plants, even though the soil may appear to be wet. Because plants can only transpire "pure" water, usable plant water in the soil solution decreases dramatically as EC increases. An actual yield reduction from irrigation with high EC water varies substantially. Factors influencing yield reductions include soil type, drainage, salt type, irrigation system and management. The amount of water transpired through a crop is directly related to yield; therefore, irrigation water with high EC_w reduces yield potential (Table 1). Beyond effects on the immediate crop is the long term impact of salt loading through the irrigation water. Water with an EC_w of only 1.15 dS/m contains approximately 2,000 pounds of salt for every acre foot of water. You can use conversion factors in Table 2 to make this calculation for other water EC levels.

Limitations for use	Electrical Conductivity
	(dS/m)*
None	≤0.75
Some	0.76 - 1.5
Moderate[1]	1.51 - 3.00
Severe[2]	≤3.00

*dS/m at 25°C=mmhos/cm
[1]Leaching required at higher range.
[2]Good drainage needed and sensitive plants may have difficulty at germination.

Table 1. General guidelines for salinity hazard of irrigation water based upon conductivity.

Salt-affected soils develop from a wide range of factors including: soil type, field slope and drainage, irrigation system type and management, fertilizer and manuring practices, and other soil and water management practice. However, the most critical factor in predicting, managing, and reducing salt-affected soils is the quality of irrigation water being used. Besides affecting crop yield and soil physical conditions, irrigation water quality can affect fertility needs, irrigation system performance and longevity, and how the water can be applied. Therefore, knowledge of irrigation water quality is critical to understanding what management changes are necessary for long-term productivity.

Electrical conductivity (EC) is the most convenient way of measuring water salinity. EC is determined as the reciprocal of the specific resistance (ohms.m) of the water sample corrected to a standard temperature, usually 25°C.The basic unit of EC in SI units is Siemens m^{-1} (previously mhos m^{-1}). Formerly water salinities were expressed in micro mhos cm^{-1}. Some useful conversions are:

1 mS m^{-1} = 0.01 m mho cm^{-1} =10 mμho cm^{-1}

e.g. a water may have EC = 2000 μmho cm^{-1} = 2 mmho cm^{-1} = 200 mS m^{-1}

Frequently EC is multiplied by a factor to obtain total soluble salts (mass/volume) as an expression of salinity. There is however no unique factor that can be applied and the factor will vary with composition and concentration. Factors found for W.A. waters vary between 5.0 and 8.5 (EC in mS m^{-1}). Generally it is more convenient to use electrical conductivity as the measure of salt content as criteria are usually published in this form. Other terms that laboratories and literature sources use to report salinity hazard are: salts, salinity, electrical conductivity (EC_w), or total dissolved solids (TDS). These terms are all comparable and all quantify the amount of dissolved "salts" (or ions, charged particles) in a water sample. However, TDS is a direct measurement of dissolved ions and EC is an indirect measurement of ions by an electrode. Although people frequently confuse the term "salinity" with common table salt or sodium chloride (NaCl), EC measures salinity from all the ions dissolved in a sample. This includes negatively charged ions (e.g., Cl^-, NO_3^-) and positively charged ions (e.g., Ca^{++}, Na^+). Another common source of confusion is the variety of unit systems used with EC_w. The preferred unit is deciSiemens per meter (dS/m), however millimhos per centimeter (mmho/cm) and micromhos per centimeter (μmho/cm) are still frequently used. Conversions to help you change between unit systems are provided in Table 2.

Component	To Convert	Multiply By	To Obtain
Water nutrient or TDS	mg/L	1.0	ppm
Water salinity hazard	1 dS/m	1.0	1 mmho/cm
Water salinity hazard	1 mmho/cm	1,000	1 µmho/cm
Water salinity hazard	EC_w (dS/m) for EC <5 dS/m	640	TDS (mg/L)
Water salinity hazard	EC_w (dS/m) for EC >5 dS/m	800	TDS (mg/L)
Water NO_3N, SO_4-S,B applied	Ppm	0.23	lb per acre inch of water
Irrigation water	acre inch	27,150	gallons of water

Table 2. Conversion factors for irrigation water quality laboratory reports. Source: Bauder *et al.*, 2011

Definitions

Abbrev.	Meaning
mg/L	milligrams per liter
meq/L	milliequivalents per liter
Ppm	parts per million
dS/m	deciSiemens per meter
µS/cm	microSiemens per centimeter
mmho/cm	millimhos per centimeter
TDS	total dissolved solids

2.2 Sodium hazard/SAR

Although plant growth is primarily limited by the salinity (EC_w) level of the irrigation water, the application of water with a sodium imbalance can further reduce yield under certain soil texture conditions. Reductions in water infiltration can occur when irrigation water contains high sodium relative to the calcium and magnesium contents. This condition is termed "sodicity," and results from excessive accumulation of sodium in soil. Sodic water is not the same as saline water. Sodicity causes swelling and dispersion of soil clays, surface crusting and pore plugging. This degraded soil structure condition in turn obstructs infiltration and may increase runoff. Sodicity therefore, causes a decrease in the downward movement of water into and through the soil, and actively growing plants roots may not get adequate water, despite pooling of water on the soil surface after irrigation. The most common measure to assess sodicity in water and soil is called the Sodium Adsorption Ratio (SAR). The SAR defines sodicity in terms of the relative concentration of sodium (Na) compared to the sum of calcium (Ca) and magnesium (Mg) ions in a sample. The SAR assesses the potential for infiltration problems due to a sodium imbalance in irrigation water. The SAR is used to estimate the sodicity hazard of the water, where:

$$SAR = \frac{Na}{0.5\ (ca + mg)} \quad \text{and all concentrations are in meq/L}$$

SAR is a measure of the tendency of the irrigation water to cause the replacement of calcium (Ca) ions attached to the soil clay minerals with sodium ions (Na). Sodium clays have poor structure and develop permeability problems. The Residual sodium carbonate (RSC) is the measure in milli equivalents per litre (meq/L) of the excess of carbonates (CO_3) and bicarbonates (HCO_3) over magnesium (Mg) and calcium (Ca). With high RSC (>1.25) there is a tendency for Ca and Mg to precipitate in the soil, thus increasing the proportion of Na and increasing the SAR of the soil solution.

Potential for Water Infiltration Problem		
Irrigation water SAR	Unlikely	Likely
	$EC_w{}^*$ (dS/m)	
0-3	>0.7	<0.2
3-6	>1.2	<0.4
6-12	>1.9	<0.5.
12-20	>2.9	<1.0
20-40	>5.0	<3.0

*Modified from Ayers and Westcot. 1994. Water Quality for Agriculture, Irrigation and Drainage Paper 29, rev. 1, Food and Agriculture Organization of the United Nations, Rome.

Table 3. Guidelines for assessment of sodium hazard of irrigation water based on SAR and $EC_w{}^2$.

2.3 pH and alkalinity
The acidity or basicity of irrigation water is expressed as pH (< 7.0 acidic; > 7.0 basic). The normal pH for irrigation water ranges from 6.5 to 8.4. High pH's above 8.5 are often caused by high bicarbonate ($HCO_3{}^-$) and carbonate ($CO_3{}^{2-}$) concentrations, known as alkalinity. Calcium and magnesium ions become insoluble due to high carbonates and bicarbonates thereby leaving sodium as the dominant ion in solution. As also described in the sodium hazard section, this alkaline water could intensify the impact of high SAR water on sodic soil conditions. The main use of pH in a water analysis is for detecting an abnormal water. The normal pH range for irrigation water is from 6.5 to 8.4. An abnormal value is a warning that the water needs further evaluation. Irrigation water with a pH outside the normal range may cause a nutritional imbalance or may contain a toxic ion. Low salinity water (EC_w < 0.2 dS/m) sometimes has a pH outside the normal range since it has a very low buffering capacity. This should not cause undue alarm other than to alert the user to a possible imbalance of ions and the need to establish the reason for the adverse pH through full laboratory analysis. Such water normally causes few problems for soils or crops but is very corrosive and may rapidly corrode pipelines, sprinklers and control equipment. Any change in the soil pH caused by the water will take place slowly since the soil is strongly buffered and resists change. An adverse pH may need to be corrected, if possible, by the introduction of an amendment into the water, but this will only be practical in a few instances. It may be easier to correct the soil pH problem that may develop rather than try to treat the water.

Lime is commonly applied to the soil to correct a low pH and sulphur or other acid material may be used to correct a high pH. Gypsum has little or no effect in controlling an acid soil problem apart from supplying a nutritional source of calcium, but it is effective in reducing a high soil pH (pH greater than 8.5) caused by high exchangeable sodium. The greatest direct hazard of an abnormal pH in water is the impact on irrigation equipment. Equipment will need to be chosen carefully for unusual water.

2.4 Chloride

Chloride is a common ion in most of the irrigation waters. Although chloride is essential to plants in very low amounts however, it can cause toxicity to sensitive crops at high concentrations. Like sodium, high chloride concentrations cause more problems. The degree of damage depends on the uptake and the crop sensitivity. The permanent, perennial-type crops (tree crops) are more sensitive. Damage often occurs at relatively low ion concentrations for sensitive crops. It is usually first evidenced as marginal leaf burn and interveinal chlorosis. If the accumulation is great enough, reduced yields result. The more tolerant annual crops are not sensitive at low concentrations but almost all crops will be damaged or killed if concentrations are amply high.

2.5 Boron

Boron, unlike sodium, is an essential element for plant growth (Chloride is also essential but in such small quantities that it is frequently classed non-essential.) Boron is needed in relatively small amounts, however, if present in amounts appreciably greater than needed, it becomes toxic. For some crops, if 0.2 mg/l boron in water is essential, 1 to 2 mg/l may become toxic. Surface water rarely contains enough boron to be toxic but well water or springs occasionally may contain toxic amounts, especially near geothermal areas and earthquake faults. Boron problems originating from the water are probably more frequent than those originating in the soil. Boron toxicity can affect nearly all crops but, like salinity, there is a wide range of tolerance among crops. Boron toxicity symptoms normally show first on older leaves as a yellowing, spotting, or drying of leaf tissue at the tips and edges. Drying and chlorosis often progress towards the centre between the veins (interveinal) as more and more boron accumulates with time. On seriously affected trees, such as almonds and other tree crops which do not show typical leaf symptoms, a gum or exudate on limbs or trunk is often noticeable. Most crop toxicity symptoms occur after boron concentrations in leaf blades exceed 250–300 mg/kg (dry weight) but not all sensitive crops accumulate boron in leaf blades. For example, stone fruits (peaches, plums, almonds, etc.), and pome fruits (apples, pears and others) are easily damaged by boron but they do not accumulate sufficient boron in the leaf tissue for leaf analysis to be a reliable diagnostic test. With these crops, boron excess must be confirmed from soil and water analyses, tree symptoms and growth characteristics.

A wide range of crops were tested for boron tolerance by using sand-culture techniques (Eaton 1944). Previous boron tolerance tables in general use have been based for the most part on these data. These tables reflected boron tolerance at which toxicity symptoms were first observed and, depending on crop, covered one to three seasons of irrigation. The original data from these early experiments, plus data from many other sources, have recently been reviewed (Maas 1984). Table 4 presents this recent revision of the data. It is not based on plant symptoms, but upon a significant loss in yield to be expected if the indicated boron value is exceeded.

Sensitive		Moderately Sensitive	Moderately Tolerant	Tolerant
0.5-0.75	0.76-1.0	1.1-2.0	2.1-4.0	4.1-6.0
Peach	Wheat	Carrot	Lettuce	Alfalfa
Onion	Barley	Potato	Cabbage	Sugar beet
	Sunflower	Cucumber	Corn	Tomato
	Dry Bean		Oats	

Source: Mass (1984) Salt tolerance of plants. CRC Handbook of Plant Science in Agriculture. B.R. Cristie (ed.). CRC Press Inc.

Table 4. Boron sensitivity of selected Colorado plants (B concentration, mg/ L*)

2.6 Sulfate
The sulfate ion is a major contributor to salinity in many of the irrigation waters. As with boron, sulfate in irrigation water has fertility benefits, and most often irrigation water has enough sulfate for maximum production for most crops. Exceptions are sandy fields with <1 percent organic matter and <10 ppm SO_4-S in irrigation water.

2.7 Nitrate
Nitrogen in irrigation water (N) is largely a fertility issue, and nitrate-nitrogen (NO_3-N) can be a significant N source as it is found in most of the wastewaters throughout the world. The nitrate ion often occurs at higher concentrations than ammonium in irrigation water. Nitrogen is a plant nutrient and stimulates crop growth. Natural soil nitrogen or added fertilizers are the usual sources, but nitrogen in the irrigation water has much the same effect as soil-applied fertilizer nitrogen and an excess will cause problems, just as too much fertilizer would. If excessive quantities are present or applied, production of several commonly grown crops may get disturbed because of over-stimulation of growth, delayed maturity or poor quality. However, these problems can usually be overcome by good fertilizer and irrigation management. The sensitivity of crops varies with the growth stage. High nitrogen levels may be beneficial during early growth stages but may cause yield losses during the later flowering and fruiting stages (Tak et al., 2010). High nitrogen water can be used as a fertilizer early in the season. However, as the nitrogen needs of the crop diminish later in the growing season, the nitrogen applied to the crop must be substantially reduced. For crops irrigated with water containing nitrogen, the rates of nitrogen fertilizer supplied to the crop should be reduced by an amount very nearly equal to that available from the water supply. Regardless of the crop, nitrate should be credited toward the fertilizer rate especially when the concentration exceeds 10 ppm NO_3-N (45 ppm NO_3^-).

2.8 Trace elements and heavy metals
A number of elements are normally present in relatively low concentrations, usually less than a few mg/l, in conventional irrigation waters and are called trace elements. They are not normally included in routine analysis of regular irrigation water, but attention should be paid to them when using sewage effluents, particularly if contamination with industrial wastewater discharges is suspected. These include Aluminium (Al), Beryllium (Be), Cobalt (Co), Fluoride (F), Iron (Fe), Lithium (Li), Manganese (Mn), Molybdenum (Mo), Selenium

(Se), Tin (Sn), Titanium (Ti), Tungsten (W) and Vanadium (V). Heavy metals are a special group of trace elements which have been shown to create definite health hazards when taken up by plants. Under this group are included, Arsenic (As), Cadmium (Cd), Chromium (Cr), Copper (Cu), Lead (Pb), Mercury (Hg) and Zinc (Zn). These are called heavy metals because in their metallic form, their densities are greater than 4g/cc. Table 6 gives the insight on the recommended maximum concentration of trace elements in irrigation water.

2.9 Guidelines

Water quality criteria can never be absolute because soils, management and drainage can influence water suitability. There is, for example a ten−fold range in the salt−tolerance of plants which gives wide scope for utilizing water of different quality. Table 5 (FAO 1985) gives broad guidelines that have been developed for the preliminary evaluation of irrigation water quality. Where a water quality parameter is in the range of increasing problems more detailed investigation is required. For instance, in the range of increasing salinity problems (0.7 to 3.0 dS m^{-1}) more concern is required in the selection of plant species and precautions are needed to minimize salt injury. Where limitations are given the application of appropriate management methods may mean that the waters are still viable. The table thus avoids rigid classification methods which can at times be misleading. They emphasize the long-term influence of water quality on crop production, soil conditions and farm management, and are presented in the same format as in the 1976 edition but are updated to include recent research results. This format is similar to that of the 1974 University of California Committee of Consultant's Water Quality Guidelines which were prepared in cooperation with staff of the United States Salinity Laboratory. The guidelines are practical and have been used successfully in general irrigated agriculture for evaluation of the common constituents in surface water, groundwater, drainage water, sewage effluent and wastewater. A modified set of alternative guidelines can be prepared if actual conditions of use differ greatly from those assumed. Ordinarily, no soil or cropping problems are experienced or recognized when using water with values less than those shown for 'no restriction on use'. With restrictions in the slight to moderate range, gradually increasing care in selection of crop and management alternatives is required if full yield potential is to be achieved. On the other hand, if water is used which equals or exceeds the values shown for severe restrictions, the water user should experience soil and cropping problems or reduced yields, but even with cropping management designed especially to cope with poor quality water, a high level of management skill is essential for acceptable production. If water quality values are found which approach or exceed those given for the severe restriction category, it is recommended that before initiating the use of the water in a large project, a series of pilot farming studies be conducted to determine the economics of the farming and cropping techniques that need to be implemented. Table 5 is a management tool. As with many such interpretative tools in agriculture, it is developed to help users such as water agencies, project planners, agriculturalists, scientists and trained field people to understand better the effect of water quality on soil conditions and crop production. With this understanding, the user should be able to adjust management to utilize poor quality water better. However, the user of Table 5 must guard against drawing unwarranted conclusions based only on the laboratory results and the guideline interpretations as these must be related to field conditions and must be checked, confirmed and tested by field trials or experience. The guidelines are a first step in pointing out the quality limitations of a water supply, but this alone is not enough; methods to overcome or adapt to them are also needed. The guidelines do not evaluate the effect of unusual or special water constituents sometimes found in wastewater, such as pesticides and organics. However, suggested limits of trace element concentrations or normal irrigation water are given in Table 6. The World Health Organization (WHO) or a local health agency should be consulted for more specific information. Laboratory determinations and calculations needed to use the guidelines are given in Table 7 along with the symbols used. Analytical procedures for the laboratory determinations are given in several publications: USDA Handbook 60 (Richards 1954), Rhoades and Clark 1978, FAO Soils Bulletin 10 (Dewis and Freitas 1970), and Standard Methods for Examination of Waters and Wastewaters (APHA 1980). The method most appropriate for the available equipment, budget and number of samples should be used. Analytical accuracy within ±5 percent is considered adequate.

Potential irrigation problem	Units	Degree of restriction on use		
		None	Slight to moderate	Severe
Salinity				
EC_W	ds/m	<0.7	0.7-3.0	>3.0
Or				
TDS	mg/l	<450	450-2000`	>2000
Infiltration				
SAR and EC_W				
0-3		>0.7	0.7-0.2	<0.2
3-6		>1.2	1.2-0.3	<0.3
6-12		>1.9	1.9-0.5	<0.5.
12-20		>2.9	2.9-1.3	<1.3
20-40		>5.0	5.0-2.9	<3.0
Specific ion toxicity				
Sodium (Na)				
Surface irrigation	SAR	<3.0	3-9	>9
Sprinkler irrigation	me/l	<3.0	>3	
Chloride				
Surface irrigation	me/l	<4.0	4-10	>10
Sprinkler irrigation	m³/l	<3.0	>4.0	
Boron (B)	mg/l	<0.7	0.7-3.0	>3.0
Nitrogen (NO_3-N)	mg/l	<5.0	5.0-30	>30
Bicarbonate (HCO_3)	me/l	<1.5	1.5-8.5	>8.5
pH		Normal range 6.5-8.0		

Table 5. Guidelines for determination of water quality for irrigation Source FAO, 1985

Element	Recommended Maximum Concentration* (mg/l)	Remarks
Al(aluminium)	5.0	Can cause non-productivity in acid soils (pH < 5.5), but more alkaline soils at pH > 7.0 will precipitate the ion and eliminate any toxicity.
As (arsenic)	0.10	Toxicity to plants varies widely, ranging from 12 mg/l for Sudan grass to less than 0.05 mg/l for rice.
Be (beryllium)	0.10	Toxicity to plants varies widely, ranging from 5 mg/l for kale to 0.5 mg/l for bush beans.
Cd (cadmium)	0.01	Toxic to beans, beets and turnips at concentrations as low as 0.1 mg/l in nutrient solutions. Conservative limits recommended due to its potential for accumulation in plants and soils to concentrations that may be harmful to humans.
Co (cobalt)	0.05	Toxic to tomato plants at 0.1 mg/l in nutrient solution. Tends to be inactivated by neutral and alkaline soils.
Cr (chromium)	0.10	Not generally recognized as an essential growth element. Conservative limits recommended due to lack of knowledge on its toxicity to plants.
Cu (copper)	0.20	Toxic to a number of plants at 0.1 to 1.0 mg/l in nutrient solutions.
F (fluoride)	1.0	Inactivated by neutral and alkaline soils.
Fe (iron)	5.0	Not toxic to plants in aerated soils, but can contribute to soil acidification and loss of availability of essential phosphorus and molybdenum. Overhead sprinkling may result in unsightly deposits on plants, equipment and buildings.
Li (lithium)	2.5	Tolerated by most crops up to 5 mg/l; mobile in soil. Toxic to citrus at low concentrations (<0.075 mg/l). Acts similarly to boron.
Mn (manganese)	0.20	Toxic to a number of crops at a few-tenths to a few mg/l, but usually only in acid soils.
Mo (molybdenum)	0.01	Not toxic to plants at normal concentrations in soil and water. Can be toxic to livestock if forage is grown in soils with high concentrations of available molybdenum.
Ni (nickel)	0.20	Toxic to a number of plants at 0.5 mg/l to 1.0 mg/l; reduced toxicity at neutral or alkaline pH.
Pd (lead)	5.0	Can inhibit plant cell growth at very high concentrations.
Se (selenium)	0.02	Toxic to plants at concentrations as low as 0.025 mg/l and toxic to livestock if forage is grown in soils with relatively high levels of added selenium. An essential element to animals but in very low concentrations.
Ti (titanium)	----	Effectively excluded by plants; specific tolerance unknown.
V (vanadium)	0.10	Toxic to many plants at relatively low concentrations.
Zn (zinc)	2.0	Toxic to many plants at widely varying concentrations; reduced toxicity at pH > 6.0 and in fine textured or organic soils.

*The maximum concentration is based on a water application rate which is consistent with good irrigation practices (10000 m3 per hectare per year). If the water application rate greatly exceeds this, the maximum concentrations should be adjusted downward accordingly. No adjustment should be made for application rates less than 10 000 m³ per hectare per year. The values given are for water used on a continuous basis at one site. Source: FAO, 1985

Table 6. Recommended maximum concentrations of trace elements in irrigation water

Parameters	Symbol	Units
Physical		
Total dissolved solids	TDS	mg/l
Electrical conductivity	Ecw	dS/m1
Temperature	T	°C
Colour/Turbidity		NTU/JTU2
Hardness		mg equiv. CaCO3/l
Chemical		
Acidity/Basicity	pH	
Type and concentration of anions and cations:		
Calcium	Ca^{++}	me/l[3]
Magnesium	Mg^{++}	me/l
Sodium	Na^+	me/l
Carbonate	CO_3^{--}	me/l
Bicarbonate	HCO_3^-	me/l
Chloride	Cl^-	me/l
Sulphate	SO_4^-	me/l
Sodium Absorption Ratio	SAR	
Boron	B	mg/l[4]
Nitrate-Nitrogen	NO_3-N	mg/l
Phosphate Phosphorus	PO_4-P	mg/l
Potassium	K	mg/l
Trace metals		mg/l
Heavy metals		mg/l

Table 7. Parameters used in the evaluation of agricultural water quality Source: Kandiah (1990)

3. Health guidelines

Guidelines for the safe use of wastewater in agriculture need to maximize public health benefits while allowing for the beneficial use of scarce resources. Achieving this balance in the variety of situations that occur worldwide (especially in settings where there may be no wastewater treatment) can be difficult. Guidelines are needed to be adaptable to the local social, economic, and environmental conditions and should be co-implemented with such other health interventions as hygiene promotion, provision of adequate drinking water and sanitation, and other healthcare measures. The *Hyderabad Declaration on Wastewater Use in Agriculture* recognises these principles and recommends a holistic approach to the management of wastewater use in agriculture. Following a major expert meeting in Stockholm Sweden in 1999, the International Water Association (IWA) on behalf of the

World Health Organization (WHO) published *"Water Quality: Guidelines, Standards and Health: Assessment of Risk and Risk Management for Water-related Infectious Disease"*. This publication outlines a harmonized framework for the development of guidelines and standards for water-related microbiological hazards (Bartram *et al.*, 2001; Prüss and Havelaar, 2001). The suggested framework involves the assessment of health risks prior to setting health targets; defining basic control approaches, and evaluating the impact of these combined approaches on public health status (Fig. 1). The framework is flexible and allows countries to adjust the guidelines to local circumstances and compare the associated health risks with risks that may result from microbial exposures through drinking water or recreational/occupational water contact (Bartram *et al.*, 2001). It is important that health risks from the use of wastewater in agriculture be put into the context of the overall level of gastrointestinal disease within a given population. The regulation of water quality for irrigation is of international importance because trade in agricultural products across regions is growing and products grown with contaminated water may cause health effects at both the local and transboundary levels. Exports of contaminated fresh produce from different geographical regions can facilitate the spread of both known pathogens and strains with new virulence characteristics into areas where such pathogens are not normally found.

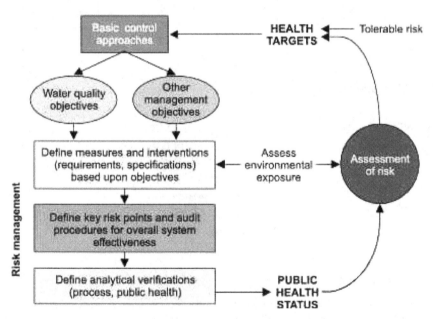

Fig. 1. Stockholm Framework for assessment of risk for water-related microbiological hazards (adapted from Bartram et al., 2001).

Effective guidelines for health protection should be: feasible to implement; adaptable to local social, economic, and environmental factors; and include the following elements:
• Evidence-based health risk assessment
• Guidance for managing risk (including options other than wastewater treatment)
• Strategies for guideline implementation (including progressive implementation where necessary).

Worldwide many different microbial standards for wastewater use in agriculture have been developed. Most guidelines lay heavy emphasis on microbial standards, but it should be recognized that other strategies for managing health risks may also be effective.

4. Conclusion

Water quality criteria for the use in agriculture should not be used without considering various interactive factors. Guidelines established by Ayres and Westcott (1976) takes a better approach to water quality evaluation. Improving the overall levels of agronomic and cultural management to improve water use efficiency can offset yield losses due to poor quality water and improve or maintain profitability. Plants respond to osmotic and ion toxicity effects with a gradually declining yield as water quality deteriorates. This enables the farmer either to accept this loss or to determine whether the marginal return to the cost incurred in avoiding salinity or toxicity problems is worthwhile. Because of the wide range of species tolerance to salinity and toxicity the type of crop being grown has a large bearing on these decisions. Occasional irrigations with poor quality water are often more beneficial than no water at all especially in a Mediterranean environment where salt build—up in the soil is prevented by heavy winter rainfall. Developing realistic guidelines for using wastewater in agriculture also involves the establishment of appropriate health-based targets prior to defining appropriate risk-management strategies. Establishing appropriate health-based targets primarily involves an assessment of the risks associated with wastewater use in agriculture, using evidence from available studies of epidemiological and microbiological risks, and risk-assessment studies. Considerations of what is an acceptable or tolerable risk are then necessary; these may involve the use of internationally derived estimates of tolerable risk, but these need to be put into the context of actual disease rates in a population related to all the exposures that lead to that disease, including other water- and sanitation-related exposures together with food-related exposure. Positive health impacts resulting from increased food security, improved nutrition, and additional household income should also be considered. Individual countries may therefore set different health targets, based on their own contexts. Strategies for managing health risks to achieve the health targets include wastewater treatment to achieve appropriate microbiological quality guidelines, crop restriction, waste application methods, control of human exposure, chemotherapy, and vaccination.

5. Acknowledgement

The author Dr. Hamid Iqbal Tak is thankful to the staff of environmental plant physiology lab for providing necessary support during the course of study and to University Grants Commission for providing financial assistance.

6. Refrences

Adriel Ferreira da Fonseca, Uwe Herpin, Alessandra Monteiro de Paula, Reynaldo Luiz Victória1, Adolpho José Melfi. (2007). Agricultural use of treated sewage effluents: agronomic and environmental implications and perspectives for brazil. *Sci. Agric. (Piracicaba, Braz.)*, 64 (2): 194-209.

Al-Jaloud, A.A., Hussain, G., Al-Saati, A.J. and Karimulla, S. (1995). Effect of wastewater irrigation on mineral composition of corn and sorghum plants in a pot experiment. *J. Plant Nutr.*, 18: 1677-1692.

APHA. 1980 American Public Health Association - Standard Methods for the Examination of Water and Wastewater, 15th Edition. APHA-AWWA-WPCF, Washington DC. 1000 p.

Asano T. and Tchobanoglous G. (1987) Municipal wastewater treatment and effluent utilization for irrigation. Paper prepared for the Land and Water Development Division, *FAO, Rome.*

Ayers, R.S. and Westcot. D.W. (1994). Water Quality for Agriculture, Irrigation and Drainage Paper 29, rev. 1, Food and Agriculture Organization of the United Nations, Rome.

Ayres, R.S. and Westcott, D.W. (1976). Water quality for agriculture. F.A.O. Irrigation and Drainage Paper No. 29 F.A.O. Rome.

Bartram, J., Fewtrell, L. and Stenström, T.A. (2001) Harmonised assessment of risk and risk management for water-related infectious disease: an overview. In: Fewtrell, L. and Bartram, J. (eds.) Water Quality: Guidelines, Standards and Health; Assessment of Risk and Risk Management for Water-related Infectious Disease. International Water Association (IWA) on behalf of the World Health Organization, London, UK, pp. 1–16.

Bauder, T. A., Waskom, R.M., Sutherland, P.L. and Davis, J. G. (2011). Irrigation Water Quality Criteria. Colorado State University. Agriculture Publication No. 0.506 http://www.ext.colostate.edu/pubs/crops/00506.html.

Bouwer H. and Chaney R.L. (1974) Land treatment of wastewater. *Advances in Agronomy,* Vol. 26. N.C. Brady (ed). Academic Press, New York.

Dewis J. and Freitas F. 1970 Physical and chemical methods of soil and water analysis. FAO Soils Bulletin 10. FAO, Rome. 275 p.

Eaton F.M. (1944). Deficiency, toxicity and accumulation of boron in plants. *J. Agric. Res.* 69:237–277.

FAO. (1985) Water quality for agriculture. R.S. Ayers and D.W. Westcot. Irrigation and Drainage Paper 29 Rev. 1. FAO, Rome. 174 p.

Feigin, A., Ravina, I. & Shalhevet, J. (1991). Irrigation with Treated Sewage Effluent. Springer-Verlag, Berlin.

Feigin, A., Vaisman, I. and Bielorai, H. (1984). Drip irrigation of cotton with treated municipal effluents: II. Nutrient availability in soil. *J. Environ. Qual.,* 13: 234-238.

Kandiah A. (1990) Water quality management for sustainable agricultural development. Natural Resources Forum. 14 (1): 22-32.

Maas E.V. (1984) Salt tolerance of plants. The Handbook of Plant Science in Agriculture. B.R. Christie (ed). CRC Press, Boca Raton, Florida.

Prüss, A. and Havelaar, A. (2001) The global burden of disease study and applications in water, sanitation, and hygiene. In: Fewtrell, L. and Bartram, J. (eds) Water Quality: Guidelines, Standards and Health; Assessment of Risk and Risk Management for Water-related Infectious Disease. International Water Association (IWA) on behalf of the World Health Organization, London, UK, pp. 43–60.

Rhoades J.D. and Clark M. 1978 Sampling procedures and chemical methods in use at the United States Salinity Laboratory for Characterizing Salt-affected Soils and Waters. Memo Report. US Salinity Laboratory, Riverside, California.

Richards L.A. (1954). Diagnosis and improvement of saline and alkali soils. USDA Agricultural Handbook No. 60, US Department of Agriculture, Washington DC. 160 p.

Tak, H. I, Inam. A and Inam, A. (2010). Effects of urban wastewater on the growth, photosynthesis and yield of chickpea under different levels of nitrogen. *Urban Water Journal,* 7(3), 187 – 195.

Vazquez-Montiel, O., Horan, N.J. and Mara, D.D. (1996). Management of domestic wastewater for reuse in irrigation. *Water Sci. Tech.,* 33: 355-362.

4

Geospatial Relationships Between Morbidity and Soil Pollution at Cubatão, Brazil

Roberto Wagner Lourenço[1], Admilson Irio Ribeiro[1],
Maria Rita Donalisio[2], Ricardo Cordeiro[2],
André Juliano Franco[3] and Paulo Milton Barbosa Landim[4]
*[1]Department of Environmental Engineering,
São Paulo States University, Sorocaba Campus, Sorocaba, SP,
[2]Department Social and Preventive Medicine, Campinas University, Campinas, SP
[3]São Carlos Federal University, Sorocaba Campus, Sorocaba, SP,
[4]Department of Applied Geology, State University of São Paulo, Rio Claro Campus, SP,
Brazil*

1. Introduction

Some diseases such as allergic asthma may be mainly caused by presence in lung of Ascaris lumbricoides (which causes ascariasis), the Ancilostoma, and Strongiloides (causing to Strongyloidiasis) that cause or exacerbate respiratory symptoms, especially in the lung causing coughing, wheezing, and dyspnea (shortness of breath). Some states of allergic contact dermatitis, are also often associated with direct contact with some type of allergen, or substance that the body identifies as dangerous, producing a rash in the local where the contact occurred. Therefore these diseases are important to public health considering the strong social component directly related to poverty and lack of primary health care in areas of high humidity, high concentration of waste and can often be transmitted by direct contact with degraded areas with contaminated soils (Kakkar and Jaffery, 2005).

Soil often acts as a filter to a large part of the impurities deposited in it. However, this capacity is limited, causing an accumulation of material resulting from atmospheric deposition of pollutants, pesticides and fertilizers (Moreira-Nodermann, 1987). Soil contamination is threat to human health and for environmental quality. Among the main pollutants of the soil, heavy metals are very dangerous when in contact with living beings (Lourenço et. al. 2010; Lourenço and Landim, 2005; Alloway, 2001; Franssen et. al. 1977). In fact, several metals are known to be carcinogens, including arsenic, chromium and nickel. (Tang et al., 1999; Winneke et al., 2002; Stein et al., 2002; Yang et al., 2003).

Studies of the spatial distribution of pollutants in air, water and soil, are traditionally carried out by different scientists in different fields of geosciences using different spatial analysis techniques in order to contribute to the understanding of the variability space of certain events that cause damage to the environment and health (Lourenço et. al., 2010; Amini et. al. 2005). Goria et. al (2009) conducted a study in four French administrative departments and highlighted an excess risk in cancer morbidity for residents around municipal solid waste incinerators. The steps to evaluate the association between the risk of cancer and the exposure to incinerators, was performed by statistical analysis and dispersion modeling using GIS. The study showed that is important to use advanced methods to better assess dose-response relationships with disease risk. Bilancia and Fedespina (2009) studied the triennial mortality rates for lung cancer in the two decades 1981–2001 in the province of Lecce, Italy. The study showed that there is a dramatic increase in mortality for both males and females. Vincenti et. al. (2009) examined the relation between exposure to the emissions from a municipal solid waste incinerator and risk of birth defects in a northern Italy community, using Geographical Information System (GIS). Among women residing in the areas with medium and high exposure, prevalence of anomalies in the offspring was substantially comparable to that observed in the population control, nor dose-response relations for any of the major categories of birth defects emerged. McGrath et al. (2004) produced maps of pollution based on the spatial distribution of Pb in Silvermines, Ireland, where the generated maps serve as valuable information on areas of risk to public health and as decision support and planning. Critto et al. (2003) used geostatistics and the main components of the distribution of chemical contaminants in the soil around a lake near Venice, Italy and evaluated their effects on health. Lin et al. (2002) used the methods to factorial kriging and indicator kriging to analyze the spatial variation of heavy metals in farmland north of Changhua, Taiwan in order to assist in monitoring for environmental remediation proposals and planning. Hills and Alexander (1989) studied surveys which presented the occurrence of leukemia near nuclear plants, and Glass et al. (1995) produced a risk map for the Lyme disease from epidemiological data and from a geographic information system, Mason (1975) presented several field studies conducted as a result of issues related to environmental determinants of cancer that has been raised after the analysis of several atlas published by the American National Cancer Institute. We can also cite important studies oral cancer (Winn et al., 1981), cancer of the bowel (brine et al., 1981), lung cancer (Ziegler et al., 1984), bladder cancer (Hoover and Strasser , 1980) and, finally, studies of associations between sources of contamination and high risk areas, including risk of childhood leukemia in areas near nuclear power plants (Diggle et al., 1990, Elliot et al., 1992) . Given the present discussion, the aim was to study the spatial correlation between the distribution of contaminants in the soil with the spatial distribution of infant morbidity in children under one year of age affected by diseases of the respiratory and intestinal tract. in the city of Cubatão, southern coast of São Paulo, Brazil.

2. Material and methods

2.1 Studied area soil sampling
The research was conducted in the city of Cubatão, southern coastal region of the State of São Paulo, Brazil. The studied area has strong industrial activities in the area with a big petroleum refinery and various chemical activities. Small factories are concentrated in the center area near of port region. There are several wastewater treatment plants in the region with high risk of pollution associated to sewage sludge and compounds.

In this study we analyzed the pollutants cadmium-Cd, lead-Pb and mercury-Hg. Cd is a trace element in various industrial uses, such as fungicides, batteries, rubber processing, production of pigments and galvanic industries, among others. Once the Pb is a toxic element and occurs as an environmental pollutant, is given its use in industrial large scale in the petroleum industry, dyes and paints, ceramics, and others. Both Cd and Pb cause serious health problems to people when exposed to them, or by eating contaminated food. Can cause problems with anemia, infections, headache, sweating (sweat) and various muscle aches. The most serious consequence of chronic exposure to Cd and Pb is cancer, especially cancer of the airways, causing pulmonary emphysema (Okada, et. al, 1997). The Hg is a metal and odorless liquid at room temperature, but when the temperature increases becomes toxic and corrosive vapors denser than air. Integrate the class of transition metals. The risk of disease is high. According to the temperature, the concentration of metallic mercury is changed and when absorbed by the human body tends to accumulate in the brain, liver and kidneys. Because of this, contamination manifests itself by acute problems in the nervous system (sensory and motor disturbances) and deficiencies of bowel function (Zavaris and Glina, 1992).

To determine total metal concentrations Cd, Pb and Hg, the soil samples were sampled with distances from 95 to 650m (Figure 1). After the soil samples were dried and conventionally decomposed by a mixture of nitric acid and concentrated hydrochloric acid according to a standardized procedure (Alloway, 2001). After that, it was weighed (approximately 2.00 g) of pre-dried soil sample that was mixed with 21 mL of 30% HCl (Suprapur) and 7 mL of 65% HNO_3 (Suprapur) in a highly pre-purified quartz vessel (200 mL). The solution was heated first to 100_C and then to 120_C. Subsequently, the samples were digested using 20 mL of concentrated HNO_3 under reflux for 3 hours. Finally, the digested samples were diluted with high-purity water to a final volume of 100 mL. Small undigested soil remainders (approximately 5%) were removed by filtration. Metal determinations were usually carried out with 1/10 dilutions of the digestion solutions. The result of soil digestion by aqua regia was assessed from five replicates and metal determinations were performed by ICP-OES (spectrometer: TJA IRIS AP, Thermo Jarrell Ash, Franklin, MA, USA).

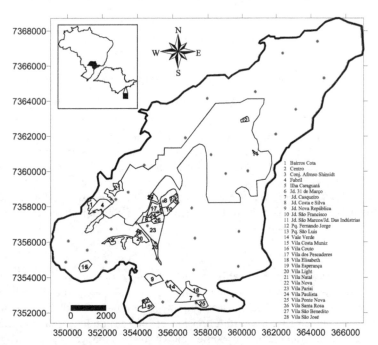

Fig. 1. Study area (read point is soils samples)

2.2 Mapping soil metal concentrations

To the mapping of total metal concentrations Cd, Pb and Hg of the soil samples it was used geostatistical methods (Van Meirvenne and Goovaerts, 2001; Webster and Oliver, 2001; Lin et. al. 2001; Romic and Romic, 2003; McGrath et. al. 2004). Geostatistics analysis methods are based in the spatial variation of data often distributed irregular, known as a regionalized variable. Therefore, for a geostatistical modeling it is used an interpolation known as kriging. The procedure is similar to that used in weighted moving average interpolation, except that the weights are derived from a variografy analysis of the data model. The weights are given by:

$$\hat{z}(X_0) = \sum_{i=1}^{n} \lambda_i \cdot z(x_i) \tag{1}$$

with $\sum_{i=1}^{n} \lambda_i = 1$. The weights λ_i are chosen so that the estimate $\hat{z}(X_0)$ has less variance σ_e^2 in relation to the sampled values compared to any other linear estimator.

The minimum variance of $[\hat{z}(X_0) - z(X_0)]$, the prediction error, or 'kriging variance' is given by:

$$\hat{\sigma}_i^e = \sum_{i=1}^{n} \lambda_i \gamma(x_i, x_0) + \phi \tag{2}$$

and is obtained when

$$\sum_{i=1}^{n} \lambda_i \gamma(x_i, x_j) + \phi = \gamma(x_j, x_0) \text{ for all } j \tag{3}$$

The $\gamma(x_i, x_j)$ is the semivariance of z between the sampling points x_i and x_j; $\gamma(x_i, x_0)$ is the semivariance between the sampling point x_i and the unvisited point x_0. Both semivariance are obtained from the fitted variogram. The semivariance ϕ is a Lagrange multiplier required for the minimization. This method is known as *ordinary kriging* and it is very well described by many authors (eg. Landim, 2003; Gringarten and Deutsch, 2001; Olea, 1999; Burrough et. al. 1997; Goovaerts, 1997; Isaaks and Srivastava, 1989; Journel and Huijbregts, 1978).

Thus, in this study, variogram models were used to analyze spatial patterns and ordinary kriging to obtain a continue surface of the distribution of soil pollutants Cd, Pb and Hg in the area.

2.3 Mapping morbidity

Morbidity was determined as the health damages to the movement of hospitalization and outpatient care of the study area. The data of the studied area were provided by the Health Brazilian Agency in the year 2007. These data were filtered to obtain only the data of hospital admissions according to the 10th revision of International Classification of Diseases (ICD-10), for hospitalizations related to some kind of disease that can be caused by direct or indirect contact with contaminated soil with high concentration of pollution.

The cases of hospitalization and outpatient care of the study area were used for construction of the discrete map by district and surface map of the morbidity distribution using interpolation of the inverse of the distance with power squared (Burrough, 2004). The equation used for Inverse Distance to a Power (IDP) is:

$$\hat{Z}_j = \frac{\sum_{i=1}^{n} \frac{Z_i}{h_{ij}^{\beta}}}{\sum_{i=1}^{n} \frac{1}{h_{ij}^{\beta}}} \tag{4}$$

$$h_{ij} = \sqrt{d_{ij}^2 + \delta^2} \tag{5}$$

where:
h_{ij} is the effective separation distance between grid node "j" and the neighboring point "i."

$\hat{Z}_j$ is the interpolated value for grid node "j";

Z_i are the neighboring points;
d_{ij} is the distance between the grid node "j" and the neighboring point "i";
β is the weighting power (the Power parameter); and
δ is the Smoothing parameter.
This procedure was carried out in order to obtain a surface continue distribution morbidity that could be compared in a space with other maps of soil pollutants.

2.4 Measures of spatial relationship between morbidity and soil pollution
To analyze the relationship between the morbidity spatial maps distribution and the soil maps was used the multiple regression spatial analysis technique. There are many cases where the variation of a variable can be explained by a number of other variables. The variables that help predict the variable of interest are called the independent variable, while the predicted variable is called dependent variable, assuming that a linear relationship exists between them. This is study used the independent variables to the soil pollution and dependent variable to the morbidity by multiple linear regression and the multiple linear regression equation is written as:

$$Y = a + b_1{}^*x_1 + b_2{}^*x_2 + b_3{}^*x_3 \tag{6}$$

where Y is the dependent variable; x_1, x_2, and x_3 are the independent variables; a is the *intercept*; and b_1, b_2, and b_3 are the *coefficients* of the independent variables x_1, x_2, and x_3, respectively. The intercept represents the value of Y when the values of the independent variables are zero, and the parameter coefficients indicate the change in Y for a one-unit increase in the corresponding independent variable.
In the multiple regression results the R represents the multiple correlation coefficients between the independent variables and the dependent variable. R squared represents the extent of variability in the dependent variable explained by all of the independent variables. The adjusted R and R squared are the R and R squared after adjusting for the effects of the number of variables.

The individual contribution of each independent variable to the individual dependent variable is express in the regression coefficients. Its significance is expressed in the form of a *t*-statistic. The *t*-statistic is the most common test used in estimating the relative success of the model and for adding and deleting independent variables from a regression model. The *t*-statistic verifies the significance of the variables departure from zero (i.e., no effect)

Multiple spatial linear regressions allow the construction of spatial maps in predicting morbidity and a map of the residual according to the linear model fitted. As the map of prediction has its variation as a function of predictor variables it can be understood as a risk map from the exposure of pollutants to the occurrence of morbidity, and the residual map as a measure of success of prediction.

3. Results and discussion

Table 1 shows the variogram parameters for soil samples after chemical analysis.

Soil attributes	Model	C_o	$C + C_o$	$C_o/C + C_o$	Range (m)	R^2
Cd	Gaussian	0.15	0.80	0.187	395	0.38
Hg	Gaussian	0.55	0.95	0.578	1100	0.30
Pb	Exponential	0.5	0.96	0.520	410	0.52

Table 1. Variogram models of heavy metals and their parameters. C_o = nugget variance, C = structural variance, and $(C + C_o)$ = sill variance.

The range values of variograms for Cd and Pb were similar and around 400m, and were lower than those for Hg (around 1100m). The Nug/Sill ratio for the Cd metal was around the 18% (Co / C + Co) showing randomness of the data that is important for good modeling variography while for the values of Hg and Pb variation is more random and unpredictable. The R2 was between 0.30 and 0.52 suggesting a good correlation between samples.

The experimental variograms of the heavy metal in soil with the fitted models are presented in Fig. 2 a-c.

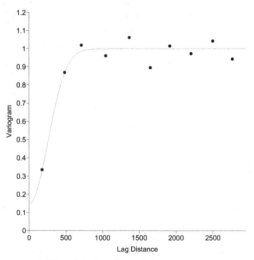

Fig. 2a. Gaussian variogram of Cd with fitted models

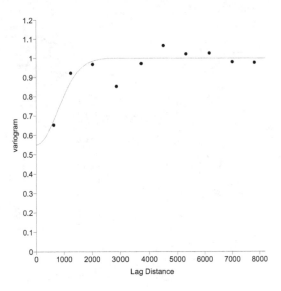

Fig. 2b. Gaussian variogram of Hg with fitted models

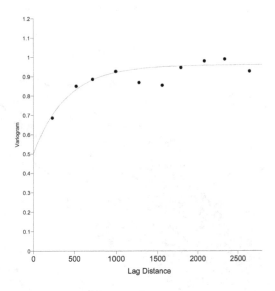

Fig. 2c. Exponential variogram of Pb with fitted models

The results showed that soil with Cd (a), Hg (b) were best fitted with the gaussian model and Pb (c) with the exponential model. The ordinary kriging technique was used here to obtain a surface of the spatial distribution of soil pollution fitted with parameters of the variogram (Figure 3).

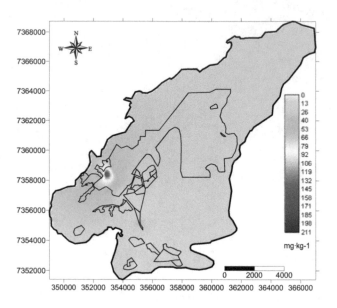

Fig. 3a. Prediction mapping of Cd concentration in soil generated by ordinary Kriging

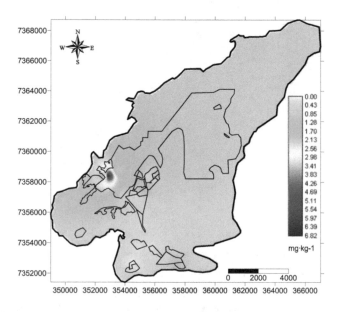

Fig. 3b. Prediction mapping of Hg concentration in soil generated by ordinary Kriging

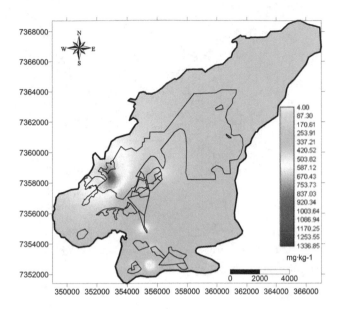

Fig. 3c. Prediction mapping of Pb concentration in soil generated by ordinary Kriging

The Figure 3 a-c presents the spatial patterns of the three heavy metals in soil from the studied area generated from their variograms. The spatial distribution maps showed similar geographical trends, especially for Cd (a) and Hg (b), with higher concentration in the west area and decreasing presence towards northeast. Meanwhile despite Pb (c) showed similar spatial trend, the intensity is higher in west area and also the southern area is emerging as an important local pollution.

3.1 Morbidity map

Two maps of the spatial distribution of cases of hospitalization and outpatient care of the study area were constructed. The first, which contains all the cases for the year 2007 divided by neighborhoods (Figure 4a) and a second map (Figure 4b) constructed with Inverse Distance to a Power technique to obtain a surface of the spatial distribution of soil pollution.

The map of Figure 4a of spatial distribution of morbidity by neighborhood showed that concentrations are localized in the west area. This area is very industrialized beyond to be place with houses of poor and low social standing.

The map of Figure 4b of spatial distribution surface of morbidity showed that concentrations are localized in three different area: a coincident with the map of Figure 4a and two others, one near the central districts and other areas closer to the south of the map. The southern sector is characterized by areas of proximity to the sea shore, with influences of the waters of the mangroves, which can be further more dangerous for people living on fishing and consumption of other foods from the sea.

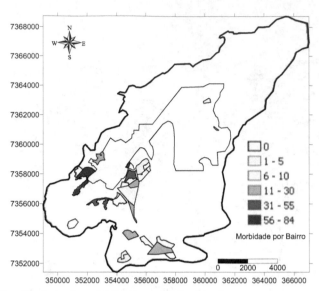

Fig. 4a. Spatial distribution of morbidity by neighborhood

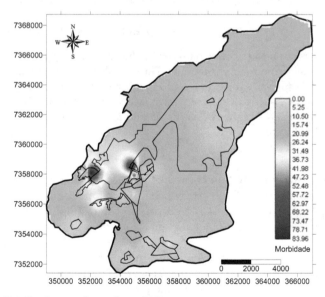

Fig. 4b. Spatial distribution surface of morbidity

3.2 Spatial relationship between morbidity and soil pollution

In order to analyze the relationship between the morbidity spatial distribution maps and the soil maps was used the spatial linear regression analysis. The Figures 5 to 7 show the regression graphs and the statistical parameters of spatial relationship.

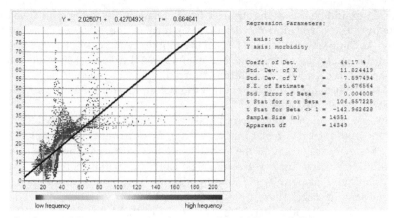

Fig. 5. Graph of spatial linear regression between Morbidity and Cd.

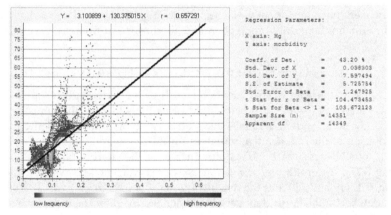

Fig. 6. Graph of spatial linear regression between Morbidity and Hg.

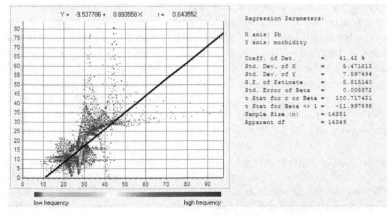

Fig. 7. Graph of spatial linear regression between Morbidity and Pb.

The regression trend line shows the stronger linear relationship to the data at soil pollution with morbidity. The correlation coefficient ("r") next to the equation tells us the same numerically. As can be seen in our data when the morbidity increases, the soil pollution also increases. In this study the correlation coefficients ranged around 0.65 indicating a strong positive relationship between soil pollution, morbidity and the coefficient of determination (r^2) around 40%, which leads us to accept that the pollutants from the soil strongly influence the morbidity in the studied area.

However, we see that all pollutants were highly correlated with morbidity and it is unclear which of them would have a greater influence on it. To determine which pollutant found in the soil has a greater influence on the variation of morbidity was performed an analysis known as multiple spatial linear regression. The multiple spatial regression analysis is an important technique that permits the investigation of the relationship of spatial variables over the same sample space. There are many cases where the variation of a variable can be explained by a number of other variables. The variables that help to predict the variable of interest are called the independent variable, while the predicted variable is called dependent variable, assuming a linear relationship existing between them.

This is study used the independent variables to the soil pollution and dependent variable to the morbidity for linear multiple regression and the linear multiple regression equation is written as:

$$\text{Morbidity} = 0.1326 + 0.4777*cd - 0.0024*Hg + 0.0027*Pb \qquad (7)$$

The regression equation shows coefficients for each of the independent variables and the intercept. The intercept (0.1326) can be thought of as the value for the dependent variable when each of the independent variables takes on a value of zero. The coefficients indicate the effects of each of the independent variables on the dependent variable. For example, if the emission of Cd increases by 100 units, increases the morbidity to 47.77% (i.e., 100 multiplied 0,4777). The multiple correlation spatial coefficient between the independent variables (ie, Cd, Hg and Pb) and the dependent variable (morbidity) was R = 0.91 and the extent of variability in the dependent variable explained by all of the independent variables was R^2 = 84%, i.e., 84% of the variance in the morbidity is explained by independent variables soil pollution.

The individual regression coefficients express the individual contribution of each independent variable to the dependent variable. The significance of the coefficient is expressed in the form of a t-statistic. The t-statistic verifies the significance of the variables departure from zero (i.e., no effect). In this study, the t-statistic has to exceed the following critical values in order for the independent variable be significant. To 99% confidence level with ∞ degrees of freedom the value is 2.57, and Cd coefficient has a t-statistic of 8.71, the Pb t-statistic is 4.75 and the Hg t-statistic is 3.66 indicating that all variables are highly significant (99%). The t-statistic is the most common test used in estimating the relative success of the model and for adding and deleting independent variables from a regression model.

Multiple linear spatial regressions allow the construction of spatial maps in predicting morbidity and a map of the residual according to the linear model fitted (Fig. 8a-b). As the map of prediction has its variation as a function of predictor variables it can be understood as a risk map from the exposure of pollutants to the occurrence of morbidity, and the residual map as a measure of success of prediction.

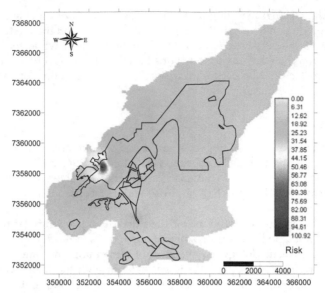

Fig. 8a. Risk map morbidity

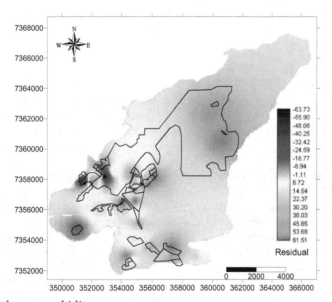

Fig. 8b. Residual map morbidity

The risk map areas (Figure 8a) with higher risk of morbidity are concentrated in the western area, coincidentally where is the greatest concentration of pollutants in soil. However, this is not the only place that appears morbidity in the area. This suggests that the variation in morbidity may have other factors, or other pollutants causing the variation of morbidity that is not being used in this study.

The analysis of the residual map (Figure 8b) is an important tool for spatial assessment of areas where the prediction obtained the best results depending on the model fitted. Usually the areas closest to the zero values have the best predictions, while the areas with the highest residual would be the worst predictions based on the model fitted. Thus, as seen in this work, the smaller residual occurs in areas with the predictions for higher morbidity risks.

4. Conclusions

The study showed that areas with high concentrations of pollutants in the soil influence the occurrence of morbidity especially those related to the intestinal tract and skin and respiratory allergies. In general, the studied area has serious problems related to the use and occupation by people with low purchasing power and, as a consequence, with little access to public health.

The proposed methodology was efficient for the purpose of showing that there is a degree of relationship between pollutants from soil and some cases of morbidity that can affect the health of people. This methodology may be useful for planning programs and management in promoting the welfare of people. This is possible through the identification of priority areas to assist people beyond the actions of government agencies to control the emission of pollutants into the environment.

Finally it is expected that the results, particularly the maps generated through the techniques of GIS, can be an important tool for urban planning and management, with main purpose to help improving the people quality of life.

5. References

Alloway, Bj (2001) *Heavy Metals In Soils*. New York: John Wiley, 339p.

Amini, M.; Afyuni, M.; Fathianpour, N.; Khademi, H.; Flühler, H. (2005) *Continuous Soil Pollution mapping Using Fuzzy Logic and Spatial Interpolation*. Geoderma 124: 223-233.

Bilancia, M.; Fedespina, A. (2009) *Geographical clustering of lung cancer in the province of Lecce, Italy: 1992–2001*. International Journal of Health Geographics 2009, 8:40 (1 July 2009).

Burrough, P.A.; Mcdonell, R. (2004) *Principles of Geographical Information Systems*. Oxford, Oxford University Press.

Burrough, P.A.; Van Gaans, P.; Hootsmans, R.J. (1997) *Continuous classification in soil survey: spatial correlation, confusion and boundaries*. Geoderma, 77: 115-136.

Critto, A.; Carlon, C.; Marcomini, A. (2003) *Characterization of contaminated soil and groundwater surrounding an illegal landfill (S. Giuliano, Venice, Italy) by principal component analysis and kriging*. Environmental Pollution.; 122:235-44.

Diggle, P. J.; Gatrell, A. C. & Lovett, A. A. (1990). *Modelling the prevalence of cancer of cancer of larynx in part of Lancashire: a new methodology for spatial epidemiology*. In: Spatial Epidemiology (R. W. Thomas, ed.), pp. 153-171. Pion: London.

Elliott, P. J.; Hills, M.; Beresford, J.; Kleinschmidt, I.; Jolley, D.; Pattenden, S.; Rodrigues, L.; Westlake, A. & Rose, G. (1992). *Incidence of cancer of the larynx and lung cancer near incinerators of waste solventes and oil in Great Britain*. The Lancet, 339:854-858.

Franssen, H.; Eijnsbergen, A.C.; Stein, A. (1997) *Use of spatial prediction techniques and fuzzy classification for mapping soil pollutants*. Geoderma 77: 243-262.

Goria, S.; Daniau, C.; Crouy-Chanel, P.; Empereur-Bissonnet, P.; Fabre, P., Colonna, M.; Duboudin, C.; Viel, J.F.; Richardson; S. (2009) *Risk of cancer in the vicinity of municipal solid waste incinerators: importance of using a flexible modelling strategy.* International *Journal of Health Geographics* 2009, 8:31 Doi:10.1186/1476-072X-8-31.

Glass G.E., Schwartz B.S., Morgan J.M., Johnson D.T., Noy P.M., Israel E. (1995). *Environmental risk factors for lyme disease identified with geographic information systems.* American Journal of Public Health 85, no. 7: 944-48.

Goovaerts, P. (1997) *Geostatistics For Natural Resources Evaluation.* Oxford University Press.

Grigoletto, J.C.; Oliveira, A.S.; Muñoz, S.I.S.; Alberguini, L.B.A.; Takayanagui, A.M.M. (2008) *Exposição ocupacional por uso de mercúrio em odontologia: uma revisão bibliográfica.* Ciência & Saúde Coletiva, 13(2), 533-542. Retrieved March 04, 2012, from http://www.scielo.br/scielo.php?script=sci_arttext&pid=S1413-81232008000200029&lng=en&tlng=pt. http://dx.doi.org/10.1590/S1413-81232008000200029.

Gringarten, E.; Deutsch, C.V. (2001) *Teacher's aide: variogram interpretation and modeling.* Mathematical Geology 33:507–534.

Hills, M.; Alexander, F. (1989) *Statistical methods used in assessing the risk of disease near a source of possible environmental pollution: a review.* J R Stat Soc A 1989; 152:353-63.

Hoover, R. N. & Strasser, P. H. (1980). *Artificial sweeteners and human bladder cancer: preliminary results.* Lancet, 1:837-840.

Isaaks, E.; Srivastava, R.M. (1989) *An Introduction to Applied Geostatistics.* Oxford University Press, Oxford.

Journel, A.G.; Huijbregts, C. (1978) *Mining Geostatistics.* Academic Press.

Kakkar, P; Jaffery, Fn (2005) *Biological markers for metal toxicity.* Environmental Toxicology and Pharmacology 19: 335–349.

Landim, P.M.B. (2003) *Statistical analysis of geological data.* Ed. UNESP. 2ª. ed. – São Paulo, Brazil.

Lin, Y.; Chang, T.; Shih, C.; Tseng, C. (2002) *Factorial and indicator kriging methods using a geographic information system to delineate spatial variation and pollution sources of soil heavy metals.* Environmental Geology; 42:900-9.

Lin, Y.P.; Chang, T.K.; Teng, T.P. (2001) *Characterization of soil lead by comparing sequential Gaussian simulation, simulated annealing simulation and kriging methods.* Environmental Geology 41:189–199.

Lourenço, R.W.; Rosa, A.H.; Roveda, J.A.F.; Martins, A.C.G.; Fraceto, L.F. (2010) *Mapping soil pollution by spatial analysis and fuzzy classification.* Environmental Earth Sciences. DOI: 10.1007/s12665-009-0190-6.

Lourenço R.W.; Landim, P.M.B. (2005) *Risk mapping of public health through geostatistics methods.* Reports in Public Health (CSP), Vol.21, no.1, p.150-160.

Mason Tj, Mckay Fw, Hoover R, Blot Wj, Fraumeni Jf Jr. (1975) *Atlas of cancer mortality for US counties 1950-1969.* Department of Health, Education and Welfare Publication No. (Nffl)75-780. Washington (DC): U.S. Government Printing Office.

Mcgrath, D.; Zhang, C.S.; Carton, O. (2004) *Geostatistical analyses and hazard assessment on soil lead in Silvermines, area Ireland.* Environmental Pollution 127:239–248.

Moreira-Nodermann, L.M. (1987) *Geochemistry and environment.* Brasilienses Geochemistry, v.1,n.1, 89-107.

Okada, I.A.; Sakuma, A.M.; Maio Franca, D.; Dovidauskas, S.; Zenebon, O. (1997). *Avaliação dos níveis de chumbo e cádmio em leite em decorrência de contaminação ambiental na região do Vale do Paraíba, Sudeste do Brasil.* Rev. Saúde Pública [serial on the Internet]. 1997 Apr [cited 2012 Mar 04]; 31(2): 140-143. Available from: http://www.scielosp.org/scielo.php?script=sci_arttext&pid=S0034-89101997000200006&lng=en. http://dx.doi.org/10.1590/S0034-89101997000200006.

Olea, R.A. (1999) *Optimum Mapping Techniques using Regionalized Variable Theory.* Kans. Geol. Survey., Series on Spatial Analysis, n.2.

Pickle, L. W.; Greene, M. H.; Ziegler, R. G.; Toledo, A.; Hoover, R.; Lynch, H. T. & Fraumeni Jr., J. F, (1981) *Colorectal cancer in rural Nebraska.* Cancer Research, 44:363-369.

Romic, M.; Romic, D. (2003) *Heavy metals distribution in agricultural topsoils in urban area.* Environmental Geology, 43:795–805.

Stein, J.; Schettler, T.; Wallinga, D.; Vallenti, M. (2002) *In harm's way: toxic threats to child development.* Journal Development Behavior Pediatric (Suppl. 1) S13–S22.

Tang, H.W.; Huel, G.; Compagna, D.; Hellier, G.; Boissinot, C.; Blot, P. (1999) *Neurodevelopment evaluation of 9-month old infants exposed to low levels of Pb in vitro: involvement of monoamine neurotransmitters.* Journal Applied Toxicology 19: 167–172.

Van Meirvenne, M.; Goovaerts, P. (2001) *Evaluating the probability of exceeding a site-specific soil cadmium contamination threshold.* Geoderma 102, pp. 63–88.

Vincenti, M.; Malagoli, C.; Fabbi, S.; Teggi, S.; Rodolfi R.; Garavelli, L.; Astolfi, G.; Rivieri, F. (2009) *Risk of congenital anomalies around a municipal solid waste incinerator: a GIS-based case-control study.* International Journal of Health Geographics 2009, 8:8 doi:10.1186/1476-072X-8-8

Webster, R, Oliver, Ma (2001) *Geostatistics for environmental scientists.* Wiley, Chichester, pp 89–96.

Winn, D.M.; Blot, W.J.; Shy, C.M.; Pickle, L.W.; Toledo, A.; Fraumeni, J.F. Jr. (1981) *Snuff dipping and oral cancer among women in the Southern United States.* N Engl J Med 304:745-749.

Winneke, G.; Walkowiak, J.; Lilienthal, H. (2002) *PCB induced neurodevelopmental toxicity in human infants and its potential mediation by endocrine dysfunction.* Toxicology 181–182, 161–165.

Yang, J.H.; Derr-Yellin, E.C.; Kodavanti, P.R. (2003) *Alterations in brain protein kinase c isoforms following developmental exposure to a polychlorinated biphenyl mixture.* Brain Res. Mol. Brain Res., 123–135.

Zavaris, C.; Glina, D.M.R. (1993) *Avaliação clínica – neuro – psicológica de trabalhadores expostos a mercúrio metálico em indústria de lâmpadas elétricas.* Revista de Saúde Pública, v. 26, n. 5 p. 356 – 365.

Ziegler, R. G.; Mason, T. J.; Stemhagen, A.; Hoover, R.; Schoenberg, J. B.; Virgo, P.; Waltman, R. & Fraumeni Jr., J. F. (1984) *Dietary carotene and vitamin A and risk of lung cancer among white men in New Jersey.* Journal of the National Cancer Institute, 73:1429-1435.

5

Assessment of Geochemistry of Soils for Agriculture at Winder, Balochistan, Pakistan

Shahid Naseem[1], Salma Hamza[2] and Erum Bashir[1]
[1]Department of Geology, University of Karachi, Karachi,
[2]Department of Geology, Federal Urdu University of Arts,
Science and Technology, Karachi,
Pakistan

1. Introduction

Pakistan basically is an agricultural country and this sector approximately contributes 22 percent of GDP. Agricultural products not only facilitate life but are also supplies as raw material to industries and a main source of foreign exchange (Pakistan Agriculture Economy and Policy [PAPE], 2009). In Pakistan, farmers are not aware of the latest development in the field of agro-sciences, although some of them take care about the soil fertility (Iqbal & Ahmad, 2011). Proper management of nutrients and micronutrients will enhance the productivity but thus far very little attention has been paid to evaluate the impact of local geology on the soils for better crop yield.

The Japanese Government in collaboration with the Department of Agriculture, Balochistan in 1993-94 has established agriculture farms in the surrounding areas of Winder Town, Balochistan to boost fruit farming in the area, necessary to fulfill consumer's demand. These farms are situated over the ophiolitic rocks of the Cretaceous age concealed by windblown sediments (Fig. 1). Pillow basalts in association with pelagic sediments are exposed along a North-South alignment in the eastern part of the area (Naseem et al., 2002). The Ferozabad Group of Jurassic age is exposed in the Mor Range and other sedimentary rocks belonging to the Cretaceous and Tertiary ages are exposed in the Pab Range, further east of study area. The dissolution of ophiolites contributes a very discrete assemblage of ions (Neal & Shand, 2002). The abundance of elements largely depends upon the nature of bedrocks, climatic conditions and mobility. The ion distribution is also influenced by the infinite complex surface and subsurface physicochemical environments (Aghazadeh & Mogaddam, 2010). The soil differs from its parent material due to interactions between the lithosphere, hydrosphere, atmosphere and the biosphere (Chesworth, 2008; Danoff-Burg, 2000). The rock in the vicinity of an agricultural field not only provides basic lithology (soil) but also contribute major and trace elements to the soil (Fig. 2). The trace elements in the soil play a dual role in agriculture. Some trace elements are essentially required for the healthy growth of plants, as well as they contribute as essential minerals to human beings. However, some trace elements are not only toxic for plants but also post a health risk to humans (Haluschak et al., 1998).

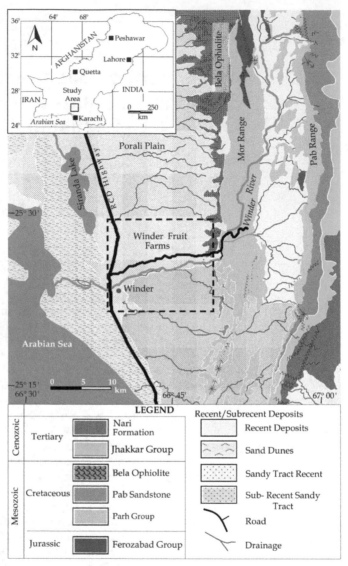

Fig. 1. Geological map of the Winder and adjoining areas (Simplified after Hunting Survey Corporation Ltd. [HSC], 1960). Main study area is marked by broken line cage. Inset showing map of Pakistan.

The majority of the soils of the study area are a mixture of sand, silt and loam. Genetically they are derived from ophiolites, sedimentary rocks and windblown sediments. The soil profile is not well developed having a poor top organic rich horizon. In the foothills of the rocky area, R-horizon is found. The soils developed in the vicinity of the ophiolites are thinner in thickness, medium to coarse grained and are light brown in colour. The western and southern portions of the study area are occupied by saline soil.

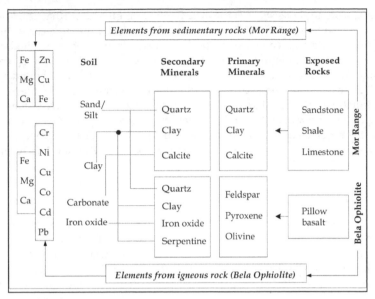

Fig. 2. Schematic diagram illustrated contribution of trace elements in soils through rocks of the study area.

Seventy-two representative soil samples were collected from Winder town and adjoining areas. The pH and electrical conductivity (EC) of soil samples were measured using a Denver Instrument Model 50. Soil organic matter was determined by the ignition method. The texture of the soils was determined with a set of sieves of different diameters. The XRD of soil samples were made using a Bruker D-8 Advance X-ray diffracto-meter, Cu and K-radiation. Estimation of major and trace elements were made through water extraction (Gupta, 2004) using AAS (Hitachi Model, Z-5000).

The objective of the study is to describe the geochemistry of the soil in relation to it potential for agriculture purposes. The major and trace elements are discussed considering its soil quality, impact of bed-rock and bioavailability of these elements in the soil. Prime irrigation qualities are elaborated to appraise its suitability for irrigation in the Winder Agriculture Farms. Possibly present study helps farmers of the area for better irrigation planning.

2. General geology

The study area lies in the southern extremity of Mor-Pab ranges, comprises of ophiolite and sedimentary rocks. The Bela Ophiolite is linked with 400km long Alpine-Himalayan Mesozoic Ophiolite Belt. It is a harzburgite sub-type ophiolite, which is common in Tethyan Domain. It has characteristics of both suprasubduction zone and mid-ocean ridge settings and is also intruded by hotspot-derived magmas (Sheth, 2008; Khan et al., 2007). It is represented by tholeiitic basalts, mafics and ultramafics. Gnos et al. (1998) divided the Bela Ophiolite into two distinct units on the basis of age and tectonic setup. The upper unit (ophiolite) is exposed in the northern part of the belt between Sonaro and Wadh. The lower unit (ophiolite accretionary wedge and trench sediments) is well exposed from Sonaro to the coast of Karachi in the south. The study area is the part of lower unit of Bela Ophiolite and intermittently exposed along the western contact of Mor Range.

Rocks of the Mor Range consists of Ferozabad Group, comprised of Kharrari, Malikhor and Anjira formations (Shah, 2009). It consists of mainly siliciclastic and carbonate rocks of Lower-Middle Jurassic age. The rocks of the Ferozabad Group were deposited on the shelf flank of a rift system resulting from the breakup of Gondwanaland and the separation of the Indian Craton from Madagascar and Africa (Gnos et al., 1997). The stratabound replacement type (MVT) deposit is confined to the allochemical limestone beds of the Kharrari and Malikhor formations. Sedimentary exhalative mineralization (Sedex) is common in the topmost Anjira Formation. Sulphide mineralization comprises sphalerite, galena, pyrite and marcasite with minor chalcopyrite (Ahsan & Mallick, 1999). The outcrops of Cretaceous sedimentary rocks (Sembar, Goru, Parh, Fort Munro, Mughal Kot and Pab formations) are present in the Pab Range (Kazmi & Abbasi, 2008).

3. Soil character and classification

The top layer of the earth's surface made up of organic material and particles of broken rocks that have been altered by chemical and environmental processes. It is generally composed of many parallel distinctive horizons (soil profile) responsible for the agricultural growth. In the study area, soils reveal variable profiles according to the geology of the area. Mostly windblown sediments are present (Fig. 3), thinning towards east, where the ophiolitic rocks are present. In the western part, windblown and fine soil is present (Fig. 3A). In the vicinity of Winder River and its Tributaries, river driven materials are common (Fig. 3B). In the foothills of rocky area, R-horizon is found. Based on the United States Department of Agriculture (USDA) and the National Cooperative Soil Survey classification system, in the study area, majority of the soils are Entisols, which are formed in the areas of very dry or cold climate, having low soil moisture and poor soil organic matter, however some of the wind-blown soils are Aridisols, which develops in very dry environments.

Fig. 3. Different types of soils present in the study area; A. Windblown silty soil, and B. Winder River channel.

Briggs et al. (1998) used sand, silt and clay fractions in the classification of agriculture soils. The information of soil through this scheme is valuable to infer texture, structure, porosity, adhesion and consistency of the soil. In addition, it largely determines the water retention and transmission properties of soils (Singh & Dhillon, 2004). The USDS sand-silt-clay triangular diagram (Fig. 4) displays 27.77% (20 out of 72) soil in the study area as sandy loam, followed by 20.83% sand, while 13.88% are silt loam. This indicates that soil of the study area is the mixture of sand and loam. Sand may drain too rapidly, while in a clay individual pore spaces are too small for adequate water holding; where clay and silt proportions are high, root penetration is difficult. Generally, loam textures are best for crop growth (Singh & Dhillon, 2004). The combination of both sandy and loamy soil is suitable for agriculture.

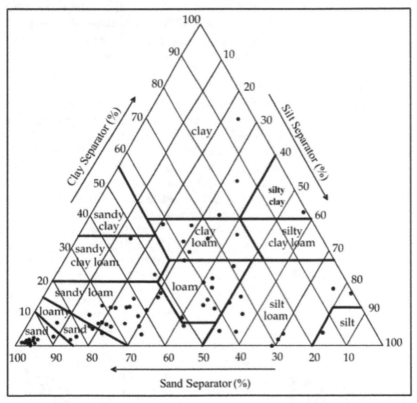

Fig. 4. Classification of soils of Winder area on sand-silt-clay ternary diagram (Class boundaries after USDA textural triangle).

4. Soil geochemistry

4.1 pH and alkalinity

The pH is one of the most important soil properties that affect the availability of elements. The majority of food crops prefer a neutral or slightly acidic soil (pH 7). Some plants however, prefer more acidic (*Manilkara zapota*) or alkaline (*Zizyphus jujuba*) conditions (Dave's, 2010). Many nutrient cations such as Zn, Fe and Cu are available for uptake by plants below pH 5.0; however at variable pH, elements show different availability. In more alkaline conditions, elements availability decreases (Sutton, 2003) and symptoms of nutrient deficiency may result, including thin plant stems, yellowing (chlorosis) or mottling of leaves and slow or stunted growth (McKenzie, 2003). In the study area, the pH range 7.06-9.20, indicates alkaline soil, which is obvious in the presence of carbonate rocks in the area.

4.2 Electrical conductivity (EC)
The accumulation of soluble salts in the soil is termed as water extractable salts and can be assessed by measuring the electrical conductivity (EC) in soil saturation extracts (Reluy et al., 2004). The EC is an important determinant of the suitability of the soil for plant growth (Gana, 2000). Singh & Dhillon (2004) classify soil with reference to salinity; soils having EC <2000 µS is termed as Non-saline, between 2000-6000 are Moderately saline, while EC >6000 are Highly saline. In the study area, only 4 samples (5.55%) are Non-saline, 44.44% samples have Moderately saline while rest of the samples are Highly saline.

4.3 Cation exchange capacity (CEC)
It is the total number of negative charges per unit weight of the soil. The quantity of positively charged ions (cations) that a clay mineral or similar material can accommodate on its negatively charged surface is expressed as milli-ion equivalent per 100g at 7pH, or more commonly as milliequivalent (meq) per 100g or cmol/kg (Troeh & Thompson, 2005). Thus, CEC is the measure of soil fertility and nutrient retention capacity. The CEC of the clay mainly depends upon the mineralogy and different members of clay can hold variable amount of cations. In the studied samples, relation between clay percent and trace elements is poor. The correlation coefficient (Fig. 5) between clay content and trace elements (Cu, Cr and Ni) exhibits negative relationship. The data also reveals that relatively high amount of elements are more in those soil samples in which clay content is between 10-20%. The distribution of trace elements is low (Table 1) in contrast to world average (Alloway, 2005). Probably the low amount is due to sandy soil and alkaline pH.

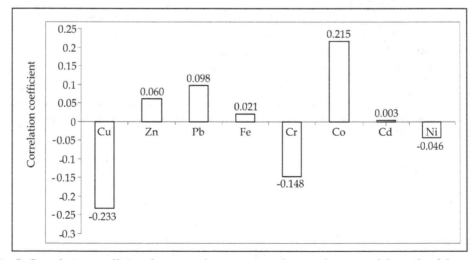

Fig. 5. Correlation coefficient between clay content and trace elements of the soils of the study area.

Element	Range in agriculture soil	World Average in agriculture soil	Average in soil of study area
Na	750-7500	6300	631
K	400-30000	8300	96
Ca	7000-500000	13700	1273
Mg	20-10000	5000	334
Fe	5000-500000	38000	28.05
Cu	1-20	13-24	2.66
Pb	3-189	32	5.12
Ni	0.2-450	20	3.12
Zn	17-125	64	1.02
Co	0.1-70	7.9	0.97
Cd	0.01-2.5	0.06-1.1	0.49

Table 1. Concentrations of selected major and trace elements in the agriculture soils (After Alloway, 2005) and their comparison with the soils of study area. All values are in mg/kg.

4.4 Water content (WC)

Water content is the quantity of water contained in the soil. It helps to maintain dissolved salts and to promote soil stratification (Hussain, 2007). It is simply express as ratio of mass of water hold in the soil (Wesley, 2010). Normally, it ranges between 15-80%. Soil with high amount of clay minerals generally contains high amount of water. It is measured by weighing a soil sample before and after it has been dried in an oven at a temperature 105°C till constant weight.

The soils of the study area have 11-74% WC with a mean value of 31%, indicating good reflection of soil texture. Nearly 14% soils have WC <15%. Mostly these soils are present in the northern barren area; however few of them are windblown soil. Soils with high WC >50% are less (6.94%) and present in the main farm area.

4.5 Soil organic matter (SOM)

Soil organic matter (SOM) is related to the productivity of a soil. Soil organic matter comprises of an accumulation of partially disintegrated and decomposed plant and animal residues and other organic compounds synthesized by the soil microbes as the decay occurs (Brady, 1990). The SOM can be divided into two major groups: Particulate Organic Matter (active fraction) and Humus (transitioning organic matter) with increasingly stable/complex compounds. The amount of SOM can vary from <1% in coarse-textured sandy soils to >5% in fertile cultivated soils. The SOM plays an important role in providing nutrients, stabilization of soil structure, water retention, cation exchange and pH buffering. The quantitative determination of SOM has high variability and ambiguities. Dry combustion and acid-dichromate methods are commonly employed for the measurement of SOM. In the present work, dry ignition method as described by Hussain (2007) is utilized. In the study area, SOM ranged between 0.35-12% with an average of 2% (Table 2). In the cultivated areas, it is high while in the rocky and wind-blown soil it is recorded low. The arid climate of the study area and sandy soil also hinder for high SOM.

	Minimum	Maximum	Average	SD	Median
pH	7.06	9.2	8.06	0.39	8
EC mS	1	22	8	5.29	5.67
TDS g/kg	1	15	5	3.68	4
Ca mg/kg	500	3507	1273	750	985
Mg mg/kg	100	2431	334	339	300
Na mg/kg	50	6875	631	1299	269
K mg/kg	13	475	96	67.87	81
Cl mg/kg	100	2747	938	691	798
SO_4 mg/kg	21	3149	803	679	700
WC %	11	74	31	13.61	28
SOM %	0.35	12	2	1.81	2
Ca:Mg	0.82	30.06	5.97	5.72	4.29
Ca:K	2.63	175	19.14	21.97	15.03
Mg:K	0.63	121	5.78	14.24	3.43
Ca epm	25	175	64	37.43	49.15
Mg epm	8	200	28	27.92	24.17
Na epm	2	299	27	56.5	11.68
K epm	0.32	12.13	2.45	1.73	2.07
ESP %	1	82	17	17.33	10.79
Cu mg/kg	0.03	8.8	2.66	2.73	1.1
Zn mg/kg	0.03	16.7	1.02	2.67	0.25
Pb mg/kg	0.08	10.55	5.12	4.14	7.55
Fe mg/kg	0.8	197.6	28.05	45.7	5.35
Cr mg/kg	0.07	9.8	4.69	2.61	4.45
Co mg/kg	0.05	2	0.97	0.38	1.05
Cd mg/kg	0.13	1.8	0.49	0.46	0.35
Ni mg/kg	0.52	13.3	3.12	2.9	1.78

ESP = Exchangeable Sodium Percentage (see Section 4.7 for definition)

Table 2. Statistical analysis of physical and chemical parameters of soils of Winder area.

4.6 Calcium (Ca) and magnesium (Mg)

Calcium improves the soil crust, structure and quality as well as reduces soil salinity, erosion and phosphorous loss. It also promotes root and leaf development, and helps to create a healthy environment for the plant growth. Calcium at optimum level will reduce disease in most plants (Dick, 2007). Imbalance of Ca will lead to tight, hardpan soils, which will restrict the flow of air and water through the soil profile (Trotter, 2011). Magnesium deficiency causes the loss of healthy green colour of leaves because of shortage of chlorophyll in plant. Magnesium is a mobile element in the plant, so the concentration of Mg usually decreases from top to bottom of plant. It also decreases according to plant maturity. Ca/Mg ratio is an important parameter to evaluate soil for agriculture purpose. A good soil has Ca:Mg between 5-8 with 6.5 as an ideal condition (Spectrum, 2011). In the presence of different rock exposures, the ratio varies. In the study area, 15 out of 72 samples i.e. 20.83% soils are within the ideal limit (5-8). Fourteen samples (19.44%) have high Ca:Mg exceeding 8. The majority of the samples 59.73% (43 samples) have low Ca:Mg (Fig. 6). This observation shows a good agreement with the geology of the area. Soils derived from ophiolites have nearly above range of ratio (Robinson et al., 1997). The present study also shows compatibility with the work of Naseem et al. (2005, 2009) carried out in the Khuzdar, north of the study area. The Ca/Mg ratio map shows that the Winder farm area has low values with few high ratio spots (Fig. 7A). The diagram shows decrease of Ca/Mg ratio in the east as well as in the west direction. High supply of Mg from ophiolitic rocks (east) and seawater from western side are mainly responsible for low Ca/Mg ratio in the study area.

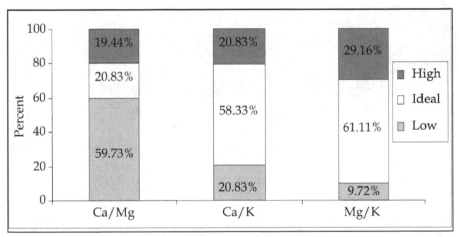

Fig. 6. Comparison of Ca/Mg, Ca/K and Mg/K ratios of soils of the Winder area, showing distribution percents of low, near ideal and high conditions.

In the study area, most of the sedimentary rocks consist of limestone such as, Parh Limestone, Malikhore and Kharrari formations. Soil becomes acidic when certain basic elements, especially Ca are removed by the percolation of rain water (Ca is very soluble in water) which carry the dissolved Ca down deep into the soil and by crops and plants removing the Ca as they grow and develop. The study area has hot and dry climate. In drier areas, the Ca remains undissolve and is not carried deep into the soil. As the Ca and some other elements build up, the soil becomes basic or alkaline (Holley, 2009).

4.7 Potassium (K) and sodium (Na)

Potassium (K) is an essential nutrient for plant growth and occupies third position after nitrogen and phosphorus. It enhances disease and drought resistance in plants. It also helps to develop fruit size, flavor and texture. Soil K is found in three forms; relatively unavailable, slowly available and readily available. Generally, 90-98% of the total K in soils is relatively unavailable in the form of primary minerals (muscovite, biotite, and feldspars); nearly 1-10% is the slowly available from clays and hardly 0.1-2% are readily available as exchangeable K (Buchholz & Brown, 1993).

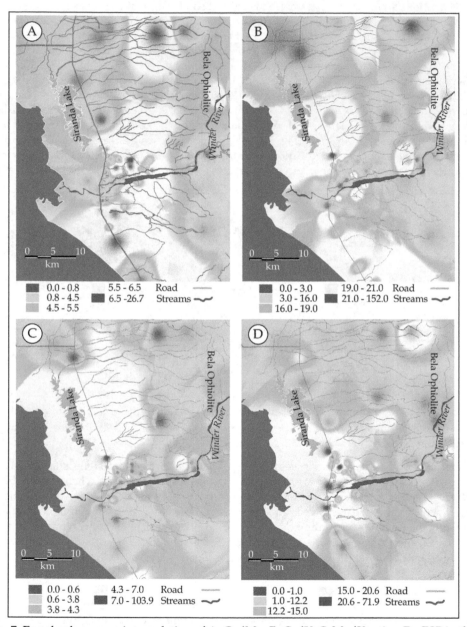

Fig. 7. Equal value range interpolation of A: Ca/Mg; B: Ca/K; C: Mg/K ratios; D: ESP in the soils of Winder and adjoining areas, based on inverse distance weighted (IDW) technique.

The level of K in the soil is termed low when it is <60mg/kg; medium (60-100) and high, when the readily available K is >100mg/kg. In the study area, nearly 29.17% soils have low K, 27.8% medium while rest have K >100mg/kg. Potassium in combination to Ca and Mg is also very significant to assessed quality of agriculture soils. The Ca/K ratio around 13:1 is approaching ideal condition. In the study area, nearly 58% samples fulfill the requirement, while the rest are either low or high (Fig. 6). The Ca/K ratio is ideal near farm areas and either side are low in K and high in Ca due to rock type (Fig. 7B). Similarly, Mg/K relationship in a ratio 2:1 is ideally fit for yielding good fruits. Here, the situation is similar, 61.11% samples are close to model ratio (Fig. 6). In and around fruit farms, the ideal condition exists with few low patches (Fig. 7C). Possibly the high Mg/K ratio is due to the presence of low K and high Mg-bearing tholeiitic basalt as one of the soil parent material in the area (Naseem et al., 1996-97). The high Ca/K ratio is due to the exposure of carbonate rocks. The other minor causes for high ratio are due to sandy soil, poor chemical weathering due to arid climate and continuous crop production since long time.

The primary problem posed by high sodium (Na) is not a toxicity hazard, but has a dispersive effect on agriculture soils, thus reduction of soil permeability and soil aeration. It can cause the clay particles to separate from each other; the particles will clog the soil pores, and cause variation in the permeability of the soil. This effect is more serious in fine textured soils than in coarse textured (Natural Resources Conservation Service [NRCS], 2007).

The amount of Na in the soil can be conveniently appraised by Exchangeable Sodium Percentage (ESP). It is calculated as follows:

$$ESP = Exchangeable \{(Na)/ (Ca + Mg + K + Na)\} \times 100$$

ESP measures the amount of soil exchange capacity occupied by Na ions and expressed as percentage. At higher ESP values, more exchangeable Na is available, and greater the potential for negative plant-soil impacts. An ESP >15% indicates that soil Na will probably limit permeability. The University of New South Wales [UNSW] (2007) has categorized soils on the basis of ESP into five classes (Table 3).

ESP Classes	ESP Values	Soil of Winder Area
Non-sodic	<6	22.2%
Sodic soil	6-10	25%
Moderately Sodic	10-15	11%
Strongly Sodic	15-25	14%
Very Strongly Sodic	>25	22.2%

Table 3. Soil classification of Winder area on the basis of ESP values (After UNSW, 2007).

The distribution of ESP in the study area illustrates high values close to Arabian Sea, which is the good source of Na in the soils (Fig. 7D). However, the farm area is safe except few high concentration zones.

4.8 Sulphate (SO_4^{2-}) and chloride (Cl^-)

Sulphur (S) is described as secondary plant nutrient and essentially required for crops. It is a major component of plant proteins and having an important role in the synthesis of chlorophyll. The deficiency of S in plants appears in the youngest leaves as pale-yellow colour (Rajendram et al., 2008). Sulphur exists in the soil predominantly as bound organic S (~94%); with small amounts (~3% each) existing as sulphate-S (SO_4-S) and extractable organic S (OS). Plant-available S is taken up by plants only in the sulfate (SO_4) form, through the roots because agricultural crops rarely respond to other types. Thus, most agriculture soil estimates plant-available S by extraction and determination of sulfate-S (Hayes, 2007).

The soils of Winder area show concentration of SO_4 in between 21-3149 with an average 803mg/kg (Table 2). According to Alloway (2005), the range of SO_4 in agriculture soil is between 90-30,000mg/kg; while ~2,000mg/kg is considered best for crop. All soil samples of the study area have low SO_4-S. The SO_4 is mainly due to sulphide mineralization in the rocks of Mor Range and to some extent from Bela Ophiolite.

Chloride (Cl) is recent addition in the list of micronutrient, needed in small quantities for crop growth. Chloride functions in photosynthesis and chemically balances the K concentration that increases in the guard cells during the opening and closing of stomata. On the other hand, high Cl may reduce the ability of roots to extract water and nutrients from the soil. Thus high Cl in the soil is major constraints to crop production (Dang et al., 2008). Chloride is taken up by plants as the Cl^- ion. In the soil, it is found as anion because it is highly soluble and most mobile.

In the study area, the rock containing appreciable amount of Cl is lacking. The soils in the surroundings of Winder show 100-2747 with an average 938mg/kg (Table 2). The Cl concentration of studied samples is much higher than world average of 100mg/kg (Flowers, 1988). Probably closeness to coastal area and arid climate is mainly responsible for elevated Cl content in the area. Additionally, it is supported from high ESP in the soil of the study area (Fig. 7D).

4.9 Trace elements

The natural concentration of trace elements in soils is a result of weathering that releases trace elements from their host minerals during soil formation (Kabata-Pendias, 1993). A close relationship between the metal content of the parent material and soils has been observed in a number of studies (Singh & Steinnes, 1994). The mobility, solubility and bioaccumulation of trace elements depend on the properties of the trace elements as well as the quality of soil, pH and other factors. The estimated amount of trace elements in the soil is also depends upon the medium of extraction. It varies for different extractors and noted minimum for simple water (Kabata-Pendias, 2004). For the present study, water extraction was used, which is close to bioavailability for plants. The bioavailability of trace elements in the soil is decisive for agriculture purposes.

The speciation of trace metals ultimately determines their bioavailability and their mobility in the soil (Pelfrêne et al., 2009). A number of trace elements (Cu, Zn, Pb, Fe, Cr, Co, Cd and Ni) are contributed through the weathering of igneous and sedimentary rocks in the study area. In order to assess the role of rock in dispersing the elements, a graph is constructed to illustrate mutual relationship (Fig. 8). It is important to note the concentration trend is quite similar except the level of enrichment. The concentration level of Cu, Zn, Cr & Co is high in the rocks and nearly 1% is transformed in the soils of the study area. Lead and Cd show relatively high proportion in the soil, while Fe and Ni are hardly half percent in the soils of the study area (Fig. 8).

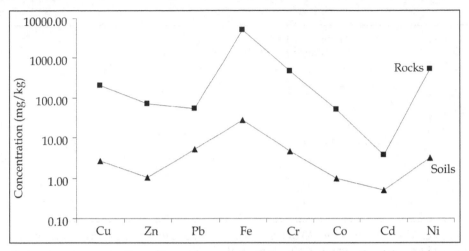

Fig. 8. Average concentration of trace elements in the rocks and the soils of the study area.

The migration of an element from rock to soil or water is termed as its mobility, i.e. the ease with which an element is move from one phase (rock) to other phase (soil). The migration can be assessed numerically from the following expression (Rose et al., 1979), modified for extractable elements of soil.

$$\text{Coefficient of Aqueous Migration } (K) = 100 \times s \ / \ t \times r$$

Where s is the concentration of the trace element in the soil (mg/kg); t is total water soluble salts (mg/kg) and r is the concentration of the trace element in the rock (%).

The assessment of trace elements indicates that Fe is immobile, Cu, Zn, Cr, Co and Ni are semi-mobile while Pb and Cd are moderately mobile elements (Fig. 9). The present situation of trace elements abundance in the soil is best demonstrated through the study of K, which is in fact influenced by the climatic condition, distance from the rock and the nature of the element. In the following paragraph, occurrence and distribution of some of the important trace elements of the soils of the study area is discussed to correlate linkage between rocks-soils-plants.

4.9.1 Chromium (Cr)

Chromium is mainly derived from the ultramafic rocks (1600-3400mg/kg), and range in the soil from 1.4-1100 mg/kg (av. 54mg/kg), depending on the soil type and other physicochemical parameters (Kabata-Pendias & Pendias, 1992). Chromium is generally found as two species Cr^{+3} and Cr^{+6} in soils. The former (Cr^{+3}) is common and serve as a micronutrient and a non-hazardous species, much less toxic than Cr^{+6} (Fendorf, 1995). The major part of the absorbed Cr is retained in the roots however it is relatively low in the leaves (5-30 mg/kg). Organic matter reduces Cr^{+6} to Cr^{+3}, which is less bioavailable to plant, because geochemically, Cr^{+3} is inert. The presence of Cr in the soil of the Winder area is best reflection of ophiolitic rocks. Its abundance is measured in the range of 0.07-9.8mg/kg (Table 2). The average value is (4.69mg/kg) beyond the world average (54mg/kg) probably due to sandy soil and organic matter in the study area.

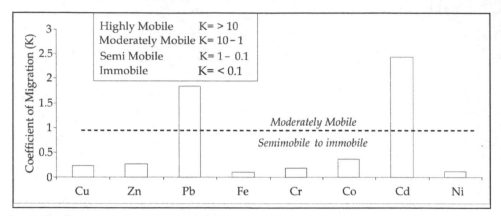

Fig. 9. Coefficient of migration (*K*) of trace elements calculated from average concentration in the rocks and soils (Categories after Rose et al., 1979).

4.9.2 Nickel (Ni)

Nickel is the most abundant (2000mg/kg) trace element of the ultramafic rocks and decreases in mafic rock (150mg/kg). Primary Ni minerals (Millerite, pentlandite, niccolite etc.) will release Ni^{2+} upon weathering into soil which is stable in the soil. In the study area, Ni is mainly derived from ophiolitic rocks and ranged between 0.52-13.3mg/kg in the soils of Winder area (Table 2). The average is lower than the world median of Ni, which is about 20mg/kg (Alloway, 2005). There is no harm of Ni in the plants through soil in the study area because the average Ni is ~3mg/kg. The present study shows good correlation matrix of Ni with Fe ($r = 0.843$) and Cd ($r = 0.788$), moderate with Zn ($r = 0.578$) and inverse ($r = -0.568$) to Pb (Table 4), reflecting origin of Ni from sulphide phase of Bela Ophiolite. Sometimes recent windblown sediments also contribute substantial amount of Ni-Cr in the soil, as it was noted by Varkouhi et al. (2006) in the neighboring country Iran.

Bioavailability of Ni decreases in alkaline pH. Furthermore, presence of Fe-Mn oxides and high organic matter also lower its availability to plants (National Research Council of Canada [NRCC], 1981). Despite the toxicity, Ni is easily translocated within plants and commonly accumulates in high quantities in leaves, exhibiting chlorosis (Kabata-Pendias & Pendias, 1992).

	Zn	Pb	Fe	Cr	Co	Cd	Ni
Cu	0.314	-0.010	0.455	0.460	0.079	0.471	-0.046
Zn		-0.316	**0.710**	0.152	0.223	**0.67**	**0.578**
Pb			**-0.567**	0.453	0.245	-0.383	**-0.568**
Fe				0.053	0.004	**0.911**	**0.843**
Cr					0.337	0.239	0.163
Co						0.09	-0.118
Cd							**0.788**

Table 4. Correlation matrix of important trace elements of soil of Winder area.

4.9.3 Iron (Fe)

Iron is the common element which may be contributed through both the different segments of ophiolites and sedimentary rocks of the study area. The water soluble Fe is in the range of 0.8-197.6mg/kg (av. 28.05). The mutual relationship among Ni-Fe-Cr in the form of triangular diagram (Fig. 10A) exhibit two separate populations, one is dominated by Cr and other by Fe. Chromium is signature for soils that developed from ultramafic rocks (Garnier et al., 2006), while high Fe is related with either mafic or sedimentary rocks.

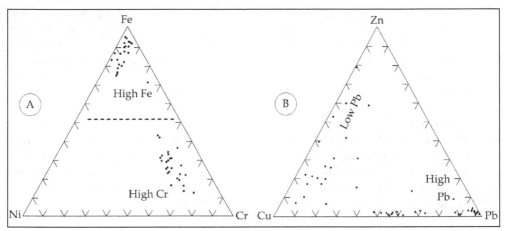

Fig. 10. Mutual relationships among A) Ni-Fe-Cr and B) Cu-Zn-Pb showing two populations of soil samples in the study area.

4.9.4 Copper (Cu)

Copper is found in relatively high levels in mafic rocks (60-120 mg/kg), which are widely exposed as pillow lavas in the study area. In the Winder area, the level of concentration of Cu in the soils is between 0.03-8.8mg/kg. The average (2.66mg/kg) is lower than world average 13-24mg/kg (Yusuf, 2007). In spite of Cu- mineralization in the ophiolitic rocks, the soils of the study area are deficient in Cu. Probably, basic pH and calcareous sandy soil reduces the mobility of Cu (Irha et al., 2009). It is an important biophile element and found as organic complex. This complexing is of great significance in maintaining adequate Cu in solution for plant use. Plants need Cu in small quantities (5-10mg/kg) for photosynthesis, and metabolism. However, Cu >100mg/kg in the soil is lethal for most common plants.

4.9.5 Zinc (Zn)

It is an essential micronutrient, required for the healthy growth of the plants. Zinc helps in chlorophyll formation and promotes formation of acetic acid in the root to prevent decaying. It motivates plant growth and prevents mottling and other disorders in the leaves (Andersen, 2000). The concentration level is 0.03-16.7; av. 1.02mg/kg in the soils of the study area (Table 2). Although, the Zn can be supply both from ophiolite and sedimentary rocks of the area, but the level is much lower than the world average (64mg/kg). Possibly the poor availability in the soil is due to long-time farming in the area.

4.9.6 Lead (Pb)

There are multi sources of Pb in the rocks of the study area, but most commonly it is derived from the ophiolites. Lead is a non essential toxic element for plants. In general, plants do not favor to absorb much Pb from soil and the translocated Pb is accumulated in the leaf. Symptoms of toxicity appear on leaf nearly in the range of 30-300mg/kg (Kabata-Pendias & Pendias, 1992). It varies between 0.08 to 10.55mg/kg in the soils of study area (Table 2). Comparing to the mean abundance in the soil (32mg/kg), the soils of the study area is quite low (av. 5.12). Despite the low abundance, Pb is one of the dominant elements in contrast to Cu and Zn (Fig. 10B). However, Cu-Zn relation in the same diagram display scattered population in which Pb is low. Probably the separate population represents two different sources. Low Pb in the soils is due to low mobility (Irha et al., 2009), especially in the arid climate. It is important to note that Pb has inverse relationship with Fe and Ni, indicating replacement with each other in the soils of Winder area (Table 4). Lead usually occurs in the soil as Pb^{2+}/Pb^{4+} and accumulates in the surface horizons of soils.

4.9.7 Cobalt (Co)

The ultramafic (100-200mg/kg) and mafic rocks (35-50mg/kg) contain elevated amount of Co and these rocks are widely exposed in the study area. The overall range for cobalt in soils on a world-wide basis is 0.1-70mg/kg (Kabata-Pendias & Pendias, 1992) and the average amount is 8mg/kg (Tisdale et al., 1985). In the Winder area soils, it ranges 0.05-2mg/kg (av. 0.97mg/kg). Despite the presence of mafic-ultramafic rocks in the area, it is measured very low. Alkaline and calcareous soil, organic matter and high Fe reduce the mobility of Co.

Biological function of Co is not very clearly understood but it is considered essential for the plants in minor quantities. The low abundance of Co in the soils of the study area possibly does not impart threat of Co in human beings through food.

4.9.8 Cadmium (Cd)

Cadmium is commonly found as sulphide mineral (Greenockite, CdS) in the mafic rock, ranges between 0.13-0.22mg/kg (Kabata-Pendias & Pendias, 1992). The weathering of rock release Cd ions, which are than accumulated in the soil in the range from 0.06-1.1mg/kg. Cadmium is phytotoxic element; it can be absorb through root and can be store in the range of 5-30mg/kg in the leaves at sub-toxic level, but it can accumulate less in the edible parts of the plant (Alloway, 1990).

The range of Cd in the soils of the study area is 0.13-1.8 with an average 0.49mg/kg, which is close to the world average 0.35 (Coskun et al., 2006). The possibilities of potential health hazards of Cd to consumers of fruits grown is minimum because the bioavailability of Cd decreases in the presence of soil carbonates and basic pH (Oluwatosin et al., 2008), as it is common in the study area. The strong correlation matrix with Fe ($r = 0.911$) and Ni ($r = 0.788$) will suggest its origin from ophiolitic rocks (Table 4).

5. Soil mineralogy

The XRD mineralogical studies of the soils of the study area revealed high quartz and calcite along with variable proportions of clay minerals and iron oxides. High quartz is not suspected from basalt country rock, so the large amounts of quartz in the soils of the study area may have contributed from windblown sediments, as indicated by Prone (2003) for the soils of Thailand. The minor mineral content of the soil is more valuable because they reflect the nature of parental rocks. These include gismondine, albite, orthoclase, anorthite, palygorskite, hercynite and rutile. Descriptions of uncommon soil minerals such as gismondine, palygorskite, chlorite-serpentine and hercynite are given below:

Gismondine is a low temperature zeolite formed by the alteration of Ca-bearing (anorthite) basaltic rocks. It is a low-silica zeolite, associated with nepheline and olivine basalt and leucite tephrite (Dyer et al., 2006; Katsuki et al., 2008). Zeolites have large vacant spaces or cages in their structures that allow space for large cations such as Na, K, Ca and Ba. The alumino-silicate structure is negatively charged and attracts the positive cations and hence the members of zeolite are capable to hold many trace elements in high quantities. It is a hydrated calcium aluminosilicate with the formula $Ca_2Al_4Si_4O_{16}\cdot 9H_2O$.

Palygorskite is a crystalline hydrated magnesium phyllosilicate ($(MgAl)_2Si_4O_{10}4H_2O$) with a fibrous habit. Its structure and absorbed cations provide it with a large specific surface area and moderate cation exchange capacity, which is very beneficial for the adsorption of heavy metals from solution (Frini-Srasra & Srasra, 2010). There are two sources of palygorskite in the soil. It is an alteration product of magnesium silicates and on the other hand it is also derived from the soil of arid regions and reported from Cretaceous-Tertiary sedimentary rocks (Daoudi, 2004). Despite the presence of both sources, the occurrence of palygorskite in the soils of the study area is less. The associated minerals in the soil indicate an igneous source.

Hercynite ($FeAl_2O_4$) is a member of the spinel group, commonly associated with ophiolite and other high grade metamorphic minerals like chromite, magnetite, ilmenite, sillimanite and andalusite (Guiraud et al., 1996). It is only reported from one locality and strengthens the idea that the soils of Winder area have influence of Bela Ophiolite.

6. Conclusion

The trace elements geochemistry of the soils of the Winder area shows relevance mainly with Bela Ophiolite of Cretaceous age along with sedimentary rocks of Mor and Pab ranges belong to Cretaceous and Tertiary ages. The Cr, Ni, Cu, Co and Cd in the soil are mainly supplied by ophiolitic rocks while high Fe and Zn can be contributed both from ophiolite and sedimentary rocks of the area. The average values of these elements are lower than world average probably due to basic pH, calcareous sandy soil and organic matter in the study area. The XRD mineralogy revealed high quartz and calcite along with variable proportions of clay minerals and iron oxides. Gismondine, albite, orthoclase, anorthite, palygorskite, hercynite, chlorite-serpentine and rutile are as minor minerals, confirming their alliance with the rocks of Bela Ophiolite.

Texturally, Entisols types of soil with some Aridisols are common. The soils on the sand-silt-clay ternary diagram are classified as sandy loam, sand and silt loam reflects the mixture of sand and loam in the study area which is suitable for agriculture. The average water content (WC) of the soils is 31%. The soil organic matter (SOM) ranged between 0.35-12% with an average of 2%; relatively high in cultivated areas and low in the rocky and wind-blown soil. The assessed Ca/Mg ratio is low while Ca/K and Mg/K ratios are ideal near farm areas while on either side these ratios vary from low to high due to rock exposures, sandy soil, poor chemical weathering and continuous crop production since long time. The distribution of ESP in the study area illustrates high values close to Arabian Sea; however, the farm areas are safe except few high concentration zones. The chloride concentration is much higher than world average while SO_4-S is low. Trace elements demonstrate Fe>Pb>Cr>Ni>Cu>Zn>Co>Cd abundance trend in the soil of the study area. The concentration level of these trace elements is low in contrast to world average, thus there is little harm suspected from the toxic trace element in the area.

The study of major and trace element geochemistry of an agriculture soil, located close to rocks exposure, is recommended for the proper management of nutrients and micronutrients as this will not only enhance the productivity and quality of the crops but also mitigate an environmental threat to humans through the food chain.

7. References

Aghazadeh, N. & Mogaddam, A.A. (2010). Assessment of Groundwater Quality and its Suitability for Drinking and Agricultural Uses in the Oshnavieh Area, Northwest of Iran. *Journal of Environmental Protection*, Vol. 1, pp. (30-400, ISSN Online: 2152-2219

Ahsan, S.N. & Mallick, K.A. (1999). Geology and genesis of Barite deposits of Lasbela and Khuzdar Districts, Balochistan, Pakistan. *Resource Geology*, Vol. 49, pp. (105-111)

Alloway, B.J. (2005). Bioavailability of Elements in soil, In: *Essential of Medical Geology, impacts of the naturak environment on public health*, Selinus, O. (ed.), pp. (347-372), Elsevier Academic Press, Amsterdam

Alloway, B.J. (1990). Cadmium, In: *Heavy Metals in Soils*, Alloway, B.J. (ed.), pp. (100-124) Blackie and Son, Ltd., Bishopbriggs, Glasgow

Andersen, A. (October, 2000). *Science in Agriculture: Advanced methods for sustainable farming*, (2nd ed.), Acres USA, ISBN 10: 0911311351, USA

Brady, N.C. (1990). *The nature and properties of soils* (10th), MacMillan Publishing Co., ISBN 0-13-852444-0, New York.

Briggs, D., Smithson, P., Addison, K. & Atkinson, K. (1998). *Fundamentals of the Physical Environment*, (2nd ed.), Routledge, ISBN 0 415 10891 8, New York

Buchholz, D.D. & Brown, J.R. (1993). Potassium in Missouri Soils, Published by University of Missouri Extension, guidelines to reprint or copy, http://extension.missouri.edu/publications/DisplayPub.aspx?P=G9185

Chesworth, W. (ed.), (2008). *Encyclopedia of soil science*, ISBN 978-1-140203994-9 Dordrecht, Netherland: Springer

Coskun, M., Steinnes, E., Frontasyeva, M.V., Sjobakk, T.E. & Demkina, S. (2006). Heavy metal pollution of surface soil in the Thrace Region, Turkey, *Environmental Monitoring and Assessment*, Vol. 119, No. 1-3, (June, 2006), pp. (545–556)

Dang, Y.P., Dalal, R.C., Mayer, D.G., McDonald, M., Routley, R., Schwenke, G.D., Buck, S.R., Daniells, I.G., Singh, D.K., Manning, W. & Ferguson, N. (2008). High subsoil chloride concentrations reduce soil water extraction and crop yield on Vertosols in north-eastern Australia. *Australian Journal of Agricultural Research*, Vol. 59, No. 4, pp. (321-330)

Danoff-Burg, J. A. (2000). The Terrestrial influence: Geology and soils, http://www.columbia.edu/itc/cerc/seeu/atlantic/restrict/modules/module10_content.html

Daoudi, L. (2004). Palygorskite in the uppermost Cretaceous-Eocene rocks from Marrakech High Atlas, Morocco, *Journal of African Earth Sciences*, Vol. 39, pp. (3-358)

Dave's Garden Copyright © 2000-2010 Dave's Garden, an Internet Brands company. All Rights Reserved http://davesgarden.com/guides/pf/b/Sapotaceae/none/none/cultivar/0/

http://davesgarden.com/guides/pf/go/174318/

Dick, C. (December 10, 2007). Calcium, more than just pH, In: *Calcium Product Incorporated,* http://blog.calciumproducts.com/posts/calcium-more-than-just-ph.cfm

Dyer, A., Heywood, B. & Szyrokyj N. (2006). Cation exchange in the synthetic gismondine-zeolite MAP, *Microporous and Mesoporous Materials,* Vol. 92, No. 1-3, (Dec. 2010), pp. (161-164), ISSN 1387-1811

Fendorf, S.E. (1995). Surface reactions of chromium in soils and waters, *Geoderma,* Vol. 67, No. 1-2, (June 1995), pp. (55-71)

Flowers, T.J. (1988), Chloride as a nutrient and as an osmoticum, In: *Advances in Plant Nutrition,* Thinker, B., Läuchli, A. (eds.), Vol. 3, pp. (55-78), Praeger, New York

Frini-Srasra, N. & Srasra, E. (2010). Acid treatment of south Tunisian palygorskite: Removal of Cd (II) from aqueous and phosphoric acid solutions. *Desalination,* Vol. 250, No. 1, (January 2010), pp. (26-34)

Gana, B.K. (2000). Effects of Agriculture on Soil Quality in Northeastern Ghana, Ph.D. dissertation, Department of Soil Science, University of Saskatchewan, Saskatoon, Canada

Garnier, J., Quantin, C., Martins, E.S. & Becquer, T. (2006). Solid speciation and availability of chromium in ultramafic soils from Niquela^ndia, Brazil, *Journal of Geochemical Exploration,* Vol. 88, No. 1-3 (Special Issue), pp. (206-209)

Gnos, E., Khan, M., Mahmood, K., Khan, A.S., Shafique, N.A. & Villa, I.M. (1998). Bela oceanic lithosphere assemblage and its relation to the reunion hotspot. *Terra Nova,* Vol. 10, No. 2, pp. (90-95)

Gnos, E., Immenhauser, A. & Peters, Tj. (1997). Late Cretaceous/early Tertiary convergence between Indian and Arabian plates recorded in ophiolites and related sediments. *Tectonophysics,* Vol. 271, pp. (1-19)

Guiraud, M., Kienast, J.R. & Ouzegane, K. (1996). Corundum–quartz-bearing assemblage in the Ihouhaouene area, (In Ouzzal, Algeria). *Journal of Metamorphic Geology,* Vol. 14, No. 6, pp. (755–761)

Gupta, P.K. (2004). *Methods in Environmental Analysis Water, Soil and Air,* Agrobio, India

Haluschak, P.W., Mills, G.F., Eilers, R.G. & Grift, S. (1998). Satus of selected trace elements in agricultural soils of Southern Manitoba. *Technical Bulletin 1998-6E,* Land Resource Unit, Brandon Research Centre, Research Branch, Agriculture and Agri-Food Canada, ISBN 0-662-27098-3

Hayes, C.F. (2007). *Test Methods for Water-Soluble Sulfate in Soils.* Portland Cement Association, PCA R&D Serial No. 3016

Holley, D. (2009). Soil Nutrients and pH Affect Plant Growth: The Availability of Minerals and Soil pH Influence Growth in Plants, In: Botany@suite101 http://www.suite101.com/content/soil-nutrients-and-ph-affect-plant-growth-a131001

Hunting Survey Co. Ltd. [HSC], (1960). *Reconnaissance geology of part of West Pakistan;* a Colombo plan-Co-operative project, Canada.

Hussain, S.S. (2007). *Pakistan Manual of Plant Ecology,* National Book Foundation, Islamabad.

Iqbal, M. & Ahmad, M. (2011). Science & Technology Based Agriculture Vision of Pakistan and Prospects of Growth, http://www.pide.org.pk/pdf/psde20agm/science%20%20technology%20based%20 agriculture%20vision%20of%20pakistan%20and%20prospects%20of%20growth.pdf

Irha, N., Steinnes, E., Kirso, U. & Petersell, V. (2009). Mobility of Cd, Pb, Cu, and Cr in some Estonian soil types. *Estonian Journal of Earth Sciences*, Vol. 58, No. 3, (September, 2009), pp. (209-214) ISSN 1736-7557

Kabata-Pendias, A. (2004). Soil-plant transfer of trace elements-an environmental issue. *Geoderma*, Vol. 122, No. 2-4, (October, 2004), pp. (143–149)

Kabata-Pendias, A. (1993) Behavioural properties of trace metals in soils. *Applied Geochemistry*, Vol. 8, No. Suppl. 2, (January 1993), pp. (3–9)

Kabata-Pendias, A. & Pendias, H. (1992). *Trace Elements in Soils and Plants* (2nd ed.), CRC Press, Inc., Boca Raton, Florida

Katsuki, K., Yoneoka, S., Mori, N., Hasegawa, M., Yamamoto, Y. & Yoshino, Y. (2008). Characterization and properties of iron-incorporated gismondine, prepared at 80°C. *Journal of Porous Materials*, Vol. 15, No. 1, (2008), pp. (35-42), ISSN 1380-2224

Kazmi, A.H. & Abbasi, I.A. (2008). *Stratigraphy and Historical Geology of Pakistan*. National Centre of Excellence in Geology, University of Peshwar

Khan, M., Kerr, A.C. & Mahmood, K. (2007). Formation and tectonic evolution of the Cretaceous-Jurassic Muslim Bagh ophiolitic complex, Pakistan: Implications for the composite tectonic setting of ophiolites. *Journal of Asian Earth Sciences*, Vol. 31, No. 2, (October, 2007), pp. (112-127)

McKenzie, R.H. (2003). Soil Fertility/Crop Nutrition, In: *Government of Alberta, Agriculture and Rural Development, Agdex 531-4*
http://www1.agric.gov.ab.ca/$department/deptdocs.nsf/all/agdex6607

Naseem, S., Bashir, E., Shireen, K. & Shafiq, S. (2009). Soil-plant relationship of Pteropyrum olivieri, a serpentine flora of Wadh, Balochistan, Pakistan and its use in mineral prospecting, *Studia Universitatis Babeş-Bolyai, Geologia, Fluorida, USA*, Vol. 54, No. 2, (Available online September 2009), pp. (33-39)

Naseem, S., Naseem, S. & Sheikh, S.A. (2005). Geochemical Evaluation of Depositional Environment of Parh Limestone, Southern Pab Range, Balochistan, Pakistan. *Proceeding of SPE/PAPG Annual Technical Conference*, Islamabad, November, 2005

Naseem, S., Sheikh, S.A., Qadeeruddin, M. & Shirin, K. (2002). Geochemical Stream Sediment Survey in Winder Valley, Balochistan, Pakistan. *Journal of Geochemical Exploration*, Vol. 76, No. 1, (July 2002), pp. (1-12)

Naseem, S., Sheikh, S.A. & Qadeeruddin, M. (1996-97). Geochemistry and tectonic setting of Gadani-Phuari segment of Bela Ophiolites, Balochistan, Pakistan. *Journal of King Abdul Aziz University, Earth Science*, Vol. 9, pp. (127-144)

National Research Council of Canada [NRCC], (1981). Effects of Nickel in the Canadian Environment. *Associate Committee on Scientific Criteria for Environmental Quality, National Research Council of Canada*, Ottawa. NRCC No. 18568

Natural Resources Conservation Service [NRCS], (2007). Soil sodium testing, In: *National Resources Conservation Service, US Department of Agriculture;*
http://www.mt.nrcs.usda.gov/technical/ecs/plants/technotes/pmtechnoteMT60/sodium_test.html

Neal, C. & Shand, P. (2002). Spring and surface water quality of the Cyprus ophiolites. *Hydrology and Earth System Sciences*, Vol. 6, No. 5, pp. (797-817), ISSN 1027-5606

Oluwatosin, G.A., Adeyolanu, O.D., Dauda, T.O. & Akinbola, G.E. (2008). Levels and geochemical fractions of Cd, Pb and Zn in valley bottom soils of some urban cities in southwestern Nigeria. *African Journal of Biotechnology*, Vol. 7, No. 19, (October, 2008), pp. (3455-3465)

Pakistan Agriculture Economy and Policy [PAEP], (2009). http://www.fas.usda.gov/country/Pakistan/Pakistan%20Agriculture%20and%20 Policy%20Report.pdf

Pelfrêne, A., Gassama, N. & Grimaud, D. (2009). Mobility of major-, minor- and trace elements in solutions of a planosolic soil: distribution and controlling factors, *Applied Geochemistry*, Vol. 24, pp. (96-105), ISSN 0883-2927

Prone, A. (2003). *L'analyse Texturale et Microstructurale des Sols: Exemple Pedologique du Nord-Est de la Thailande*. Publications de l' Universite de Provence (in French)

Rajendram, G., Ghani, A., Waller, J., Kear, M., Watkinson, J., Benge, K. & Wheeler, D. (2008). Total Sulphur: A Better Predictor of Sulphur Deficiency in Pastoral Soils. In: *Occasional Report No. 21*. Currie, L.D. & Yates, L.J. (eds.), FLRC, Massey University, Palmerston North, New Zeland

Reluy, F.V., de Paz Becares, J.M., Hernandez, R.D.Z. & Sánchez Díaz, J. (2004). Development of an equation to relate electrical conductivity to soil and water salinity in a Mediterranean agricultural environment. *Australian Journal of Soil Research*, Vol. 42, No. 4, pp. (381-388)

Robinson, B.H., Brooks, R.R., Kirkman, J.H., Gregg, P.E. H. & Alvarez, H.V. (1997). Edaphic influences on a New Zealand ultramafic ("serpentine") flora: a statistical approach. *Plant Soil*, Vol. 188, No. 1, (January 1997), pp. (11-20)

Rose, A.W., Hawkes, H.E. & Webb, J.S. (1979). *Geochemistry in mineral exploration*, (2nd edition). Academic press, London

Shah, S.M.I. (2009). *Stratigraphy of Pakistan*, Memoir of the Geological Survey of Pakistan, Quetta

Sheth, H.C. (2008). Do major oxide tectonic discrimination diagrams work? Evaluating new log-ratio and discriminant-analysis-based diagrams with Indian Ocean mafic volcanics and Asian ophiolites. *Terra Nova*, Vol. 20, pp. (229–236)

Singh, J. & Dhillon, S.S. (2004). *Agricultural Geography*, (3rd edition), Tata McGraw-Hill Publishing Company Limited, ISBN 0070532281, New Delhi

Singh, B.R. & Steinnes, E. (1994). Soil and water contamination by heavy metals, In: *Soil Process and Water Quality*, Lal, R. & Stewart, B.A. (eds), pp. (233–270), Lewis Publisher, Boca Raton, FL, USA. 233.270

Spectrum Analytic Inc, (2011). Calcium Basics, http://www.spectrumanalytic.com/support/library/ff/Ca_Basics.htm

Sutton, M. (2003). Organic farming technical summary, In: *OFTS No 38, SAC* http://www.sac.ac.uk/mainrep/pdfs/ofts38traceorganicvegetables.pdf

Tisdale, S.L., Nelson, W.L. & Beaton, J.D. (1985). *Soil Fertility and Fertilizers*. (4th edition), MacMillan Publishing Co. New York

Troeh, F.R. & Thompson, L.M. (2005). *Soil and Soil Fertility* (6th edition), Blackwell Publishing

Trotter, D. (2011). *Calcium: The Soul of Soil and Plant Health*, The Dollar Stretcher, Inc. Bradenton FL 34280

University of New South Wales (UNSW), 2007. Exchangeable Sodium Percentage (ESP), Terra GIS,

http://www.terragis.bees.unsw.edu.au/terraGIS_soil/sp_exchangeable_sodium_
    percentage.html
Varkouhi, S., Lasemi, Y. & Kangi, A. (2006). Geochemical evaluation of toxic trace elements
    in Recent Wind Driven Sediments of Zahedan Catchment Area, *Proceedings of the
    5th WSEAS International Conference on Environment, Ecosystems and Development,*
    Venice, Italy, Nov. 20-22, 2006.
Wesley, L.D. (2010). *Fundamentals of soil mechanics for sedimentary and residual soils,* John
    Willey & Sons Inc. ISBN 13: 9780470376263
Yusuf, K.A. (2007). Sequential extraction of lead, copper, cadmium and zinc in soils near Ojota
    Waste Site. *Journal of Agronomy,* Vol. 6, No. 2, (2007) pp. (331-337), ISSN1812-5379

Part 2

Managing Irrigation of Crops

6

Influence of Irrigation, Soil and Weeding on Performance of Mediterranean Cypress Seedling in Nursery

Masoud Tabari
Tarbiat Modares University,
Iran

1. Introduction

Mediterranean cypress, an evergreen softwood species, is well-known as *Cupressus sempervirens* L. var. *horizontalis* Mill. (Gord). It is an indigenous species of Mediterranean Europe and western Asia, including, Cyprus, Crete, Turkey, Syria, Saudi Arabia, Caucasian and Iran, resembling the Mediterranean climate (Sabeti, 2004). It is belonged to warming period and interglacial climate. Plant structure has a lot of the Mediterranean elements (Sagheb-Talebi et al., 2003). In Iran among the fifth-fold native coniferous trees (*Juniperus communis, J. Sabina, J. polycarpus, Biota orientalis*) it plays a unique role in restoration of deforested area of the Mediterranean zones (Rezaei, 1992); and the most important natural habitats of this species are found in northern regions and in special bio-geographical stations, as well as in warm valleys subjected to the southern winds (Mossadegh, 1996). The best developed Mediterranean cypress forests are found in the Chalous valley, and then, in the valley of Sefid-Roud (Roudbar). As small and isolated areas, it distributes in other parts of the country, particularly in west and south, in provinces of Lorestan, Sistan, Fars, Khozestan and Kohkilouye-Booyerahmad (Zare, 2002). Some scientists believe that the Mediterranean and semi-Mediterranean climates occurred in some parts of Iran are the major reason for distribution of this tree. The components of the flora of these forests include many Mediterranean elements (Asadolahi, 1991).

Although some conifers show better performance in more fertile beds (substrate or medium) but Mediterranean cypress does not care the nutritional quality of soil and is able to survive in poor and dry habitats. It is a resistant tree to hard conditions and in some parts of northern Iran particularly in Marzan-Abad, Roudbar and Manjil it exists on low nutrient, superficial and calcareous soils. It has been known as a low nutrient demanding species; its strong root system makes it able to easily establish on steep slope, rock and cliff. It tolerates well high dry; on rich soil it grows fast and on moist soil the root system forms shallow (Bolandian, 1999).

Since 1980, in dry slopes and semi-arid zones of northern Iran, more than 20.000 ha, have been afforested with this species. This is while that the increased destruction of Hyrcanean forests and the need for development of afforestation and green space in some capable regions have been given rise to producing its seedlings in forest nurseries. In 1996 the seedling production in communal nurseries was 96 million, with majority of this species. In the nurseries, due to weed competition, drought stress and poor nutrient soil, mortality rate is high and growth is low in the first growing season. Thus, the seedlings should be retained in nursery for next year; consequently, the seedling production would be costly. Generally, in nurseries, increase of seedling production, decreases of mortality rate and costs (especially for weeding and irrigation) are of main importance. Hence, good elaboration of inhibiting factors of growth and establishment and careful understanding of the ecological requirements of species are necessary (Krasowski et al. 2000).

Generally, regarding to increasing the afforestation with Mediterranean cypress in north of country, research on growth performance of seedling production of this species is inevitable (Tabari & Saeidi, 2007). Regarding to above mentioned, this question arises whether poor soils like sandy soils, longer periods of watering and weeding can decrease the nursery cost and improve the production efficiency. Hence, this research aims to determine the best treatment for better growth and higher seedling production by testing the different treatments of soil, irrigation and physical weed control.

2. Materials and methods

2.1 Sampling

In April 2008 some seeds of Mediterranean cypress (*Cupressus sempervirens*) collected from some elite trees, following 24 hours saturation in water, were sown in 108 plots 1 m × 1 m in an open area of Shahrposht nursery (Nowshahr city, north of Iran,). Based on metrological census of the study site (After Tabari et al. 2002), mean annual precipitation is 1100 mm, mean annual temperature, mean max. temperature of the warmest month and mean min. temperature of the coldest month is 16.4 °C, 30 °C and 3.7 °C, respectively. Number of dry days of year (xerothermique index) is 55 and Pluviothermique index (Q_2) is 143.6. By the classification of Emberger (1932), the climate of nursery is humid with cool winters. Seeds were kept in a suitable place and their purity and viability determined in the laboratory. Twenty seven seedlots (plots) 1 × 1 m were established (seedlot spacing = 50 cm, corridor width =100 cm) with substrates prepared from soils mentioned in Table 1. In early spring 2009, in each seedlot 400 sound seeds, moistened for 24 hours, were sown in 4 rows and covered with a thin (about 3 mm) sand layer. For controlling fungus, Agrotis and larvae, Capton and Diazinon (56%) and Malathion solutions were used. Also, slug was trapped using the Sevin toxin (Toxin 1 kg+Rice bran 12 kg as paste) and Alderin toxin (toxin 400 g in 220 liter water) was applied for controlling the syringe.

Soil type	Sand (%)	Silt (%)	Clay (%)
Loam-Clay (A)*	21	41	38
Sand (B)	83	9	8
Sand-Loam-Clay (C)	47	27	26
Loam (D)	42	49	9

* Common nursery soil

Table 1. Soil type and texture components (%) of substrates for growing C. *sempervirens* seedlings

In early May, following the seeds germination, watering was carried out at one day intervals, except in rainy days, as long as the seedlings did not become woody and reach 5 cm in height. Until commencement of the applied treatments, weeding was accomplished thrice in order to successful establishment of the seedlings produced. Watering treatment began since mid-June and continued to mid-October (4 months) with 4-, 8- and 12-day intervals. During the investigation period three weeding levels (20-, 30- and 40-day) were done manually, without using herbicides, and it was tried to prevent the removal of newly-grown seedlings. After the seed germination, during the care periods no fertilizer was added into the nursery substrate.

2.2 Measurement

The germinated seedlings in the experimental plots were counted and measured in late autumn (after ending the weeding and irrigation periods). In order to determine the differences among means (in different soil treatments, irrigation and weeds control), and the interactions of these factors Duncan's test and General Linear Multivariate (GLM)' test were performed, respectively. In fact, this research was done as factorial test in split-split plot design with 4 soil, 3 irrigation and 3 weeding treatments whereas in each soil one irrigation or weeding was repeated in 9 experimental plots. Statistical analyses were conducted using the SAS software (SAS Institute Inc., Cary, NC, USA).

3. Results

3.1 Seedling number

The statistical analysis, after square root transformation of data, showed that the frequency of survived seedlings varied in different soils (d.f.=3, F=16.028, P=0.036). The number of seedlings in loam-clay (A), sand (B), sand-loam-clay (C) and loam (D) were 28.6, 37.8, 48.9 and 31.3 individuals in m², respectively. In fact, sand-loam-clay soils (C) showed higher efficiency as compared to the other soil textures (Table 2).

By square root transformation it become evident that the number of seedlings varied in different irrigation regimes (d.f.=2, F=50.616, P=0.000). Hence, watering at intervals of 8, 4 and 12 days produced 54.5, 31.8 and 16.8 individuals/m2, respectively (Table 2). The mean number of seedlings in 20-, 30- and 40-day weeding treatments was 34, 36.7 and 32.4 individuals/m², respectively, showing a non-significant difference (d.f.=2, F=0.529, P=0.638) (Table 2). As a whole, soil and irrigation treatments separately influenced the efficiency of seedling production, but the interaction of these factors on these characteristics was not significant (Table 3)

3.2 Shoot growth

The analysis displayed that shoot growth varied in different soil textures (d.f.=3, F=2.963, P=0.000). Like frequency, shoot growth in sand-loam-clay (C) was the highest in comparison to the other soils types. Among the four soil texture types, shoot growth in sandy soil (B) was poorest (Table 2). Also, shoot growth was different in irrigation regimes (d.f.=2, F=19.616, P=0.000); it was greatest in 8-day irrigation interval followed by 12- and 4-day irrigation intervals (Table 2).

Shoot growth varied among the weeding levels (d.f.=2, F=2.751, P=0.045), so that the growth was greatest in 20-day weeding and was intermediate in 30-day weeding (Table 2). Soil, irrigation, and weeding, separately and simultaneously (except weeding × irrigation interactions) affected shoot growth (Table 3).

Treatment		Number (m²)	Shoot growth (cm)
Soil type	Loam-Clay (A)	28.6 ± 4.9 b	26.6 ± 0.7 b
	Sand (B)	37.8 ± 4.9 b	23.2 ± 0.6 c
	Sand-Loam-Clay (C)	48.9 ± 6.5 a	31.0 ± 0.6 a
	Loam (D)	31.3 ± 2.9 b	26.5 ± 0.7 b
Watering regimes	4-day interval	31.8 ± 2.6 b	24.4 ± 0.6 c
	8-day interval	54.5 ± 3.3 a	29.2 ± 0.5 a
	12-day interval	16.8 ± 1.3 c	27.1 ± 0.6 b
Weeding regimes	20-day interval	34.0 ± 2.9 ns	27.6 ± 0.6 a
	30-day interval	36.7 ± 3.2 ns	27.3 ± 0.5 ab
	40-day interval	32.4 ± 2.9 ns	25.8 ± 0.5 b

For each treatment, different letters in column show significant differences.
ns, is non-significant.

Table 2. Mean number and shoot growth (± sd) of C. *sempervirens* in different soil types, watering regimes and weeding regimes

	Number		Shoot growth	
Sources	F	P	F	P
Soil	4.229	0.008 **	35.193	0.000 ***
Watering	20.824	0.000 ***	25.954	0.000 ***
Weeding	0.577	0.564 ns	6.0.18	0.003 ***
Soil × Watering	1.214	0.309 ns	6.530	0.000 ***
Soil × Weeding	0.841	0.542 ns	3.585	0.002 ***
Watering × Weeding	0.827	0.512 ns	2.716	0.029 *
Soil × Watering × Weeding	1.499	0.145 ns	5.354	0.000 ***

ns= non-significant, *, significant at 5%, **, significant at 1%, ***, significant at 0.1%

Table 3. Analysis of Variance for number and shoot growth of C. *sempervirens* in different treatments

4. Discussion

The results of this research showed that soil texture influenced the survival and growth rates of C. *sempervirens* seedlings, whereas sand-loam-clay (C) was the suitable substrate for raising the seedlings. It indicated that with adding sand into nursery soil (loam-clay), the soil probably became a little light and *Cupressus* seedlings could spread their roots easier and attain higher growth and survival. Hassan et al (1994), working on *C. sempervirens* seedlings, showed that the mixed soil with sand, i.e. sand+clay+sponge (1:1:3) produced the highest growth rate. In our research, sandy soils (B) gave rise weak performance in *Cupressus* seedlings. In contrast, the common nursery soil (A) produced better growth as compared to sandy soil (B). Tabari et al (2007) also considered the low growth of *C. sempervirens* in sandy soil in comparison with medium-textured soil and soils containing organic matter. The similar finding was observed in report of Kiadaliri (2002) on the 2-year transplanted seedlings of *C. sempervirens*.

Accordingly, our results showed on sand-loam-clay soils (C), which normally have the higher nutrients contents in comparison with sandy soils (B), the seedling performance was more outstanding. Shahini (1996) found the better growth of pot-planted *Cupressus* seedlings in soils containing organic matter (30% peat moss). The similar finding was detected in work Ahmadloo et al. (2009) on *C. sempervirens and C. arizonica*, Román et al. (2003) on *Pistacia lentiscus, Pinus halepensis, Picea sitchensis* species, Khasa et al. (2005) on conifer seedlings, and Tsakaldimi (2006) on *Pinus halepensis* seedlings. Significant effects of nutrient and organic matter on the increased growth of other needle-leaved species have been observed in findings of Samuelson (2000), Salifu and Timmer (2001) and Blevins et al. (2006), too. This is seemingly due to proper aeration and water content in soil and easiness in absorption of nutritional elements by plant (Shibu et al., 2006). Indeed, with increasing organic matter in soil, plant is stimulated for nutrient absorption, metabolism activity and better growth (Tichy and Phuong, 1975).

Our study also showed the positive effect of watering on survival and growth of *Cupressus* seedlings, whereas the most frequent and the highest shoot length occurred in 8-day irrigation. No difference of growth was detected in findings of Johnson (1990), working with *Pinus taeda* and *P. elliottii* seedlings. In the present research, watering with shorter interval (4-day) showed a weakened function compared with medium interval (8-day). This may be due to high growth of herbaceous species and their competitive effect causing shorter growth and higher mortality for *Cupressus* seedlings.

Numerous studies have addressed how the performance of planted or naturally established woody seedlings are affected by herb competition (Morris et al., 1993; Caldwell, et al., 1995; Rey Benayas et al., 2002, Jose et al., 2003) and other limiting factors. However, their interactive effects on the performance may have complex interactions receiving little attention. Weeds compete with seedlings for resources (especially water and nutrient) and influence negatively growth (Davis, 1999; Lof et al., 1998; Duplisis et al., 2000; Lhotka and Zakzek., 2001; Mirzaei et al., 2007) but they also diminish radiation and may increase winter low temperatures at the ground level, indirectly effecting the seedling establishment and growth (Rey Benayas et al., 2005). In the current study, weeding and its interaction with soil and also with watering had a positive effect on shoot growth, a finding in line with Kolb and Steiner (1990) and Lorimer et al. (1994). In contrast to Neary et al (1990), who illustrated that early growth of *Pinus taeda* and *P. elliottii* did not reduce with weed control; in our study weed control significantly affected growth, but no did seedling production. Weed control improved shoot growth where the watering interval was 20 days. The minimum shoot growth was detected where the seedlots were removed from herbaceous species in 40-day intervals.

5. Conclusion

According to the findings of this investigation, although the better watering treatment for *C. sempervirens* seedling is 4-day interval but the 8-day interval can be recommended without that the growth and production efficiency is decreased. This is while that water consumption can be economized, too. Regarding to the little differences of growth and the importance of seedling production affected by weeding, for saving the laborer costs, 30-day weeding can be as well applied rather than 20-day one. Likewise, due to the low differences of seedling production and shoot growth presented by soil treatments, loam soil (D) and even sand soil (B) can be suggested for growing *C. sempervirens* seedlings in forest nurseries, if weed control and watering are utilized well.

6. Acknowledgments

This research was carried out as a research project and financially supported by Tarbiat Modares University in Iran. Author would like also to acknowledge Forestry Department of the Tarbiat Modares University for technical and scientific assistance. Likewise, many thanks go to Dr. Mohammad Reza Poormadjidin and to the experts of General Office of Nowshahr Natural Resources for the helps provided during the course of the field work.

7. References

Ahmadloo, F.; Tabari, M., Yousefzadeh, H., Rahmani, A. & Kooch, Y. (2009). Effect of soil on growth and performance of *Cupressus arizonica* and *C. sempervirens* seedlings. *Sciences and technologies of Agriculture and Natural Resources*, 13, 48, 437-448 (In Persian)

Asodolahi, F. (1991). Plantation and forest nursery. Forest and Rangeland Oraganization, *Afforestation and Park Boreau*, Chalous. (In Persian)

Blevins, L.L., Prescott, C.E. & Annette, V.N. (2006). The roles of nitrogen and phosphorus inincreasing productivity of western hemlock and western red cedar plantations on northern Vancouver Island. *Forest Ecology and Management*, 234, 116-122

Bolandian, H. (1999). *Knowing the Forest*. Imam Khomeini International Press. 240p (In Persian)

Caldwell, J.M.; Sucoff, E.I. & Dixon, R.K. (1995). Grass interference limits resource availability and reduces growth of juvenile red pine in the field. *New Forests*, 10, 1-15

Davis, M.A.; Wrage, K.J., Reich, P.B., Tjoelker, M.G., Schaeffer, T. & Muermann, C. (1999). Survival, growth, and photosynthesis of tree seedlings competing with herbaceous vegetation along a water–light–nitrogen gradient. *Plant Ecology*, 145, 341–350

Duplissis, J.; Yin, x. & Banghman, M.J. (2000). Effects of site preparation, seedling quality and tree shelters on planted northern red oak, University of Minnesota, Staff Paper Service, No: 141, 29p.

Emberger, L. (1932). Sur une formule climatique et ses applications en botanique. *La Meteorologie*, No. 92-93, Paris, France.

Hassan, H.A.; Mohamed. S.M., Abo. El. Ghait. Em. & Hammad. H.H. (1994). Growth and chemical composition of *Cupressus sempervirens* L. seedlings in response to growing media. *Annals of Agricultural Science*, Moshtohor, Egypt, 32, 1, 497-509

Johnson, J.D. (1990). Dry matter partitioning in loblolly and slash pines: Effects of fertilization and irrigation. *Forest Ecology and Management*, 30, 1-4, 147-157

Jose, S.; Merritt. S. & Ramsey, C.L. (2003). Growth, nutrition, photosynthesis and transpiration responses of longleaf seedling to light, water and nitrogen. *Forest Ecology and Management*, 180, 1-3, 335-344

Khasa, D.P.; Fung, M. & Logan, B. (2005). Early growth response of container-grown selected woody boreal seedlings in amended composite tailings and tailings sand. *Bioresource Technology*, 96, 857-864

Kiadaliri, SH. (2002). Determining the most proper soil treatment to produce the *Cupressus sempervirens* potted seedlings. *M.Sc. Project*, Natural Resources faculty, Tarbiat Modares University, Noor, Iran, 25p. (In Persian)

Kolb, T.E. & Steiner, K.C. (1990). Growth and biomass partitioning of northern red oak and yellow poplar seedlings: effect of shading and grass root competition. *Forest Sciences*, 36: 34-44

Krasowski, MJ. and Elder, RJF. (2000). Opportunities for improvements to reforestation success. *Extension note*, 43. *Ministry of Forest Research Program.*

Lhotka, J.M. & Zaczek, J. (2001). The use of soil scarification to enhance oak regeneration in a mixed oak bottomland forest of southern Illinois, *Proceeding of the Eleventh Biennial Southern Silvicultural Research* (USA), 392-395

Lof, M., Gemmel, P., Nilsson. U. & Welander, N.T. (1998). The influence of site preparation on growth of *Quercus robur* L. seedling in a southern Sweden clear cut and shelterwood, *Forest Ecology and Management*, 109, 241-249

Lorimer, C.G.; Chapman, J.W. & Lambert, W.D. (1994). Tall understorey vegetation as a factor in the poor development of oak seedlings beneath mature stands. *Journal Ecology*, 82: 227-237

Mirzaei, J.; Tabari, M. & Daroodi, H. (2007). Early growth of *Quercus castaneifolia* seedlings as affected by weeding, shading and irrigation. *Pakistan Journal of Biological Sciences*.10, 15, 2430-2435

Morris, L.A.; Moss, S.A. & Garbett, W.S. (1993). Competitive interference between selected herbaceous and woody plants and *Pinus taeda* L. during two growing seasons following planting. *Forest Sciences*, 39, 166–187

Mossadegh, A. (1996). *Silviculture*. Tehran University Press, 481p. (In Persian)

Neary, D.G.; Rockwood, D.L., Comerford, N.B., Swindel, B.F. & Cooksey, T.E. (1990). Importance of weed control, fertilization, irrigation and genetics in slash and loblolly pines: Early growth on poorly drained spodozole. *Forest Ecology and Management*, 30, 1-4, 271-281

Rey Benayas, J.M., Loʹpez-Pintor, A., Garcıʹa, C., de la Caʹmara, N., Strasser, R. & Goʹmez Sal, A., (2002). Early establishment of planted *Retama sphaerocarpa* seedlings under different levels of light, water and weed competition. *Plant Ecology*, 159: 201–209

Rey Benayas, J.M., Navarro, J., Espigares, T., Nicolau, J.M. & Zavala, M.A. (2005). Effect of artificial shading and weed mowing in reforestation of Mediterranean abandoned cropland with contrasting *Quercus* species. *Forest Ecology and Management*, 212, 302-314

Rezaei, A. (1992). Ecological study of natural stands of *Cupressus sempervirens* in north of Iran, *M.Sc. Thesis*, Gorgan University, Iran, 220p. (In Persian)

Román, R.; Fortún C., García López De Sá M E. & Almendros, G. (2003). Successful soil remediation and reforestation of a calcic regosol amended with composted urban waste. *Arid Land Resources Management*, 17, 297-311

Sabeti, H. (2004). *Trees and Shrubs of Iran*. Yazd University Press. 812p. (In Persian)

Sagheb-Talebi, Kh.; Sadjedi, T., Yazdian, F. (2003). *Forests of Iran*. Technical Publication, 339, 28p.

Salifu, K.F. & Timmer, V.R. (2001). Nutrient retranslocation response of *Picea mariana* seedlings to nitrogen supply. *Soil Science Society of America Journal*, 65, 905-913

Samuelson, L.J. (2000). Effects of nitrogen on leaf physiology and growth of different families of loblolly and slash pine. *New Forests*, 19, 95-107

Shahini, GH. (1996). Seedling production with composts of *Azola* and left-over plants, *M.Sc. Thesis*, Tarbiat Modares University, Iran, 153p. (In Persian)

Shibu, M.E.; Leffelaar, P.A., Van Keulen, H. & Aggarwal, P.K. (2006). Quantitative description of soil organic matter dynamics. A review of approaches with reference to rice-based cropping systems, *Geoderma*, 137: 1 – 18.

Tabari, M. & Saeidi, H.R. (2007). Restoration of deforested areas by cypress seedling in southern coast of Caspian Sea (north of Iran). *Ekoloji*, 17, 67, 60-64

Tabari, M.; Djazirei, M.H., Asadolahi, F. & Mir-Sadeghi, M.M.A. (2002). Forest communities and environmental requirements of *Fraxinus excelsior* in forests of north of Iran. *Pajouhesh-va-Sazandegi*, 15, 2, 94-103 (In Persian)

Tabari, M.; Saeidi, H.R., Alavi-Panah, K., Basiri, R. & Poormadjidian, M.R. (2007). Growth and survival response of potted *Cupressus sempervirens* seedlings to different soils. *Pakistan Journal of Biological Sciences*, 10, 8, 1312-1309

Tichy, V.; & Phuong, H.K. (1975). On the character of biological effect of humic acids. *Humus Planta*, 6, 379-382

Tsakaldimi, M. (2006). Kenaf (*Hibiscus cannabinus* L.) core and rice hulls as components of container media for growing (*Pinus halepensis* M.) seedlings. *Bioresource Technology*, 97, 19, 1631-1639

Zare, H. (2002). Native and exotic needle-leaved species in Iran. *Forest and Rangeland Research Institute*, 550p. (In Persian)

7

Developing Crop-Specific Irrigation Management Strategies Considering Effects of Drought on Carbon Metabolism in Plants

Silvia Aparecida Martim[1], Arnoldo Rocha Façanha[2]
and Ricardo Enrique Bressan-Smith[3]
[1]*Instituto de Biologia, Departamento de Ciências Fisiológicas,
Universidade Federal Rural do Rio de Janeiro*
[2]*Centro de Biociência e Biotecnologia, Laboratório de Biologia Celular e Tecidual,
Universidade Estadual do Norte Fluminense – Darcy Ribeiro*
[3]*Centro de Ciências e Tecnologia Agropecuária, Laboratório de Fisiologia Vegetal,
Universidade Estadual do Norte Fluminense – Darcy Ribeiro*
Brazil

1. Introduction

Development of appropriate irrigation managements in order to produce crops of high quality without water waste is a relevant theme worldwide in the face of the global environmental change and the related perspectives for a future increment in dryness (IPCC, 2001). Drought stress imposes adverse effects on plant yield and productivity affecting mainly leaf and root growth, stomatal conductance, photosynthetic rate and biomass gain (Blum, 1998). Based on studies carried in the last decade, it is clear that plants perceive and respond quickly to minima modifications in water status by means of a series of cellular, physiological and molecular events developing in a parallel feature (Chaves et al., 2009). The duration, intensity and rate of evolution of forced drought are responsible to modulate the various levels of response.

All living organisms depend directly or indirectly from the energy of photosynthesis, the only process that can absorb the energy from the sun to oxidize water, release oxygen and to reduce carbon dioxide forming carbohydrate. On that basis, how decreasing cellular water content affects the photosynthetic process is an interesting issue for present and future debates (Krieg, 1983; Chaves, 1991; Lawlor and Cornic, 2002; Flexas and Medrano, 2002; Chaves et al, 2009). Photosynthesis consists of a set of integrated reactions submitted to a range of conditions under environmental and genetic control. Under field conditions, it is accepted that the decrease in photosynthesis in response to moderate water stress in soil and/or atmosphere is firstly due to stomata closure (Chaves et al., 2002, 2003). However, the extent and the nature of the restriction of carbon assimilation in leaves under water stress occur, mainly due to stomata or non-stomata limitations, is still in debate (Tzara et al. 1999; Cornic, 2000, Lawlor and Cornic, 2002; Flexas et al., 2004).

Changes in plant growth elicited by low water availability have also been associated to modulations on the balance between photosynthesis and respiration. These processes are intimately related (Bartoli et al., 2005; Flexas et al., 2006; Martim et al 2009). In general, from the total carbon incorporated in carbohydrate by photosynthesis, more than 50% is spent in respiratory rates; nevertheless, this relationship may alter under drought conditions. Respiration is an essential metabolic process that generates not only Adenosine Tri-Phosphate (ATP) but several carbon skeletons - metabolites that are used in many synthetic processes essential for growth and maintenance of the cell homeostasis, including under stress conditions (MacCabe et al., 2000; Bartoli et al., 2000). Remarkably, mitochondria are involved in several metabolic process concerned in cell adaptation to abiotic stress, like photorespiratory cycle (Kramer, 1995), proline metabolism which accumulates under hyperosmotic stress (Kiousue et al., 1996), programmed cellular death (Rhoads et al., 2006), themselves and the cell defense from a excess Reactive Oxygen Species (ROS) particularly under salinity and drought (Alsher et al., 1997).

Other implications to plant development in response to drought is the altered partitioning of carbon between the leaf and others organs (Chaves, 1991). The amount of carbon available for storage, maintenance and translocation is determined by key regulatory process in source leaves. Under water deficit, there is a strong reduction in levels of inactive osmotically solutes (starch) and increase in active osmotically solutes (soluble sugars) and as a consequence the osmotic potential decreases, contributing to the maintenance of leaf water status (Pelleschi et al., 1997; Pinheiro et al., 2001; Yang et al., 2001). Sucrose plays a crucial role in plant growth and development not only as a key molecule in energy transduction and storage, but also because there is increasing evidence that sucrose or several metabolites derived from it may function as regulators of cellular metabolism (Smeekens and Rook, 1997). Moreover, the sugars play a substantial role during plant growth and development under abiotic stress, especially drought. Like a hormone, sugars can act as primary messengers and regulate signals that control the expression of several genes involved in sugar metabolism. Nevertheless, the use of sucrose as carbon source and energy depends on their hydrolysis into hexoses catalyzed by sucrose synthase or invertase. Invertase (EC 3.2.1.26, ®-fructofuranosidase) is a hydrolase present in diverse isoforms with different biochemical properties and subcellular localization (Sturm, 1996). The invertases have a key important role in several cellular processes like phloem loading/unloading, defense response to abiotic and biotic stresses and cellular turgor recover (Sturm and Tang, 1999).

Agriculture represents 70% of freshwater expenditure worldwide, and this percentage rises above 90% in arid countries (WRI, 2005). Therefore, the water availability for irrigation is scarce in many areas due to aridity increment in the world and competition for water by households, agriculture and industry (IPCC, 2007). Consequently, there are urgent needs to expand the use of irrigation management that improves high yield, quality and water use efficiency (WUE). Two techniques are widely used and investigated for this proposal: Deficit irrigation (DI) and Partial root-zone drying (PRD). Deficit irrigation is a strategy in which water is applied during drought-sensitive growth stages and water restraint is employed to drought-tolerant phenological stages, commonly the vegetative stages and the terminal ripening period. Partial root-zone drying (PRD) is innovation deficit irrigation (DI), which the water is provided only to one side of the root system leaving the other part to dry a certain water potential of the soil before rewetting by shifting irrigation to the dry side. As a consequence, the abscisic acid produced by roots in the dehydrating side is the chemical signal sent to the shoots and leaves via xylem that invariably reduces the stomatal conductance, transpiration and vegetative growth. At the same time, roots of the watered side maintain a favorable plant water status (Dry and Loveys, 1999; Dry et al., 2000, 2001; Chaves and Oliveira, 2004).

This chapter will focus the changes occurred in carbon metabolism from carbohydrate production (photosynthesis) until consumption (respiration) including the partitioning (sugar metabolism), in response to drought. Such a information will be discussed as a new approach to develop crop-specific irrigation managements.

2. Production and consumption of carbon under low water availability

From all processes that contribute to plant development, cell growth and photosynthesis are the firstly affected by drought. When the availability of water in the soil is scarce or vapor deficit pressure of the atmosphere is high, plants firstly respond by decreasing the stomatic aperture. As the main role of stomata is to perform gas exchange, this strategy has two consequences, prevent the water loss and limit CO_2 diffusion. Consequently, this might cause the reduction of the photosynthetic rates in drought conditions. Nevertheless, the extension and the nature of the diminished carbon assimilation in water stressed leaves are due to stomatic or non-stomatic limitations, a theme still under discussion (Tzara et al., 1999; Cornic, 2000; Lawlor and Cornic, 2002; Flexas et al., 2004).

Photosynthetic rate is gradually diminished with progressive reduction in relative water content (RWC). The complexity of photosynthesis is based on the activity of Rubisco (Ribulose 1,5 Biphosphate Carboxylase/Oxygenase) per unit leaf, the rate of RuBP (Ribulose Bisphosphate) resynthesis (hence on capture of photosynthetically active radiation (PAR)) and on the CO_2 supply, given by stomatic conductance (g_s) and the ambient CO_2 concentration (C_a).

The CO_2 availability to the chloroplast is intrinsically dependent of the ambient CO_2 concentration and the pathway for diffusion between air and carboxilation site, mainly stomatic conductance in gas phase (g_s) and mesophilic conductance (g_m) in liquid phase. In hydrated leaves and a saturated environment with CO_2 and light the maximum rate of photosynthesis is denominated potential photosynthesis (Apot) (Lawlor and Cornic, 2002). Moreover, when relative water content lowering, to obtain the Apot is necessary Cc to saturated Rubisco and sufficient Ca to suppress the barriers imposed by stomatic and mesophilic conductances. In the first moment, when RWC decreased and the photosynthetic rate diminished, a high CO_2 concentration ({50 a 150 mL L⁻¹} de 5 a 15%) can restore A to values near Apot, and consequently, the A is unaffected. Afterwards, the decreased photosynthesis cannot be restored by high CO_2 concentrations showing that Apot is impaired by metabolic factors. Parallel to this, photosynthesis fall gradually as RWC decreases showing that Apot is progressively inhibited and the effects of stomatic conductance are diminished. This response built from photosynthetic rate x internal concentration of CO_2 are termed response type I and Type II (Lawlor and Cornic, 2002).

Some species such *Rhamnus alatemus*, *Rhamnus ludovici-salvatoris*, *Nicotiana sylvestris*, *Phaseolus vulgaris* and *Vitis vinifera* have their photosynthetic rate mainly limited by stomatic conductance at the beginning of drought development. As drought severity evolves, activity of Rubisco is impaired and the levels of RuBP diminished, consisting as the main limiting factors for photosynthesis (Bota et al., 2004). In another way, plants like grasses show the carbon assimilation and quantic efficiency of photosystem II ($\sqrt{}$ PSII) decreased quickly with RWC decline (Ghannoum et al., 2003). For these plants, the application of 2500 $\lceil$ L L⁻¹ of CO_2 had no effect in photosynthetic rate and $\sqrt{}$PSII, showing that biochemical factors are limiting the photosynthetic process. In sunflower leaves, inhibited synthesis of RuBP caused by lower ATP content and not the CO_2 diffusion is the limiting factor to the photosynthesis (Tzara et al., 1999). Linear photosynthetic electron transfer consisting of Photosystem II (PSII) and and Photosystem I (PSI), both capable to convert light energy into chemical energy as ATP and reducing power NADPH, whereas PSI cyclic electron transfer (CET) is merely involved in ATP synthesis. ATP and NADPH from thylakoid linear electron transport are employed to chloroplast CO_2 or O_2 reduction. The electron flow in photosynthetic membranes is highly regulated and strongly dependent from substrate availability and redox level of the transfer cycle.

A little reduction in soil water content caused different degree of inhibition in photosynthetic rate and increase of non-photochemical quenching of chlorophyll a fluorescence in grapevine (Flexas et al., 1999). Under moderate drought the correlation between CO_2 assimilation and electron transfer was maintained, however under severe drought a strong inhibition of photosynthesis broke the mentioned correlation. The degree to which water stress impair the linear electron transport on thylakoid and the partitioning of electrons between produced in photosystem II (PSII) and consumed by acceptors depend on leaf water potential. In sunflower leaves, the chloroplast activity begins to reduce at the same leaf water potential that causes the stomatal closure and the electron transport begins to limit photosynthesis at leaf water potential below -1.1 MPa (Keck and Boyer, 1974). To investigate the linear photosynthetic electron transfer under drought conditions, it is necessary to consider two issues: the photosynthetic apparatus resistance and the substrate availability to chloroplast reactions. The carbon reduction by photosynthetic cycle is so far the main consumption of the electrons from water oxidation. So, when carbon assimilation decreases in consequence of a lower RWC, there is an increase in reactive oxygen species (ROS) due to strong reduction in LPET and consequently increase the electron transfer to oxygen. In the absence of alternative cycles to use the electrons this is extremely damaging to the cells. When the photosynthesis decreases in response to lowering RWC, the most important way to use the electrons is the photorespiration, dark respiration and Mehler reactions (Cornic and Briantais, 1991). In general, a large portion of the total flow of electron is used by photorespiration that plays a key role in redox regulation until LPTE-reducing by low carbon assimilation (Haupt-Herting and Fock, 2000, 2002).

It is clear that drought induces several alterations in photosynthetic rate, often arisen as part of an overall response to desiccation which involve specific gene expression. In some cases, these alterations are consequence to self-resistant of the photosynthetic apparatus. The decrease in Ci caused metabolic change which limits photosynthesis reversibly being the photosynthetic apparatus damaged with advancement of drought (Cornic, 2000). However, under filed conditions is usually accepted that the decrease in photosynthetic rate in response to moderate water stress (RWC 70-75%) is in first instance due to closed stomata (Chaves et al., 2002, 2003). In general reductions in transpiratory rate are associated with the stomatal aperture occurred at low water potential. In this condition the water use efficiency (WUE) instantly defined as the ratio A_N/E can be enhanced to maintain cellular metabolism and plant survival. This event can be mostly observed in plants drought tolerant like grapevine (Pou et al., 2008; Martim et al., 2009), alfafa (Erice et al., 2011) and drought-tolerant maize (Hund et al., 2009). Mediterranean plants with different growth pattern showed increase in leaf WUE with the initial decrease of water availability, but WUE decreased with lower water availability (Medrano et al., 2009). According to the authors, this pattern is explained by the relative changes of photosynthesis and stomatal conductance as soil water deficit progress. This physiological process gives the plants the strategy of optimize carbon assimilation and minimize water losses related to decrease in g_s which only causes a modest photosynthetic reduction. The later decrease in WUE of plants can be explained by metabolic impairment of the photosynthetic apparatus.

Focusing on concept of water use efficiency to the amount of water transpired relative to the amount applied with emphasis on soil factors that influence this ratio, any factor that restricts the root system reduces WUE. However, it is hardly known whether transpiration or WUE is mostly affected by a long period of soil compaction. The degree of consumption and loss of water is directly related with plant productivity like observed to wheat and maize where soil compaction reduced grain yields by reducing soil water storage and/or crop WUE (Radford et al 2001). Soil compaction involves an increment in soil bulk density and connected with this are increment in soil strength and decrease in air permeability and hydraulic conductivity. In modern agriculture, the majority of soil compaction is caused by vehicular traffic (Flower and Lal, 1998). Soil compaction originating from anthropogenic or natural causes exerts an enormous impact on the establishment, growth and yield of crops in tropical regions. The capacity of intake water via roots located in deep wet soil layers is a key factor determining transpiration under soil dried conditions. Soil compaction affects root growth by increase mechanical resistance, lowering water availability and oxygen diffusion or by restricting nutrient supply. In regions where the soil resistance to penetrometer exceeding approximately 0.5 MPa, roots experiences great difficult to penetrate (Young et al. 1997) and they stop growing when soil resistance reaches 3.6 MPa (Masle, 1999). In contact with compact soil, root deepening is delayed and roots tend to have a clumped spatial arrangement causing water stress even in wet soil, due to an increase in resistance to the soil-root water flux. However, the extent to which WUE is diminished when root system employed their energy exploring complex channels rather than more direct paths toward water and nutrients is still unclear.

In limited availability of water, plant productivity is determined by amount of water available as well as water use efficiency. In agriculture, absorption and transpiration by the crop canopy determine the flow of water and CO_2 and therefore canopy photosynthetic water productivity and biomass water productivity. Shifting the focus from leaf to canopy there is an additional feature that should be considered because the approach is on a land area basis instead of leaf area basis. The radiation capture by a crop is intrinsically dependent of leaf area, generally evaluated by the leaf area index (LAI) on the arrangement of the leaves within the canopy as well as on the angle and intensity of incident radiation. Besides the solar radiation, plant density and the stage of vegetative growth determinate radiation interception. Moreover, of the total captured solar radiation just the portion that is photosynthetically active (PAR) is effective in CO_2 assimilation, while the whole spectrum is used for transpiration. At the canopy level the CO_2 assimilation of many crops usually doesn't reach light saturation meaning that a linear response to irradiance is observed (Lowerse, 1980; Asseng and Hsiao, 2000). Then, any change in the quantity of radiation captured by canopy would affect at the same way CO_2 assimilation and transpiration. Nevertheless, at the canopy level, the CO_2 and water vapor share the transport pathway and energy source. A difference in that sensible heat flux can either add or remove energy for transpiration from the canopy independent of radiation. Therefore, the sharing of radiative energy source is a critical and often dominant factor in connecting CO_2 assimilation and transpiration rates at the canopy level.

When cultivated under non-stress conditions the maximum yield is defined as yield potential. WUE is mainly discussed in terms of plant production rather than gas exchange. At this framework, yield under drought conditions can be determined by genetic factors controlling yield potential and/or drought resistance and/or WUE. Photosynthesis is the primordial process to plant because the carbon compounds synthesize with input of energy from sunlight will be employed in several metabolic reactions. So, under stress conditions there should be a rearrangement at the cellular, molecular, biochemistry and physiological level to maintain the integrity of the photosynthetic apparatus. Photosynthetic apparatus is considered resistant to drought and its structure is not affected by inhibition of photosynthesis imposed by reductions in stomatal conductance which provide quickly responses of leaves to changes in environment water status. There is a number of studies showing the resistance of photosystem II (PSII) to soil desiccation (Abreu and Munné-Bosch, 2008; Munné-Bosch et al. 2009; Chernyad'ev, 2009; Georgieva et al. 2010; Ibáñez et al. 2010; Hura et al. 2011). This fact can be evidenced by the ability to recover of photosynthetic rate measured after re-irrigation of plants exposed to water scarcity. The dehydration of Haberlea plants leading the inhibition of quantum efficiency of PSII which was due to decreased efficiency of both excitation capture by open PSII reaction centres and, mainly, of photochemical quenching. In addition the CO_2 assimilation decreased sharply as a result of dehydration featuring no net assimilation seven days after dryness. But, seven days after rehydration of stressed plants the CO_2 assimilation was comparable to the irrigated plants (Georgieva et al. 2010). Table 1 shows that in stressed plants of Cabernet Sauvignon photosynthetic rate (A_N), stomatal conductance (g_s), transpiratory rate (E) and internal carbon concentration (C_i) declined with drought progress reaching the lowest values at the severe stress. However, gas exchange and leaf water potential measured 48 hours after rehydration of stressed plants presented recovery potential which A_N and C_i showed values very close to control.

Treatment	A_N (μmol CO_2 m^{-2} s^{-1})	g_s (mol m^{-2} s^{-1})	E (mmol m^{-2} s^{-1})	C_i (ppm)	WUE	A_N/C_i
Irrigated	11.9	0.8	14	304	0.8	0.04
Mild-stress	8.8	0.1	5.4	234	1.6	0.04
Severe-stress	1.8	0.01	194	0.6	3.3	0.01
Re-irrigation	11.1	0.3	11.2	294	1.0	0.04

Table 1. Photosynthetic rate (A_N), stomatal conductance (g_s), transpiration (E), internal CO_2 concentration (C_i), water use efficiency (WUE) and carboxylation efficiency (AN/Ci) measured in Cabernet plants under different water availability.

Changes in plant growth elicited by low water availability have also been related to modulations on the balance between photosynthesis and respiration. These processes are intimately related, and in the last decade has been increasing interest in the interaction between them (Bartoli et al., 2005; Flexas et al., 2006; Martim et al 2009). Photosynthesis is a complex process and to achieve optimal rates must occur the interaction of chloroplasts with cytosol and other organelles such a mitochondria. Of the total carbon assimilated in photosynthesis, commonly more than 50% is lost in respiration required for plant growth and maintenance; however, this relationship may change under drought. For instance, while the photosynthetic rate may decrease up to 100% becoming totally damaged in severe drought conditions, the respiration rate may either increase (Bartoli et al., 2005; Martim et al., 2009) or decrease (Huang and Fu, 2000; Galmes et al., 2007) but may never become fully damaged. Respiration is an essential metabolic process that generates not only ATP but several other metabolites that are used in many synthetic processes essential for growth and maintenance of the cell homeostasis, including under stress conditions (MacCabe et al., 2000; Bartoli et al., 2000).

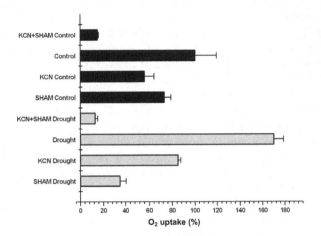

Fig. 1. Respiratory activity of Cabernet Sauvignon leaves. Oxygen consumption was monitored in young leaves of plants either Irrigated (control) or expressed to drought treatment (D) and following the addition of 1 mM KCN, 10 mM SHAM or both.

A special feature of plant cell respiration is the presence of an alternative pathway which drains electrons from the ubiquinone pool without involvement of the cytochrome oxidase (COX) (Brownleader et al., 1997). The mitochondrial alternative oxidase (AOX) apparently reduces molecular oxygen to water in a single four-electron transfer step (Day et al., 1991; Moore and Siedow, 1991). This alternative pathway is nonphosphorylating, resistant to cyanide and antimicine and inhibited by salicilhydroxamic acid (SHAM) and n-propyl gallate (nPG) (Schonbaum et al., 1971; Siedow and Grivin, 1980). Studies focusing respiration rate in response to drought is particularly lower than those about photosynthetic rate. In general, the respiratory pathway decreases during a drought period due to decrease photosynthetic activity and growth. However, this behavior is species dependent but the respiration can increase particularly under severe stress (Flexas et al., 2005; Ghasghaie et al., 2001). After experiencing a drought period, plants may require a increasing in respiratory rate (Kirschbaum, 1987). In wheat leaves Bartoli et al (2005) observed that respiratory rate increased 41% when RWC decreased to 75% in relation to 97.2% in hydrated leaves, this respiratory increase was due to alternative oxidase that doubled the oxygen consumption. Data of our previous research showed that in *Vitis vinifera* L. cv. Cabernet Sauvignon, the respiratory rate of leaves increased around 74% 12 days after water suspension. This increase was due mostly to an enhancement of the alternative pathway activity as observed in leaves treated with 1 mM potassium cyanide (KCN) an effective COX inhibitor (Table 2).

At the same time, stomatal conductance, assimilation rate and internal CO_2 concentration in leaves of stressed plants decreased 98%, 83%, and 21%, respectively, along with the progress of the water deficit treatment. So, it is worth emphasizing that photosynthesis also requires interactions of chloroplasts with the mitochondria to attain optimal rates (Hoefnagel et al. 1998; Padmasree and Raghavendra, 1999; Dutilleul et al. 2003; Noctor et al. 2004; Nunes-Nesi et al. 2008). In this context, it is likely that the stimulation of the AOX pathway observed in grapevine might represent an important response protecting the photosynthetic machinery under drought conditions.

There are several studies showing that the alternative oxidase increase in plants under some environmental conditions (Padmasree and Raghavendra, 1999; Bartoli et al., 2005; Yoshida et al., 2007; Feng et al., 2008). The increment in alternative oxidase under stress conditions represent a important response to protection of the photosynthetic apparatus against the harmful effects of energy excess. In despite this, the electron partitioning between COX and AOX is modified at the same values of stomatic conductance limiting photosynthesis. At the saturating photosynthetic photon flux density (PPFD) AOX functioning keeps the photosynthetic electron transport chain more oxidized in Broad beans plants, where the AOX inhibition decreased the O_2 evolution rates and quantum efficiency of PSII (ϕII) (Yoshida et al. 2006). The authors observed that AOX inhibition by SHAM induced a clear decreased in photochemical quenching (qP) and AOX inhibition by nPG lead the prompt increase in non-photochemical quenching (qN) after onset of irradiation. Moreover, AOX inhibition diminished the O_2 evolution and ϕII even at low PPFD and cause imbalance of operating efficiencies between two photosystems which suggest that even at low PPFD the AOX is essential for optimal photosynthesis. In a posterior work, Yoshida et al. (2007) propose that the major physiological function of AOX is to serve as an electron sink that prevents over-reduction of the photosynthetic apparatus and thereby mitigates photoinhibition under excess PPFD. So, AOX activity in plant mitochondria becomes an essential accessory for stabilizing the autotrophic system of higher plants.

Respiratory capacities						
Treatment	Vt (μmol O_2 g^{-1} DW h^{-1})	Vcyt (μmol O_2 g^{-1} DW h^{-1})	V_{AOX} (μmol O_2 g^{-1} DW h^{-1})	V_{Cyt}/ V_{AOX} (ratio)	V_{Cyt}/ Vt (%)	V_{AOX} /Vt (%)
Irrigated	46.9 ± 4.6	23.1 ± 2.9	15.2 ± 2.1	1.5	49	32
Drought	90.8 ± 5.7	31.3 ± 1.8	60.7 ± 5.3	0.5	34	67

Table 2. Measurement of the capacities of respiratory pathways in Cabernet Sauvignon leaves. Rates were calculated from data of figure 1. Residual respiration (O_2 uptake in the presence of 1 mM KCN + 10 mM SHAM was subtracted from all values. Vt is the rate of O_2 uptake in the absence of inhibitors, V_{Cyt} is the capacity of the cytochrome pathway estimated as inhibition in the presence of 10 mM SHAM, V_{AOX} is the capacity of the AOX pathway estimated as inhibition in the presence of 1mM KCN. All measurements were made in the fully expanded leaf. Values are means ± SEM of three to four replicates.

Beyond the request of interaction of chloroplast with the mitochondria to obtain the greatest photosynthetic rate, it is now well established that mitochondrial electron transport is necessary to optimize photosynthesis (Padsmaree and Raghavendra, 1999; Dutilleul et al.

2003; Noctor et al., 2004). Working with wheat Bartoli et al. (2005) showed that under drought the mitochondrial AOX pathway is up-regulated, and that is capable to maintaining photosynthetic electron transport under drought either by direct consumption of reducing power, by sustained production of CO_2, or by allowing other power-consuming process to operate unabated. Thus, the nonphosphorylating pathways may function as a mechanism for plant photo-protection, but the components of this mechanism have not been characterized in detail. There are studies which that the AOX can efficiently respond to changes in the light environment such as ones developed by Svensson and Rasmusson (2001) and Finnegan et al. (1997) where AOx gene is induced by light in potato and soybean, respectively. Intrinsic interaction between photosynthesis and the respiratory chain has been observed in studies using specific inhibitors or mutants. In environments of high light and CO_2 concentrations, better photosynthesis requires a continuous flow of metabolites through the calvin cycle and cytosolic sucrose synthesis. According Krömer et al. (1988, 1993) in barley protoplasts a low concentration of oligomycin (inhibitor of mitochondrial phosphorylation) diminished the photosynthetic rate, the Adenosine Tri-Phosphate/Adenosine Di-phosphate (ATP/ADP) ratio and the Triose Phosphate/3-Phosphoglyceric Acid (TP/3-PGA) ratio. Some research refer that respiratory ATP is consumed by sucrose synthesis leading to an optimal rate of photosynthesis. Antimycin A (inhibitor of complex III) and SHAM (inhibitor of AOX) decreased the photosynthetic rate in pea mesophyll protoplast. The presence of antimycin A incresead TP/3-PGA ratio more than malate/oxaloacetic acid ratio, where in presence of SHAM the last one was strongly increased (Pdsmaree and Raghavendra, 1999). These findings suggest that the cytochrome pathway sustains the triose-phosphate export and the alternative pathway sustains the oxidation of malate in the light.

In leaves, the type and strength of the metabolic interconnections between chloroplast and mitochondria is broadly dependent on chloroplastidial nicotinamide adenine dinucleotide phosphate (NADPH) and ATP production driven by light energy. In turn, both ATP and NADPH are the driving force for carbon uptake, nitrogen assimilation and photorespiration. Already in the dark or in heterotrophic tissues the relationship between the plastids and mitochondria are distinct because the mitochondria is the main source of energy (Hoefnagel et al. 1998). However, independent of tissue type the mitochondria always preserve their basic function of being the powerhouse of the cell with the capacity for mitochondrial energy production being ubiquitous among contrasting tissues and developmental stages. Furthermore, another relevant point to be considered is the effect of water stress in electrons partitioning between cytochrome oxidase and alternative oxidase and its consequence in the ATP production. The mitochondrial electron transport chain (ETC) was more important than oxidative phosphorilation to optimize photosynthesis, particularly under low CO_2 concentrations (Padmasree e Raghavendra, 1999). The importance of ETC mitochondrial to photosynthetic process was verified by reduced photosynthetic rate 20 a 30% in tobacco plants with impaired complex I (Dutilleul et al., 2003). In absence of complex I the contribution of ETC mitochondrial to glycine oxidation is reduced increasing extramitochondrial drain causing an increment in chloroplastidial reducing molecules and photosynthetic inhibition.

A main role of AOX is to balance the necessities of carbon metabolism and mitochondrial electron transport (Vanlerberghe and McIntosh, 1997) and under stress conditions its operation allows the TCA cycle to continue providing carbon skeletons for metabolism and synthesis of compatible solutes (Mckenzie and McIntosh, 1999). At the inner membrane of mitochondria AOX short cuts electron transport by transferring electrons directly from reduced ubiquinone to oxygen consequently the ATP production is diminished. Hence, part of the source for energy coupling is "wasted" as heat. On the other hand, AOX ignore adenylate and Pi control, and, under a high-energy charge, AOX assist to avoid incomplete reduction of oxygen to water as a source for reactive oxygen species. Thereby, AOX activity enables high turnover rates of carbon skeletons in the cytosol and the citric acid cycle at lower productivity levels of harmful reactive oxygen. Indeed, as photosynthesis decreased under water stress, an excess of reducing power is frequently generated and thus over-reduction of photosynthetic electron chain may result in oxidative burst. Thus, the AOX has been the focus of many studies in plant respiratory metabolism under several environmental stresses, mainly because an increase in AOX capacity might contribute to controlling the formation of reactive oxygen species (ROS) (Wagner, 1995; Popov et al., 1997; Maxwell et al., 1999; Umbach et al., 2005). At least part of the drought effects on plant physiology is related to ROS formation, such as superoxide ($O2^-$), hydrogen peroxide (H_2O_2), hydroxil radicals ($\cdot OH$) and singlet oxygen ($^1O_2$) (Li and Staden, 1998). These ROS may initiate destructive oxidative process such as lipid peroxidation, chlorophyll bleaching, protein oxidation, and damage to nucleic acids (Scandalios, 1993). Water stress invariably decreases the photosynthetic rate and the intensity of this effect influences the capacity of different species to cope with the drought, which also depends on the duration of stress and plant genetic background (Kaiser, 1987; Chaves, 1991, Chaves et al., 2002). Generally, when respiration rate decreases upon drought conditions the photosynthesis and growth requirements are further affected. Nevertheless, this behavior seems to be somewhat species dependent, and respiratory rates can also increase, particularly under severe drought (Gashghaie et al., 2001; Flexas et al., 2005).

Has long be known that the mitochondria are the main source of cellular ROS, and a number of environmental stress that increase ROS production in plants also leads to an increase alternative pathway respiration. Culture of plant cells and fungi showed an increased in AOX activity in response the addition of hydrogen peroxide to the culture medium (Wagner, 1995; Vanlerberghe and McIntosh, 1996). Furthermore, essays with isolated soybean and pea mitochondria showed that additions of AOX inhibitors like as SHAM and nPG induce H_2O_2 production (Popov et al. 1997). Under abiotic stress plants unavoidably undergo a disruption of cellular homeostasis with predictable consequences for the functioning of mitochondria, including their ability to regulate cellular energy status to cope with unfavorable conditions and during recovery. Soon, due the key role of mitochondria in plant cells, one might expect that cells with impaired mitochondria should not be able to survive stress. As discussed previously, respiration is generally affected by drought, but in a lesser extent than photosynthesis which fits well with the essential role of mitochondria. In water-stressed wheat seedlings pretreated with 1 mM SHAM there was more generation of ROS than seedling either subjected to drought or SHAM treatment alone did (Feng et al. 2008). In addition, 1 mM SHAM did not significantly change the activity of peroxidases in drought leaves but inhibited most of AOX activity, evidencing that AOX inhibition lead to additional ROS production under drought conditions.

As far as water stress is an issue, it is interesting to focus at models that show adaptation to dryness, such as drought-resistant species and desiccation-tolerant organisms such as resurrection plants and orthodox seeds. In the meantime, mitochondrial biology in refer to water stress has been extensively studied in durum wheat a drought-tolerant cereal (Bartoli et al. 2005; Pastore et al. 2007). Analyzing messenger ribonucleic acid (mRNA) expression of

AOX in Arabidopsis Clifton et al. (2006) show that five Aox genes are expressed with organ and development regulation, suggesting regulatory specialization of AOX genes members. Moreover, studying genes coexpressed with AOXs in response to various treatments that modify mitochondrial functions and/or in plants with altered AOX levels reveals that this gene set encodes more functions outside the mitochondrion than in mitochondria itself (Clifton et al. 2006). Despite this, the authors concluded that this have a role in reprogramming cellular metabolism in response to constant changing environment encountered by plants. The changes in the efficiency of energy coupling (amount of total respiration and AOX) induced by environmental stress are proposed to keep growth rates constant even a change to more unfavourable conditions (Hansen et al. 2002). The AOX pathway is regarded less efficient in energy conservation, so the authors propose that less-efficient metabolism in terms of energy coupling might be more efficient in terms of growth stability.

The optimizing metabolic efficiency for adaptive regulation of growth and development seems to be partly regulated by differential activity of AOX pathway. Order to verify the potential role of AOX as marker for stress tolerance a schematic global strategy for future experimentation on AOX was proposed by Arnholdt-Schmitt et al. (2006). Firstly, the system analyses and ecophysiological modeling should be carried out at the whole plant level to determine the importance of identifiable yield determining parameters as a basis for molecular research. This work has to be carried out at the species levels and should regard the interaction between developmental stages and the environment. Stress adaptation may results of sustaining growth by maintaining homeostase, down-regulating growth by avoid nutrient imbalance or up-regulating growth by induced secondary root growth and root hair formation. To breeding a growth adaptation efficient is a function of its effect on yield stability and in accordance of modeling; responsive tissues and cells should be identified parameters and will then be available for AOX analyses.

3. Changes in sugar metabolism in response to low water availability

As discussed in the previous section, drought alters the production and consumption of photoassimilates, so not surprisingly, the carbon partitioning between leaves itself and others plants organs will also be affected. The partitioning of photoassimilate is the result of a coordinated set of processes of anabolism and transport between source/sink and is under the control of genetic, environment and development factors (Chaves, 1991). The amount of carbon that will be available for storage, maintenance and transport is determined by regulatory processes in the sources leaves. Part of this process, referred as metabolic control of the of triose-phosphate export from the chloroplast for the synthesis of sucrose in the cytosol, it is reasonably comprehensible (Stitt and Quick, 1989). Sucrose, a main product of photosynthesis, is the major form of translocated carbon and the most important substrate for sink metabolism. In full expanded leaves, the carbon is split between leaf itself and the whole plant, already in mature leaves, the majority of carbon is translocated to another parts of plant. In accordance to Huber and Huber (1996) sucrose contents in leaf is determined by several factors including the rate of photosynthesis, the partitioning of photosynthetic carbon between starch and sucrose, the rate of sucrose hydrolysis and the rate of sucrose export. The sucrose contents in leaves can imply the existing availability of carbon assimilates for growth and development, since this sugar is both, the principal and the preferred form of photoassimilate for transport to sink organs (Liu et al. 2004).

Under water stress is a strong decreased in starch content, inactive osmotically, and concomitant increase in soluble sugars, osmotically active, which provide the lowering the osmotic potential and contributes to maintenance of leaf water status (Pelleschi et al. 1997; Pinheiro et al. 2001; Yang et al. 2001). Moreover, according Bray (1997) sugars are energy and carbon sources required for defense response and water stress adaptation, hence, the higher supply of these molecules is necessary for plant survival in this condition. When reserve polysaccharides are mobilized, often has sucrose as a product of hydrolysis. The drought modifies the carbon partitioning between starch and sucrose synthesis and the sucrose/starch ratio can increase 2 or 3 fold (Quick et al. 1989, 1992; Vassey and Sharkey, 1989). Plants of grapevine cv. Cabernet Sauvignon exposed to severe stress showed decreases in leaf starch levels with fall in ¬w, so, as a consequence of increased synthesis *versus* starch breakdown higher levels of sucrose were expected. However, the drought leads to reductions of 32% in the levels of sucrose, but, the sucrose/starch ratio increased from 0.54 to 0.69 in control and dryness plants, respectively (table 3). This increase being more directly related to starch reduction than sucrose synthesis itself. Moreover, the decrease in leaf water potential reduced 18% in the SPS activity and increased 60% in the vacuolar acid invertase activity (Figure 2). In leaves, sucrose content is dependent on its synthesis catalyzed by sucrose phosphate synthase (SPS) and its breakdown catalyzed by invertases. The employ and allocation of sucrose for different pathways and different cellular compartments are greatly dependent on the physiological and biochemical requirements of tissues: 1) driven to glicolytic way and tricarboxylic acid cycle to produce ATP and NADH; 2) employed in biosynthesis of primary metabolites essential for growth and development; 3) converted in polymers such starch, triacylglycerides or polypeptides for long-term storage or 4) converted in secondary compounds enabling the plants to produce with predators, pests and environment changes. Many of these processes can occur simultaneously in the same compartment, so the allocation of sucrose for such events requires a mechanism of precise control. The use of sucrose as a source of carbon and energy depends on their breakdown into hexoses, and in plants either sucrose synthase (SS) or invertase catalyze this reaction. Invertase (EC 3.2.1.26, β-fructofuranosidase) is a hidrolase and cleaving sucrose into glucose and fructose, whereas SS (EC 2.4.1.13) is a glycosyl transferase that in presence of Uridine Diphosphate (UDP) cleaving sucrose into UDP-glucose and fructose (Sturm, 1999). Plants have two classes of invertase, which differ by their optimum pH, being alkaline invertase (AI) with maximal activity at pH 7.0 located in the cytoplasm, and the acid isoforms with optimum activity at pH around 5.0 and at least two subcellular locations. Most plants have at least two isoforms of vacuolar invertase, which accumulate as soluble proteins (soluble acid invertases) in the lumen of this acidic compartment and several isoforms of extracellular invertase that are ionically bound to the cell wall (Sturm, 1999).

Treatment	Starch ($\mu g\ g^{-1}$ FW)	Sucrose ($\mu mol\ g^{-1}$ FW)	Glucose ($\mu mol\ g^{-1}$ FW)
Irrigated	0.4	300	254
Drought	0.1	164	153

Table 3. Starch, sucrose and glucose levels measured in Cabernet Sauvignon under severe water stress.

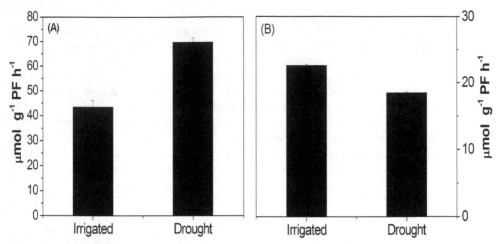

Fig. 2. Invertase activity (A) and sucrose phosphate synthase activity measured in cabernet sauvignon under severe stress.

The role of invertases is extremely important for many cellular processes, such as for example, their involvement in phloem loading and unloading, the involvement in plant defense response to stresses and recovery of turgor for cell expansion (Sturm and Tang, 1999). Sugars and invertases are crucial for reproductive development under drought. In several parts of the world, reproductive crops are the mainstays of agriculture being large extensions devoted to grain, fruit, nuts, flowers and so on. During the production cycle of plants, it is vitally to maintain these structures, especially under environment unfavorable conditions. In virtually all crops, water stress can trigger a smaller floral development or induce aborts when occurred around the time of pollination. Even if occurs the rehydration by rain or irrigation management, there is no resumption of floral growth and the abortion will irreversible. The result is fewer grains or fruits, but, if dryness is delayed until later in reproduction, abortion may not occur and, instead, floral and fruit development may diminish reversible (Saini and Westgate, 2000; Boyer and Westgate, 2004). After phloem unloading in sinks, sucrose is translocated to recipient cells through either the apoplastic pathway (cell wall matrix) or symplastic pathway (via plasmodesmata). The apoplastic pathway is effective at various points during development of most of reproductive organs where CWIN is characteristically expressed to hydrolyse extracellular sucrose. The cell wall invertase (CWIN) and vacuolar invertase (VIN) are associated to the earliest phases of flower development and the activity supplies hexoses to developing anthers and ovaries before pollination occurs. Recipient cells in sinks receive the hexose released from CWIN activity through hexose transporter (HT) and co-expressinon o CWIN and HT has been observed in many systems (Roitsch and Gonzalez, 2004).

Tissue elongation can be attributed to both process cell division and cell expansion, and it is believed that vacuolar invertase plays a role in cell expansion. But, it is difficult to obtain direct evidence for this, mainly due the multi-cellular nature of plant tissues where a given cell type is often deeply embedded and thus not readily accessible for harvest and measurement. It is believed that the invertase act in cell expansion by promoting the osmotic regulation (Roitsch and Gonzales, 2004; Sergeeva et al., 2006). This is possible because through its activity invertase hydrolyzing sucrose into two hexoses which doubles the osmotic contribution thus favoring water influx to drive cell expansion. The invertase activity in response to water stress present different patterns, which vary according to the intensity of the stress and the species studied. According to Zinselmeier et al. (1995), the increased in soluble and insoluble invertase activity during pollination and grain development of maize was blocked by water stress. This was correlated with low levels of reducing sugars, increased in sucrose contents, starch depletion and growth inhibition of ovarian culminating in abortion. However, there was strong accumulation of hexoses correlated with induction of vacuolar invertase activity in leaves of maize exposed to water stress (Peleschi et al. 1997). These results show that the response of invertase to water stress is specific to tissue and organ. In addition, in leaves of Lupinus albus L. the activity of vacuolar and cytosolic invertases and soluble sugar levels increased 6 days after water suspension, being the magnitude of the increased more pronounced in leaf blade (Pinheiro et al. 2001). In hydrated maize plants except cell wall invertase, the activity of other enzymes responsible for hydrolysis of sucrose is higher in sink than in source leaves, being none detected or only marginal effect of water stress on the activity of CWIN, CIN and SS in both organs (Kim et al. 2000). But, the water stress has caused a remarkable increase in VIN in mature leaves, leaf sheath and primary roots. In maize under drought, the most pronounced activity of VIN was due Ivr2 gene expression (Peleschi et al. 1999; Kim et al. 2000), while none of the other genes encoding both the soluble (Ivr1) and insoluble (Incw1, Incw2, Incw3, Incw4) form were expressed or modulated (Kim et al. 2000).

The buildup of sucrose and hexoses in source leaves may have several consequences, such as change in cellular osmotic potential and modulation of the expression of various genes of carbohydrate metabolism. According Sturm and Tag (1999) sucrose and its hydrolysis hexoses have crucial role in several signal transduction pathways. Because are sessile, plants have few alternative to survive and/or acclimate to environment changing, then are highly sensitive and responsive. Thus, the genes regulation by carbohydrates represents a valuable mechanism for adaptation to environment failures (Smeekens, 2000; Rolland et al. 2006). The sugar concentration in plant tissues varies over wide range, which typically exceeds that found in other homeostatic systems, such as the blood of mammals. This provides the plants with broad spectrum of signs and great capacity for adjustment. The effects of carbohydrate availability on the expression of specific genes enable and amplify the possible interference of more immediate metabolic control. In general way decreases in carbohydrate levels increase gene expression of photosynthetic process (Sheen, 1990; Krapp et al; 1993; Ehness, 1997) and storage mobilization/translocation (Ho et al. 2001; Conde et al. 2006). On the other hand, high levels of carbohydrate lead the expression of storage and consumption genes (Roitsch et al. 1995; Atassanova et al. 2003).

Abscisic acid (ABA), a important component of several signal transduction pathway of stress response led increases in invertase activity when exogenously applied in soybean green beans (Ackerson, 1985) and maize leaves exposed to drought (Trouverie et al. 2003). Abscisic acid induces not only increases in VIN activity but also the expression of Ivr2 in leaves and root maize (Trouverie et al. 2004). Some methodical and striking research of the sugar-ABA interface have been carried out recently and hexose-based signals originating from sucrose cleavage are implicated in regulation of ABA biosynthetic genes (Gazarrani and McCourt, 2001). Like the several role exerted by the sucrose, eg. nutrition, osmoregulation and signaling, plant invertase may have different functions. Often, this enzyme hydrolyzes sucrose into hexoses providing fuel for respiration as well carbon and energy for the synthesis of diverse compounds. The cleavage of sucrose into glucose and fructose causes a marked increase in osmotic pressure of cells, suggesting a possible role of invertase in cell elongation and plant growth (Sturm and Tag, 1999). Thus, in water stress conditions the invertase can act as mediators in the process of osmotic adjustment required for the survival of plant.

4. Water-saving irrigation and plant metabolism

It is known that many crops have high water requirements and supplemental irrigation is necessary to achieve the optimal production rates. According to actually scenario of climate changing and future predictions the demand for irrigation will increase considerably in years to alleviate the consequences of these changes. Nevertheless, as a consequence of global climate changes and environment pollution, water for agriculture has been limited in many regions (FAO, 2003). Because of this, water resources saving and increasing agricultural productivity per unit of water are becoming of a strategic importance for many countries. Irrigation is applied for preventing water deficits that reduce crop yield. The crop water use has two major components related to water loss: evapotranspiration (ET) that represents evaporation losses from the soil and crop and other losses that result from the distribution of water to the land. An important fact is that all irrigation water contains salts and, as water evaporates, salts concentrate in soil profile and must be displaced below the root zone before they reach a concentration that limits crop yield. Salt leaching is achieved by the movement of water applied in excess of evapotranspiration and some of water losses are unavoidable and are needed to maintain the salt balance. So, this effect can be minimized with efficient irrigation methods and by appropriate management. The evaporation from crop canopies is strongly attached with carbon assimilation (Steduto et al. 2009), so diminishing ET without deleterious effects in crop yield is so difficult. A restraint in water supply that decreases transpiration under the rate dictated by the evaporative demand of the atmosphere is paralleled by a reduction in biomass gain.

Thus, improving management irrigation is most likely the best option in most agricultural systems for increasing the efficiency of water use (Steduto et al. 2007). Deficit irrigation (DI) and partial root-zone drying (PRD) are water-saving irrigation strategies. The practice of water supply below the ET requirements is termed deficit irrigation (DI), where the irrigation is reduced relative to that needed to meet maximum ET and the mild stress has minimal effects on the crop productivity (English, 1990; English and Raja, 1996). Thus, water employed for irrigation can be reduced and the water saved can be diverted for another uses. In areas of frequent water scarceness and long drought periods the application of DI is a common practice; however, an effective use of DI requires prior knowledge of specific crop-growth stages showing tolerance to water stress. Moreover, the decreasing in the water improvement for irrigation to an area requires a lot of adjustments in agricultural system. PRD is a further refinement of DI; which involves irrigating only half of the root zone leaving the other to dry to a predetermined level before alternating irrigation. The PRD practice can save up to 50% of irrigation water with only a marginal yield reduction in tomato (Kirda et al. 2004). So, accumulated evidence has demonstrated that, given a same amount of irrigation water, PRD is superior to DI in terms of yield maintenance and increase WUE (Dodd 2009; Wang et al. 2010). Wang et al. (2009) reported that in potatoes plants total N content leaf layer was significantly higher for PRD than for the DI and full irrigated (FI).

However, the increment of tuber dry weight was similar for all the irrigation treatments; as a result the WUE was similar for the PRD and DI treatments and which was significantly greater than that for FI treatment.

Results of Mingo et al. 2004 showed that PRD tomato plants achieved biomass equivalence with FI plants but in PRD plants translocation of dry matter from leaves and stems was increased which led increase of root biomass by 19%. Has been shown that the exposure of roots to soil drying and soil re-watering increases root growth (Laing et al. 1996) and that PRD increase root growth of grapevine plants (Dry et al. 2000). The conservative vision of drought is that soil drying induces restriction of water supply and this outcome in a sequential reduction of tissue water content, growth and stomatal conductance. Definitely this is the case, and in some events changes occurred in leaf physiology are more closely associated to the changes in soil water content (Passioura, 1988). This sort of reaction demands that the plants have certain mechanism for sensing the soil water content and regulating stomatal aperture and leaf growth accordingly. This might involve transfer of chemical signal from the roots to shoots through xylem and such control has been called chemical signaling (Jones, 1980). Both practices, PRD and DI led the ABA-based root-to-shoot chemical signaling regulating stomatal conductance and leaf expansion growth thereby increasing WUE (Dodd, 2007; Wang et al. 2010). But, earlier studies indicated that in a similar soil water deficit, PRD can intensify ABA signaling relative to the DI treatment resulting in better control of plant water loss causing further improvement of WUE (Dodd, 2007; Wang et al. 2010). PRD can stimulate root growth and maintain a constant ABA signaling to regulate shoot physiology; whereas plants under DI some of the roots in dry soils for long period may die and signaling may diminish and shoot water deficits may occur (Kirda et al 2004; Davies and Hartung, 2004).

Beyond saving water and improvement WUE, various researches have demonstrated that PRD may improve fruit quality (dos Santos et al. 2003; Zgebe et al. 2006; Treeb et al. 2007; Jensen et al. 2010). Although the physiological reasons for such event remain unclear, it is suggested that improved crop nutrient status may partly be responsible for the higher crop quality under the PRD treatment (Shahnazari et al. 2008). The authors predicted that PRD led to less mineral N left in the soil and the crop showed a clear "stay green" phenomenal late in the season, and might have contributed to the higher marketable tuber yield. In agreement, Wang et al. (2009) observed that PRD significantly increased the total N content in leaves, stems, and tubers of potato plants compared with FI and DI treatments. Another crop, like maize and wheat has also been denoted an increase in N uptake under PRD irrigation (Kirda et al. 2005; Li et al. 2005). Nevertheless, PRD does not improve N uptake in plants of maize (Hu et al. 2009) and tomato (Topcu et al. 2007) not being clarified the reasons for this discrepancy.

The PRD technique has been applied in several studies related to distinct culture such potato, tomato, soybean, wheat, olive and so on. But, to Vitis vinifera L plants which the production is frequently dependent on irrigation the technique is now undergoing extensive commercial trials. The main effects of PRD in grapevine are that water use efficiency is increased, vegetative vigour is reduced while crop yield and berry size are not significantly reduced. The reduction in canopy density can result in better light penetration to the bunch zone and a consequent improvement in grape quality (Dry et al. 1996). So, the concept to use PRD as a tool to manipulate water deficit responses in this way had its origin in the observation of Loveys and During (1984) who the root-derived ABA signal was important to grapevine stomatal conductance. Thereafter, Gowing et al. (1990) showed that split-root plants could be used to show that many of the effects of water stress could be explained in terms of the transport of chemical signals from root to shoot without changes in water relations. The management of irrigation allows manipulating vegetative development if both wet and dry root zones could be maintained (Loveys, 1991). The necessary chemical signals would derive from the dry roots and water supplied from the wet roots would prevent the development of severe water deficit. In a pot experiment where a comparision was made between partial and complete rootzone drying, partial drying resulted in an 80% reduction in stomatal conductance an 60% in leaf ABA content whereas when a similar reduction in gs occurred in response to drying the entire root system foliar ABA increased 5-fold (Stoll et al. 2010). According Loveys et al. (2000) the changes in ABA content of grapevine roots exposed to drying soil not necessary translate to an equivalent change in leaf ABA. The amount of ABA accumulated in leaves may not be important in determining stomatal conductance but the levels of ABA in the xylem is the important factor that control stomatal aperture (Jia and Zhang, 1999). Both, the high concentration of ABA in the xylem and the increase in xylem sap pH in PRD vines may contribute to greater transpiration efficiency and regulation of stomatal conductance (Stoll et al. 2010).

One of the greatest strengths of PRD over other forms of deficit irrigation is the better control of vegetative growth without reduction in fruit yield and quality (Santos et al. 2003). Frequently, this control is associated to reduce total leaf area index in response to PRD application (Santos et al. 2003; Liu et al. 2006). So, a lowest green leaf area will not only reduce transpiration but can also reduce photosynthesis due to less light capture. But, in some studies the PRD treatment have not influenced the photosynthetic rate (Aganchich et al. 2009; Melgar et al. 2010). As discussed previously in this chapter water deficits can affect photosynthesis and reduce carbon photoassimilates in source leaves. The response of plant growth is closely related to carbon supply and allocation within its organs, and under water deficit the export of photoassimilates from source leaves to sink organs is reduced and the competition for assimilates among the organs is modified. As a result, carbon assimilation, partitioning and plant growth changes in response to water deficit, since the growth of any specific sink organ especially reproductive organ, depends mostly on the carbon assimilates available to the organ. Opposed to a simple water deficit treatment, DI scheduling is often applied by withholding water specifically in periods when organ of the plant with economic value (eg. Fruit, grain, root...) is less sensitive to water deprivation. Although, the responses of carbon allocation to water stress are well clarified, there is still a lack of knowledge about how PRD and DI interfere allocation patterns of carbon assimilates.

Deficit irrigation is more frequent applicable in tree crops and vines because economic returns be higher in fruit trees than in field crops. When tested in tomatoes, PRD saved up 50% of the irrigation water and caused a marginal reduction in yield with reduced leaf area index and vegetative growth suggesting that photosynthesis assimilates were predominately partitioning to fruit growth so that significant yield reduction was prevent (Kirda et al. 2004). According to yours findings the authors suggest that PRD can be viable and beneficial option with the conventional DI to avoid crop-yield reductions when and if there is water shortage. Moreover, high crop yield can be maintained under water scarcity if the PRD is applied, so are still needed more studies to test yield responses of other horticultural, field and tree crops with a high-irrigation-water requirement. However, when applying the deficit irrigation is important to know the low, medium and drought-tolerant growth stage to optimize the reduction of water that can be deployed. In various cases PRD is capable to maintain a elevated yield as well as higher quality and thereby a elevated water saving per produced unit than DI. Higher local soil water flow might be present and higher nitrogen mineralization occurred under PRD as compared with DI influencing the transport of nutrients to the root surfaces (Jensen et al. 2010).

Although it has shown positive results for some crops, many tests are required with the practice of PRD does occur before its implementation on a commercial scale. So far, the most valued characteristics in studies of PRD and DI applied are WUE and the gas exchange and in some cases the crop yield and quality. However, there is a great gap about the behavior of sugar partitioning between organ source and organ sink as well as activity of key enzymes of this process, in cellular respiration and in other processes related to cellular energy in response to DI and PRD applied. Even though there studies showing few or no modification on gas exchange in plants exposed to DI or PRD failure to elucidate the physiological mechanisms responsible to this. Given the fact that production (photosynthesis) partitioning (sugar metabolism) and consumption (respiration) are the mainstays to plant growth and development, studies of such processes in plants exposed to DI or PRD is extremely important to support the recommendation of this technique.

5. Conclusion

Actually, more than told the global climate change has been experienced by all living organisms, especially the plants because they are sessile. In this scenario a major concern in the plant science community is the aridity increasing ever more reducing water availability for agriculture practices. The plants respond quickly to reduced water potential in soil and atmosphere by altering several metabolic reactions, physiological process and genes expression, where the first one is the decrease in stomatal aperture. Photosynthesis and respiration are closely related process, and obviously the changes in plant growth elicited by reduced water availability have been associated to modulations between the production and consumption of carbon. Furthermore, another strong implication to plant development inherent drought is the impaired partitioning of carbon between the leaf and others organs. So, before this situation the development of suitable irrigation managements in order to maintain crop yield without impaired quality and avoidance water waste is an important issue worldwide. Some research employing irrigation management like deficit irrigation and partial root-zone drying techniques has been widely investigated with promising results but somewhat conflicting. Although, the major works employing DI and PRD technique were performed in perennial crops, with a few works in horticultural crops. Moreover, the results obtaining by these irrigation management are mainly related to quality and water use efficiency, being required studies on the parameters of production (photosynthesis), consumption (respiration) and partitioning (sugar metabolism) of carbon.

6. References

Abreu, M.E. & Munné-Bosch, S. (2008). Salicilic acid may be involved in the regulation of drought-induced leaf senescence in perennials: A case study in field-grown Salvia officinalis L. plants. *Environmental and Experimental Botany*, 64, 105-112.

Ackerson, R.C. (1985). Osmoregulation in Cotton in response to water stress. III. Effects of phosphorus fertility. *Plant Physiology*, 77, 309-312.

Aganchich, B. Wahbl, S., Loreto, F., & Centritto, M. (2009). Partial root zone drying: regulation of photosynthetic limitations and antioxidant enzymatic activities in young olive (*Olea europaca*) saplings. *Tree Physiology*, 29, 685-696.

Alscher RG, Donahue JL, & Cramer CL. (1997). Reactive oxygen species and antioxidants: relationships in green cells. *Physiologia Plantarum*, 100, 224–233.

Arnholdt−Schmitt, B., Costa, J.H. & de Melo, D.F. (2008). AOX − a functional marker for efficient cell reprogramming under stress? *Trends In Plant Science*, 11, 281-287.

Asseng, S., & Hsiao TC (2000). Canopy CO_2 assimilation, energy balance, and water use efficiency of an alfalfa crop before and after cutting. *Field Crops Research*, 67, 191–206.

Atassanova, R., Leterrier, M., Gaillard, C., Agasse, A., Sagot, E., Coutos-Thévenot, P. & Delrot, S. (2003). Sugar-regulated expression of a putative hexose transport gene in grape. *Plant Physiology*, 131, 326-334.

Bartoli, C.G., Gomez, F., Gergoff, G., Gulaét, J.J. & Puntarulo, S. (2005). Up-regulation of the mitochondrial alternative oxidase pathway enhances photosynthetic electron transport under drought conditions. *Journal of Experimental Botany*, 56, 1269-1276.

Bartoli, C.G., Pastori, G.M. & Foyer, C.H. (2000). Ascorbate biosynthesis in mitochondria is linked to the electron transport chain between complexes III and IV. *Plant Physiology*, 123, 335-343.

Bartoli, CG, Pastori, GM & Foyer, CH (2000). Ascorbate biosynthesis in mitochondria is linked to the electron transport chain between complexes III and IV. *Plant Physiology*, 123, 335-343.

Blum, A (1998). Improving wheat grain filling under stress by stem reserve mobilization. *Euphytica*, 100, 77-83.

Bota, J., Medrano, H. & Flexas, J. (2004). Is photosynthesis limited by decreased rubisco activity and RuBP content under progressive water stress? *New phytologist*, 162, 671-681.

Boyer, J.S., & Westgate, M.E. (2004). Grain yields with limited water. *Journal of Experimental Botany*, 55, 2385–2394.

Bray, E.A. (1997). Plant responses to water deficit. *Trends in Plant Science*, 2, 48-54.

Brounleader, M.D., Harbone, J.B. & Dey, P.M. (1997). Carbohydrate metabolism: Primary metabolism of monosaccharides. Chapter 3. In: DEY P.M. & HARBORNE J.B. Eds. *Plant Biochemistry*, 111-140.

Chaves, M.M., Maroco, J.P., & Pereira, J.S. (2003). Understanding plant response to drought: from genes to the whole plant. *Functional Plant biology*, 30,239-264.

Chaves, M.M., & Oliveira, M.M., (2004). Mechanisms underlying plant resilience to water deficits: prospects for water-saving agriculture. Water-saving in Agriculture Special issue. *Journal of Experimental Botany*, 55, 2365–2384.

Chaves, M.M., Pereira, J.S., Rodrigues, M.L., Ricardo, C.P.P., Osório, M.L., Carvalho, I., Faria, T., & Pinheiro, C. (2002). How plants cope with water stress in the Field: photosynthesis and growth. *Annals of Botany*, 89, 907-916.

Chaves, MM, Flexas, J. & Pinheiro, C. (2009). Photosynthesis under drought and salt stress: regulation mechanisms from whole plant to cell. *Annals of Botany*, 103, 551-560.

Chaves,M.M. (1991). Effects of water deficits on carbon assimilation. *Journal of Experimental Botany*, 42, 1-16.

Chernyad'ev, LI.I. (2009). The protective action of cytokinins on the photosynthetic machinery and productivity of plants under stress (review). *Applied Biochemestry and Microbiology*, 45, 351-362.

Clifton R, Millar AH, & Whelan J. (2006). Alternative oxidases in Arabidopsis: a comparative analysis of differential expression in the gene family provides new insights into function of nonphosphorylating bypasses. *Biochimica et Biophysica Acta*, 1757, 730–741.

Conde, C., Agasse, A., Glissant, D., Tavares, R., Gerós, H.; & Delrot, S. (2006). Pathways of glucose regulation of monosaccharide transport in grape cells. *Plant Physiology*, 141, 1563-1577.

Cornic, G. (2000). Drought stress inhibits photosynthesis by decreasing stomatal aperture – not by affecting ATP synthesis. *Trends in Plant Science*, 5,187-188.

Cornic, G., & Briantis, J.M. (1991). Partitioning of photosynthetic electron flow between CO_2 and O_2 reduction in a C_3 leaf (*Phaseolus vulgaris* L.) at different CO_2 concentration and during drought stress. *Planta*, 183, 178-184.

Davies, W.J., & Hartung, W., (2004). Has extrapolation from biochemistry to crop functioning worked to sustain plant production under water scarcity? In: *Proceeding of the Fourth International Crop Science Congress*, 26 September–

Day, DA, Dry, IB, Soole, KL, Wiskich, JT, & Moore, AL (1991). Regulation of alternative pathway activity in plant mitochondria. *Plant Physiology*, 95, 948-953.

Dodd IC (2007). Soil moisture heterogeneity during deficit irrigation alters root-to-shoot signaling of abscisic acid. *Functional Plant Biology*, 34, 439–448.

Dodd, IC (2009). Rhizposphere manipulations to maximize 'crop per drop' during deficit irrigation. *Journal of Experimental Botany*, 60, 1–6.

Dos Santos, T.P., Lopes, C.M., Rodrigues, L., de Souza, C.R., Maroco, J.P. Pereira, J.S., Silva, J.R. and Chaves, M.M (2003). Partial rootzone drying: effects on growth and fruit quality of field-grown grapevines (Vitis vinifera L.). Functional Plant Biology, 30: 663-671.

Dry, P.R., Loveys, B.R., Botting, D. and During, H. (1996) Effects of partial root-zone drying on grapevine vigour, yield, composition of fruit and use of water. Proceedings of the 9th Australian Wine Industry Technical Conference, 126-131.

Dry, P.R., & Loveys, B.R., (1999). Grapevine shoot growth and stomatal conductance are reduced when part of the root system is dried. *Vitis*, 38, 151–156.

Dry, P.R., Loveys, B.R., & During, H., (2000). Partial drying of rootzone of grape. I. Transient changes in shoot growth and gas exchange. *Vitis* 39, 3–7.

Dry, P.R., Loveys, B.R., McCarthy, M.G., & Stoll, M., (2001). Strategic management in Australian vineyards. *Journal International Sciences de la Vigne et du Vin*. 35, 129–139.

Dutilleul, C, Driscoll, S, Cornic, G, De Paepe, R, Foyer, CH & Noctor, G (2003). Functional mitochondrial complex I is required by tobacco leaves for optimal photosynthetic performance in photorespiratory conditions and during transients. *Plant Physiology*, 131, 264-275.

Ehness, R., Ecker, M., Dietmute, E., Godt, E. & Roitsch, T. (1997). Glucose and stress independently regulate source and sink metabolism and defense mechanism via signal transduction pathways involving protein phosphorylation. *The Plant Cell*, 9, 1825-1841.

English M, & Raja SN (1996). Perspectives on deficit irrigation. *Agricultural Water Management*, 32, 1–14.

English M. (1990). Deficit irrigation. I. Analytical framework. *Journal of Irrigation and Drainage Engineering*, 116, 399–412.

Erice,G, Louahlia, S, Irigoyen, JJ, SAnches-Diaz, M, Alami, TI & Avice, J-C (2011). Water use efficiency, transpiration and net CO_2 exchange of four alfafa genotypes submitted to progressive drought and subsequent recovery. *Environmental and Experimental Botany*, 72, 123-130.

FAO, 2003. AQUASTAT–Global information system on water and agriculture. Global Map of Irrigation Areas Serbia. See: http://www.fao.org/ES/ess/index.en.asp.

Feng, H, Li, H Li, X, Duan, J, Liang, H, & Zhi, D, Ma, J (2007). The flexible interrelation between AOX respiratory pathway and photosynthesis in rice leaves. *Plant Physiology and Biochemistry*, 45, 228-235.

Finnegan, P.M., Soole, K.L., & Umbach, A.L., (2004). Alternative mitochondrial electron transport proteins in higher plants. In: Day, D.A., Millar, A.H., Whelan, J. (Eds.), *Plant Mitochondria: From Genome to Function*. Kluwer Academic Publishers, Dordrecht, pp. 163–230.

Flexas, J. & Medrano, H. (2002). Drought-inhibition of photosynthesis in C_3 plants: Stomatal and non-stomatal limitations revisited. *Annals of Botany*, 89, 183-189.

Flexas, J., Bota, J., Clifre, J.; Escalona, J.M., Galmes, J.; Gulias, J., El-Kadri Lefi; Martinez-Cantellas, S.F.; Moreno, T.; Ribas-Carbo, M.; Riera,D.; Sampol, B. & Medrano, H. (2004). Understanding down-regulation of photosynthesis under water stress: future prospects and searching for physiological tools for irrigation management. *Annals Applied Biology*, 144, 273-283.

Flexas, J., Escalona, J.M. & Medrano, H. (1999). Water stress induces different levels of photosynthesis and electron transport rate regulation in grapevines. *Plant, Cell and Environment*, 22, 39-48.

Flexas, J.; Galmés, J.; Ribas-Carbo, M. & Medrano, H. (2005). The effects of water stress in plant respiration. Chapter 6. In: H lambers, M. Ribas-Carbo (Eds) *Plant respiration: from cell to ecosystem*, vol 18. Advances in photosynthesis and respiration: Series . p. 85-93.

Flexas, J.; Ribas-Carbo, M.; Bota, J.; Galmés, J.; Henkle, M.; Martinez-Cañellas, S, S. & Medrano, H. (2006). Deceased rubisco activity during water stress is not induced by decreased relative water content but related to conditions of low stomatal conductance and chloroplast CO_2 concentration. *New Phytologist*, 172, 73-82.

Flowers, M.D., & Lal, R., (1998). Axle load and tillage effects on soil physical properties and soybean grain yield on a mollic ochraqualf in northwest Ohio. *Soil Tillage Research*, 48, 21–35.

Galmés, J., Medrano, H. & Flexas, J. (2007). Photosynthetic limitations in response to water stress and recovery in Mediterranean plants with different growth forms. *New Phytologist*, 175, 81-93.

Gazarrani, S. & McCourt, P. (2001). Genetic interactions between ABA, ethylene and sugar signaling pathways. *Current opinion in Plant Biology*, 4, 387-391.

Georgieva, K., Sárvári, E. & Keresztes, A. (2010). Protection of thylakoids against combined light and drought by a luminal substance in the resurrection plant *Harberlea rhodopensis*. *Annals of Botany*, 105, 117-126.

Ghannoum, O., Conroy, J.P., Driscoll, S.P.; Paul, M.J., Foyer, C.H. & Lawlor, DW (2003). Nonstomatal limitations are responsible for drought-induced photosynthetic innhibition in four C_4 grasses. *New Phytologist*, 159, 599-608.

Ghashghaie, J, Duranceau, M, Badeck, FW, Cornic, G, Adeline, MT, & Deleens, E (2001). $\delta^{13}C$ of CO_2 respired in the dark in relation to $\delta^{13}C$ of C of leaf metabolites comparison between *Nicotiana sylvestris* and *Helianthus annuus* under drought. *Plant, Cell and environment*, 28, 834-849.

Gowing, G.J., Davies, W.J., & Jones, H.G. (1990). Regulated deficit irrigation of Cabernet Sauvignon grapevines. *Australian New Zealand Wine Industry Journal*, 5, 131-133.

Hansen, L.D., Church, J.N., Matheson, S., McCarlie, V.W., Thygerson, t., Criddle, R.S. & Smith, B.N. (2002) Kinetics of plant growth and metabolism. Thermochimestry Acta 388: 415-425.

Haupt-Herting, S. & Fock, H.P. (2000). Exchange of oxygen and its role in energy dissipation during drought stress in tomato plants. *Physiologia Plantarum*, 110, 489-495.

Haupt-Herting, S. & Fock, H.P. (2002). Oxygen Exchange in relation to carbon assimilation in drought stressed leaves during photosyntheis. *Annals of Botany*, 89, 851-859.

Ho, S.L., Chao, Y.C., Tong, W.F. & Yu, S.M. (2001). Sugar coordinately and differentially regulates growth-and stress-related gene expression via a complex signal transduction network and multiple control mechanisms. *Plant Physiology*, 125, 877-890.

Hoefnagel, MHN, Atkin, OK, & Wiskich, JT (1998). Interdependence between chloroplasts and mitochondria in the light and the dark. *Biochimica et Biophysica Acta*, 1366, 235-255.

Huber, S.C. & Huber, J.L. (1996). Role and regulation of sucrose-phosphate synthase in higher plants. *Annual Review Plant Physiology and Plant Molecular Biology*, 47, 431-444.

Hund, A, Ruta, N & Liedgens, M (2009). Rooting depth and water use efficiency of tropical maize inbred lines, differing in drought tolerance. *Plant Soil*, 318, 311-325.

Hura, T., Hura, K. & Grzesiak, M. (2011). Soil drought applied during the vegetative growth of triticale modifies the physiological and biochemical adaptation to drought during the generative development. *Journal of Agronomy and Crop Science*, 197, 113-123.

Ibáñez, H., Ballester, A., muñoz, R. & Quiles, M.J. (2010). Chlororespiration and tolerance to drought, heat and high illumination. *Journal of Plant Physiology*, 167, 732-738.

IPCC Fourth Assessment Report (2007). Climate Change 2007. United Nations Intergovernmental Panel on Climate Change (IPCC).

Jensen, C.R., Battilani, A., Plauborg, F., Psarras, G., Chartzoulakis, K., Janowiak, F., Stikic, R., Jovanovic, Z., Li, G., Qi, X., Liu, F., Jacobsen, S.E. & Andersen, M.N. (2010). Deficit irrigation based on drought tolerance and root signaling in potatoes and tomatoes. *Agricultural Water Management*, 98, 403-413.

Jia, W.S. & Zhang, J.H. (1999). Stomatal closure is induced rather by prevalling xylem abscisic acid than by accumulated amount of xylem-derrived abscisic acid. *Physiologia Plantarum*, 106, 268-275.

Jones, H.G, (1980). Interaction and integration of adaptive responses to water stress: the implication of an unpredictable environment. In: *Adaptation of plants to water and high temperature stress*. Turner, N.C., P.J. Kramer, Ed. Wiley, New York, 353-365.

Kaiser, WM (1987). Effects of water deficit on photosynthetic capacity. *Physiologia Plantarum*, 71, 142-149.

Keck, R.W. & Boyer, J.S. (1974). Chloroplast response to low leaf water potentials. III. Differing inhibition of electron transport and photophosphorylation. *Plant Physiology*, 53, 474-479.

Kim, J.Y., Mathé, A., Guy, S., Brangeon, J., Roche, O., Chourey, P.S., & Prioul, J.L. (2000). Characterization of two members of the maize gene family. Ivr2 and Incw4, encoding cell-wall invertases. *Gene*, 245, 89-102.

Kirchbaum, M.U.F. (1987). Water stress in Eucalyptus pauciflora: comparison of effects on stomatal conductance with effects on the mesophyll capacity for photosynthesis, and investigation of a possible involvement of photoinhibition. *Planta*, 171, 466-473.

Kirda C, Topcu S, Kaman H, Ulger AC, Yazici A, Cetin M, & Derici MR (2005). Grain yield response and N-fertilizer recovery of maize under deficit irrigation. *Field Crops Research*, 93, 132–141.

Kirda, C., Cetin, M., Dasgan, Y., Topcu, S., Kaman, H., Ekici, B., Derici, M.R. & Ozguven, A.I. (2004). Yield response of greenhouse grown tomato to partial root drying and conventional deficit irrigation. *Agricultural Water Management*, 69, 191-201.

Kiyosue T, Yoshiba Y, Yamaguchi-Shinozaki K, & Shinozaki K. (1996). A nuclear gene encoding mitochondrial proline dehydrogenase, an enzyme involved in proline metabolism, is upregulated by proline but downregulated by dehydration in Arabidopsis. *The Plant Cell* 8, 1323–1335.

Kramer,P.J., & Boyer, J.S. (1995). Water relations of plants and soils. New York: Academic Press, 1995, 495 p.

Krapp, A., Hofmann, B., Schàfer, C. & Stitt, M. (1993). Regulation of the expresion of rbcS and other photosynthetic genes by carbohydrate: a mechanism for the "sink regulation" of photosynthesis? *The Plant Journal*, 3, 817-828.

Krieg, N.D. (1983). Photosynthetic activity during stress. *Agricultural Water Management*, 7, 249-263.

Krömer, S., Malmberg, G., & Gardeström, P., (1993). Mitochondrial contribution to photosynthetic metabolism. A study with barley (Holdeum vulgare L.) leaf protoplasts at different light intensities and CO_2 concentrations. *Plant Physiology*, 102, 947–955.

Krömer, S., Stitt, M., & Heldt, H.W., (1988). Mitochondrial oxidative phosphorylation participating in photosynthetic metabolism of a leaf cell. *FEBS Letters*, 226, 352–356.

Laing, J., Zhang, J., & Wong, M.H., (1996). Effects of air-filled porosity and aeration on the initiation and growth of secondary roots of maize (Zea mays). *Plant Soil*, 186, 245–254.

Lawlor, D.W. & Cornic, G. (2002). Photosynthetic carbon assimilation and associated metabolism in relation to water deficits in higher plants. *Plant, Cell and Environment*, 25, 275-294.

Li Z, Zhang F, & Kang S (2005). Impacts of the controlled roots divided alternative irrigation on water and nutrient use of winter wheat. *Trans CSAE*, 21, 17–21.

Li, L, & Staden, JV (1998). Effects of plant growth regulators on the antioxidant system in callus of two maize cultivars subjected to water stress. *Plant Growth regulation*, 24, 55-66.

Liu F, Jensen CR, & Andersen MN (2004) Drought stress effect on carbohydrate concentration in soybean leaves and pods during early reproductive development: its implication in altering pod set. *Field Crop Research*, 86, 1–13.

Louwerse, W. (1980). Effects of CO_2 concentration and irradiance on the stomatal behavior of maize, barley, and sunflower plants in the field. *Plant Cell Environment*, 3, 391–398.

Loveys, B.R. (1991). What use is a knowledge of ABA physiology for crop improvement? In: *Environment plant biology*. Physiology and biochemistry of abscisic acid. Oxford: Bios Scientific Publishers, 245-250.

Loveys, B.R. Stoll, M., Dry, P.R. & McCarthy, M.G. (2000). Using plant physiology to improve the water use efficiency of

Loveys, B.R. & During, H. (1984). Diurnal changes in water relations and abscisic acid in field-grown *Vitis vinifera* cultivars. II. Abscisic acid changes under semi-arid conditions. *New phytologist*, 97, 37-47.

MacCabe, TC, Daley, D, & Whelan, J (2000). Regulatory, developmental and tissue aspects of mitochondrial biogenesis in plants. *Plant Biology*, 2, 121-135.

Mackenzie, S. & MacIntosh, L. (1999). Higher plant mitochondria. *The Plant Cell*, 11, 571-585.

Martim, SA, Santos, MP, Peçanha, AL, Pommer, C, Campostrini, E, Viana AP, Façanha, AR & Bressan-Smith, R (2009). Photosynthesis and cell respiration modulated by water deficit in grapevine (*Vitis vinifera* L.) cv. Cabernet Sauvignon. *Brazilian Journal of Plant Physiology*, 21, 95-102.

Masle J (1999). Root impedance: sensing, signaling and physiological effects. In: *Plant responses to environmental stresses from phytohormones to genome reorganization*. HR Lerner (editor) pp.475-495, Marcel Dekker Inc. New York.

Maxwell, DP, Wang, Y & McIntosh, L (1999). The alternative oxidase lowers mitochondrial reactive oxygen production in plant cell. *Proceedings National Academy Science*, 96, 8271-8276.

Medrano, H, Flexas, J & Galmés, J (2009). Variability in water use efficiency at the leaf level among Mediterranean plants with different growth forms. *Plant Soil*, 317, 17-29.

Melgar, J.C., Dunlop, J.M. & Syvertsen, J.P. (2010). Growth and physiological responses of the citrus rootstock Swingle citrumelo seedlings to partial rootzone drying and deficit irrigation. *Journal of Agricultural Science*, 148, 593-602.

Mingo, D.M., Theobald, J.C., Bacon, M.A., Davies, W.J., & Dodd, I.C., (2004). Biomass allocation in tomato (Lycopersicon esculentum) plants grown under partial rootzone drying: enhancement of root growth. *Functional Plant Biology*, 31, 971–978.

Moore, AL, & Siedow, JN (1991). The regulation and nature of the cyanide resistant alternative oxidase of plant mitochondria. *Biochimestry and Biophysical Acta*, 1058, 121-140.

Munné-Bosch, S., Falara, V., Pateraki, I., López-Carbonell, M., Cela, J., & Kanellis, A.K. (2009). Physiological and molecular responses of the isoprenoid biosynthetic pathway in a drought-resistent Mediterranean shrub, Cistus Crectus exposed to water deficit. *Journal of Plant Physiology*, 166, 136-145.

Noctor, G, Dutilleul, C, De Paepe, R, & Foyer, CH (2004). Use of mitochondrial electron transport mutants to evaluate the effects of redox state on photosynthesis, stress tolerance and the integration of carbon/nitrogen metabolism. *Journal of Experimental Botany*, 55, 49-57.

Nunes-Nesi, A., Sulpice, R., Gibon, Y. & Fernier, A.R. (2008). The enigmatic contribution of mitochondrial function in photosynthesis, *Journal of Experimental Botany*, 59, 1675-1684.

Padsmareee, K & Raghavendra, AS (1999.) Importance of oxidative electron transport over oxidative phosphorylation in optimizing photosynthesis in mesophyll protoplasts of pea (*Pisum sativum* L.). *Physiologia Plantarum*, 105, 546-553.

Passioura, J.B., (1988). Root signals control leaf expansion in wheat seedlings growing in drying soil. *Australian Journal of Plant Physiology*, 15, 687–693.

Pastore, D., Trono, D., Laus, M.N., Di Fonzo, N. & Flagella, Z. (2007). Possible plant mitochondria involvement in cell adaptation to drought stress. A case study: durum wheat mitochondria. *Journal of Experimental Botany*, 58, 195-210.

Pelleschi, S., Guy, S., Kim, J.Y., Pointe, C., Mathé, A., Barthes, L., Leonardi, A., & Prioul, J.L. (1999). Ivr2, a candidate gene for a QTL of vacuolar invertase activity in maize leaves. Gene-Specific expression under water stress. *Plant Molceular Biology*, 39, 373-380.

Pelleschi, S., Rocher, P. & Prioul, J.L. (1997). Effect of water restriction on carbohydrate metabolism and photosynthesis in mature maize leaves. *Plant, Cell and Environment*, 20, 493-503.

Pinheiro, C., Chaves, M.M. & Ricardo, C.P. (2001). Alterations in carbon and nitrogen metabolism induced by water deficit in the stems and leaves of *Lupins albus* L. *Journal of Experimental Botany*, 52, 1063-1070.

Popov, VN, Simonian, RA, Skulachev, VP & Starkov, AA (1997). Inhibition of the alternative oxidase stimulates H_2O_2 production in plant mitochondria. *FEBS Letters*, 416, 87-90.

Pou, A, Flexas, J, Alsina Mdel mar, Bota, J, Carambula, C, Herralde, F, Galmés, J, Lovisolo, C, Jimenez, M, Ribas-Carbo, M, Rusjan, D, Sechi, F, Tomàs, M, Zsófi, Z & Medrano, H (2008). Adjustments of water use efficiency by stomatal regulation during drought and recovery in the drought-adapted Vitis hybrid Richter-110 (V. berlandieri x V. ruprestris). *Physiologia Plantarum*, 134, 313-323.

Quick, W.P., Chaves, M.M., Wendler, R., David, M.M., Rodrigues, M.L., Pereira, J.S., Leegood, R. & Stitt, M. (1992). Effects of water stress on photosynthetic carbon metabolism in four species grown in Field conditions. *Plant Cell and Environment*, 15, 25-35.

Quick, W.P., Siegl, G., Neuhaus, E., Feil, R. & Stitt, M. (1989). Short-term water stress leads to a stimulation of sucrose synthesis by activating sucrose phosphate synthase. *Planta*, 177, 535-547.

Radford, BJ, Yule, DF, McGarry, D & Playford, C (2001). Crop responses to applied soil compaction and to compaction repair treatments. *Soil & Tillage Research*, 61, 157-166.

Rhoads DM, Umbach AL, Subbaiah CC, & Siedow JN. (2006). Mitochondrial reactive oxygen species: contribution to oxidative stress and interorganellar signaling. *Plant Physiology*, 141, 357–366.

Roitsch, T., & Gonza´ lez, M.C. (2004). Function and regulation of plant invertases: sweet sensations. *Trends Plant Science*, 9, 606–613.

Roitsch, T., Büttner, M., & Godt, D.E. (1995). Induction of apoplectic invertase of chenopodium rubrum by D-glucose and a glucose analog and tissue-specific expression suggest a role in sink-source regulation. *Plant Physiology*, 108, 285-294.

Rolland, F., Gonzales-Baena, E. & Sheen, J. (2006). Sugar sensing and signaling in plants: conserved and novel mechanisms. *Annual Review of Plant Biology*, 57, 675-709.

Saini, H.S., & Westgate, M.E. (2000).. Reproductive development in grain crops during drought. *Advances in Agronomy*. 68, 59–96.

Scandalios, JG (1993). Oxygen stress and superoxide dismutase. *Plant Physiology*, 101, 7-12.

Schonmaum, GR, Bonner, WD, Storey, BT, & Bahr, JT (1971). Specific inhibition of the cyanide-insensitive respiratory pathway in plant mitochondria by hydroxamic acids. *Plant Physiology*, 47, 124-128.

Sergeeva, L.I., Keurentjes, J.J., Bentsink, L., Vonk, J., van der Plas, L.H., Koornneef, M., & Vreugdenhil, D. (2006). Vacuolar invertase regulates elongation of Arabidopsis thaliana roots as revealed by QTL and mutant analysis. *Proceedings of National Academy Science*, 103, 2994–2999.

Shahnazari A, Ahmadi SH, Laerke PE, Liu F, Plauborg F, Jacobsen S-E, Jensen CR, & Andersen MN (2008). Nitrogen dynamics in the soil-plant system under deficit and partial root-zone drying irrigation strategies in potatoes. *European Journal of Agronomy*, 28, 65–73.

Sheen, J. (1990). Metabolic repression of transcription in higher plants. *The Plant Cell*, 2, 1027-1038.

Siedow, JN & Grivin, ME (1980). Alternative respiratory pathway. Its role in seed respiration and its inhibition by propyl galalate. *Plant Physiology*, 65, 669-674.

Smeekens, S & Rook, F (1997). Sugar sensing and sugar-mediated signal transduction in plants. *Plant Physiology*, 115, 7-13.

Smeekens, S. (2000). Sugar-induced signal transduction in plants. *Annual Review Plant Physiology and Plant Molecular Biology*. 51, 49-81.

Steduto, P., Hsiao, T.C., & Fereres, E., (2007). On the conservative behavior of biomass water productivity. *Irrigation Scence*, 25, 189–207.

Steduto, P., Hsiao, T.C., Raes, D., & Fereres, E., (2009). AquaCrop – the FAO crop model to simulate yield response to water: I. Concepts and underlying principles. *Agronomy Journal*, 101: 426–437.

Stitt, M. & Quick, W.P. (1989). Photosynthetic carbon partitioning: its regulation and possibilities for manipulation. *Physiologia Plantarum*, 77, 633-641.

Stoll, M., Loveys, B. & Dry, P. (2000). Hormonal changes induced by partial rootzone drying of irrigated grapevine. *Journal of Experimental Botany*, 51, 1627-1634.

Sturm A (1996). Molecular characterization and functional analysis of sucrose-enzymes in carrot (Daucus carota L.). *Journal of Experimental Botany*, 47, 1187–1192.

Sturm, A. & Tang, G.Q. (1999). The sucrose-cleaving enzymes of plants are crucial for development, growth and carbon partitioning. *Trends in Plant Science*, 4, 401-406.

Svensson, A.S., & Rasmusson, A.G., (2001). Light-dependent gene expression for proteins in the respiratory chain of potato leaves. *Plant Journal*, 28, 73–82.

Topcu S, Kirda C, Dasgan Y, Kaman H, Cetin M, Yazici A, &Bacon MA (2007). Yield response and N-fertiliser recovery of tomato grown under deficit irrigation. *European Journal of Agronomy*, 26, 64–70.

Treeb, M.T., Henriod, R.E., Bevington, K.B., Milne, D.J. & Storey, R. (2007). Irrigation management and rootstock effects on navel orange (Citrus sinensis L. Osbeck) fruit quality. *Agricultural Water Management*, 91, 24-32.

Trouverie, J., Chateau-Joubert, S., Thévenot, C., Jacqemot, M.P., & Prioul, J.L. (2004). Regulation of vacuolar invertase by abscisic acid or glucose in leaves and roots from maize plants. *Planta*, 219, 894-905.

Trouverie, J., Thévenot, C., Rocher, J.P., Sotta, B., & Prioul, J.L. (2003). The role of abscisic acid in response of a specific vacuolar invertase to water stress in the adult maize leaf. *Journal of Experimental Botany*, 54, 2177-2186.

Tzara, W, Mittichell, VJ, driscoll, SD, & Lawlor, DW (1999). Water stress inhibits plant photosynthesis by decreasing coupling factor and ATP. *Nature*, 401, 914917.

Umbach, AL, Fiorani, F & Siedow, JN (2005). Characterization of transformed *Arabidopsis* with altered alternative oxidase levels and analysis of effects on reactive oxygen species in tissues. *Plant Physiology*, 139, 1806-1820.

Vanlerberghe, G.C., & McIntosh, L. (1997). Alternative Oxidase: From gene to function. *Annual Review of Plant Physiology and Plant Molecular Biology*, 48, 703-734.

Vassey, T.L. & Sharkey, T.D. (1989). Mild water stress of Phaseolus vulgaris plants leads to reduced starch synthesis and extractable sucrose phosphate synthase activity. *Plant Physiology*, 89, 1066-1070.

Wagner, AM, & Krab, K (1995). The alternative respiration pathway in plants? Role and regulation. *Physiologia Plantarum*, 95, 318-325.

Wang, H., Liu, F., Andersen, M.N. & Jensen, C.R. (2009). Comparative effects of partial root-zone drying and deficit irrigation on nitrogen uptake in potatoes (*Solanum tuberisum* L.). *Irrigation Science*, 27:443-448.

Wang, H., Liu, F., Andersen, M.N. & Jensen, C.R. (2009). Comparative effects of partial root-zone drying and deficit irrigation on nitrogen uptake in potatoes (*Solanum tuberosum* L.). *Irrigation Science*, 27, 443-448.

Wang, Y., Liu, F., Nerrgaard, A., Jensen, L.S., Luxhoi, J. & Jensen, C.R. (2010). Alternate partial root-zone irrigation induced dry/wet cycles of soils stimulate N mineralization and improve N nutrition in tomatoes. *Plant soil*, 337, 167-177.

WRI (2005). World Resources Institute: Freshwater resources.

Yang, J., Zhang, J., Wang, Z. & Zhu, Q. (2001). Activities of starh hydrolitic enzymes and sucrose-phosphate synthase in the stems of rice subjected to water stress during grain filling. *Journal of Experimental Botany*, 52, 2169-2179.

Yoshida, K, Terashima, I & Noguchi, K (2007). Up-regulation of mitochondrial alternative oxidase concomitant with chloroplast over-reduction by excess light. *Plant Cell Physiology*, 48, 606-614.

Yoshida, K., Terashima, I. & Noguchi, K. (2006). Distinct roles of the cytochrome pathway and alternative oxidase leaf photosynthesis. *Plant Cell Physiology*, 47, 22-31.

Young IM, Montagu K, Conroy J, & Bengough AG (1997). Mechanical impedance of root growth directly reduces leaf elongation rates of cereals. *New Phytologist*, 135, 613-619.

Zegbe, J.A., Benboudian, M.H., & Clothier, B.E. (2006). Yield and fruit quality in processing tomato under partial rootzone drying. *European Journal Horticultural Science*, 71, 252-258.

Zinselmier, C., Westgate, M.E., Schussier, J.R. & Jones, R. (1995). Low water potential disrupts carbohydrate metabolism in Maize (*Zea mays* L.) ovaries. *Plant Physiology*, 107, 385-391.

Growth Characteristics of Rainfed/Irrigated *Juniperus excelsa* Planted in an Arid Area at North-Eastern Iran

Masoud Tabari and Mohammad Ali Shirzad
Tarbiat Modares University,
Iran

1. Introduction

Many dry regions of the world due to shortage enough water resources are lacking vegetation cover. This is while that the severe deforestations happened during short time has decreased their surface area in recent decades. Using the drought-tolerant species for rehabilitation of these regions and compensation of water deficit is promising (Tabari et al., 2011) Although, in dry regions the survival rate of planted seedlings managed as rainfed is low and watering causes enhanced survival and growth (Koroori and Khoshnevis, 2000; Lichter, 2000; Shirzad, 2009), because of low and irregular rainfall, long dry period, insufficient water resources, only drought tolerant tree species are able to recompense the scarceness of vegetation cover (Boers, 1994; Kozlowski, 1987). These species owing to the high capacity of water storage at stem and branch, and the good rooting in soil are able to well bear the drought status and to overcome the drought stress (Oliet et al., 2002; Tabari et al., 2011). Since the vast regions of world have dry climate, strategies for perfect use of the water resources and the drought-resistant species, which are able to highly benefit from the soil moisture, can be the main targets of plantation in such regions (Oliet et al., 2002; Sanchez-Coronado et al., 2007).

Juniperus excelsa is distributed in vast areas of Irano-Touranian growing regions (Zare, 2001). It tolerates high dryness and coldness and is able to restore deforested areas in mountain arid and semi-arid zones (Hampe and Petit, 2010). Of course, except *Juniperus excelsa,* other conifers have been also investigated for rehabilitation of such lands. Baquedano and Castill (2006), Oliet et al. (2002) and Castro et al. (2005) state that *Pinus halepesis* seedling is successfully able to overcome drought stress and it is a good species for plantation in arid and semi-arid areas. Khosrojerdi et al. (2008) and Ghasemi (1996) in the order put forward *J. excelsa* and *J. polycarpos* for plantation in semi-arid regions of Iran. Likewise, plantation with *J. phoenicea* in Mediterranean semi-arid regions of Jordan, and with *J. scopulorum* in semi-arid regions of northeastern United State has been reported by Alrababah et al. (2008) and Bjugstad and Ardell (1984), respectively.

Because water resources in arid and semi-arid regions of Iran is a serious obstacle for plantation development, so researches to apply suitable methods with the drought-tolerant species is imperative. Thus, the researches necessary in order to assessing the primary establishment of *J. excelsa* in arid zones of the country is unavoidable. Although, in the country few researches with *J. polycarpos* and *J. excelsa* have been reported in semi-arid zones (Ghasemi, 1996; Khosrojerdi et al., 2008; Shirzad and Tabari, 2010), but none on *Juniperus* planted in an arid zone was reported. This investigation is aimed to determine establishment and growth characteristics of *J. excelsa* seedlings under rainfed/irrigated conditions in a region with 231 mm annual precipitation and 6 months vital dry period.

2. Materials and methods

This investigation was conducted in southern elevations of Mashhad city located in northeastern Iran (59° 27′ E, 36° 30 N′, 1450 m a.s.l.). Based on synoptic station of Mashhad, the mean annual precipitation is 231 mm and the region benefits from a dry climate. The dry season is 6 months, starting from mid-May and lasting in mid-October. In this research 270 three-year old seedlings (1+2) of *J. excelsa* were planted on a natural soil in the site study. Experiment was made as factorial with randomized completely design, with three irrigation levels including control (rainfed), 20-days interval irrigation and 40-days interval irrigation. Watering was made 15 lit/period in spring and summer. In order to inhibit drought stress the rainfed seedlings were watered once at plantation time. The planting distance was 4 m, planting depth 40-50 cm and mean width of pits 80 cm (Photos 1 and 2).

Experiment was done for three years. At the end of each growing season, survival, crown width and total height (with meter, Photo 3) and stem collar base (with digital apparatus) of seedlings were measured and the increments for each period calculated. Soil pH and EC were 7.9 and 0.9 milmos/cm², respectively. Phosphorous and Potassium were 2 and 96 ppm, respectively. Nitrogen, using the Kjeldahl method, was 0.09% and Co₃Ca and Carbon, using Walkey-Black method, were 0.43 and 32.1, respectively. The soil texture was sandy loam. Analysis data was conducted by SPSS. Quantitative factors as increments of total height, crown width and stem collar base followed determining normality was conducted by Kolmogorow-Smironov test and equality data with Levene test. For comparison of means one-way Anova and Duncan tests were used. Survival was transferred into the normal data by Arc sin (Zar, 1999).

Photo 1. A view of site study (front) and air pollution (background) of Mashhad city

Photo 2. A part of *J. excelsa* plantation on a mountain hill (h= 1450 m) in the arid region of south of Mashhad city

Photo 3. Height measurement of *J. excelsa* in year 3 after plantation

3. Results

The results of the first, second and third years indicated that only in the first year height growth was significantly affected by irrigation treatment (Table 1). The greatest height growth was detected in 20-days interval irrigation and the least in rainfed condition (Table 2). Height growth, stem collar base and survival rate differed with irrigation (Table 1). Comparison of means in second year revealed that height growth was greater in 20-days interval irrigation than 40-days interval irrigation and control, and no significant difference was found between 40-days interval irrigation and control. Stem collar base was greatest in 20-days interval irrigation and least in rainfed status. Survival rate was highest with seedlings irrigated in intervals of 20 and 40 days. The results of third year showed that increments of stem collar base and crown width were affected by irrigation treatment (Table 1). The greatest and least increments stem collar base were found in 20-days interval irrigation and control, respectively. Crown width increment was greatest in 20-days interval irrigation (Table 2).

Characteristics measured	Year 1			Year 2			Year 3		
	d.f.	F	P	d.f.	F	P	d.f.	F	P
Height growth (cm)	2	7.3	0.02*	2	13.5	0.006**	2	0.4	0.69 ns
Stem collar base growth (mm)	2	0.7	0.53 ns	2	16.9	0.003**	2	8.8	0.02*
Crown width growth (cm)	2	0.6	0.56 ns	2	3.1	0.12 ns	2	15.2	0.005**
Survival (%)	2	1.8	0.24 ns	2	5.6	0.04*	2	3.4	0.10 ns

* Significant at level of 95% probability, ** Significant at level of 99% probability, ns Non significant

Table 1. One-Way Anova of Characteristics measured of *J. excelsa* seedlings affected by irrigation in different years

	Irrigation treatment	Height growth (cm)	Stem collar base growth (mm)	Crown width growth (cm)	Survival (%)
	20-day	7.2 ± 0.7a	0.7 ± 0.2	3.5 ± 0.3	93.33 ± 3.8
Year 1	40-day	3.4 ± 0.8b	0.5 ± 0.1	3.6 ± 0.5	75.28 ± 2.1
	Rainfed	2.3 ± 1.2	0.4 ± 0.1	2.9 ± 0.6	69.65 ± 15.2
	20-day	14.6 ± 0.3a	5.1 ± 0.4a	9.4 ± 0.1	74.87 ± 5.9a
Year 2	40-day	9.5 ± 0.9b	3.9 ± 0.1b	7.3 ± 0.5	74.67 ± 4.8
	Rainfed	5.1 ± 2.0b	2.5 ± 0.3c	6.5 ± 0.8	47.77 ± 9.0b
	20-day	19.1 ± 2.4	8.6 ± 0.1a	17.1 ± 1.0a	74.67 ± 7.4
Year 3	40-day	15.4 ± 5.4	7.3 ± 0.5ab	10.5 ± 1.3b	74.33 ± 8.7
	Rainfed	14.7 ± 5.1	5.6 ± 0.6b	10.3 ± 0.5b	47.67 ± 8.9

In each column, different letters are significant among irrigation treatment of each year

Table 2. Comparison of means (± sd) of Characteristics measured of *J. excelsa* seedlings affected by irrigation in different years

4. Discussion

The results of the present investigation at the end of the first, second and third years showed survival rate of seedlings irrigated in intervals of 20 days was 99.3, 74.87 and 74.67%, respectively. In intervals of 40 days it was 75.28, 74.67 and 74.33, respectively in mentioned years. Alrababah et al. (2008) in plantation with *J. phoenicea* found that watering reduced soil moisture stress and enhanced survival rate. They stated that at the end of the first and second growing season survival was 42% and 32%, respectively. Castro et al. (2005), working with *Pinus sylvestris*, with irrigation in intervals of 10 days (2 liter/seedling) observed that survival rate was 30% and 22% an the end of years 1 and 2, respectively.

Likewise, Bjugstad and Ardell (1984) in semi-arid region of Wyoming situated in north-eastern of United State found the significant effect of soil moisture on establishment of *Prunus americana, Pinus ponderosa* and *J. scopulorum* irrigated as dripping. As a matter of fact, the 5 years finding of mean survival rate of three species was ~ 29%. According to their idea the dry summer, particularly in initial years was a limiting factor for seedling establishment. As a whole, although in the dry regions, survival and growth rate of rainfed planted seedlings is mainly low, under such a condition watering causes increase of these characteristics, but owing to low and irregular precipitation, lacking sufficient water resources and long dry period only drought-tolerant species are able to recompense the rareness of vegetation cover (Boers, 1994; Kozlowski, 1987). However, because of deficient water resources, successful establishment of water-managed plantations would not be secured. Therefore, using the low-moisture demand species or species able to establish at water-lacking status can remove to a great extent criticizes of drought in plantations of such regions (Oliet et al., 2002; Sanchez-Coronado et al., 2007).

Rainfed planted *J. excelsa* seedlings, especially in years 1 and 2, due to deprived rooting and low moisture and nutrient uptake showed higher sensibility to environmental stresses, particularly to drought. So, watering regimes induced enhanced survival. In this respect, it can be stated that because the summer drought in arid zone is a factor affecting growth and establishment of plantations, therefore watering causes decreased the soil moisture stress and increased the establishment and survival rates of seedlings, especially in the primary years (Bjugstad and Ardell, 1984; Garcia, 2001; Maria et al., 2002; Castro et al., 2005). This is while, that in year 3 watering regime did not raise the survival rate. This may me because of appropriate distribution of rooting for soil moisture uptake and also suitable adaptation of seedlings with environmental conditions in the third year after plantation (Tabari et al., 2011). As a matter of fact, the fairly suitable survival of rainfed planted seedlings at the end of years 1, 2 and 3 (69.65, 47.77 and 47.67%) confirms that *J. excelsa* seedling benefits from the ecological adaptation in this arid area and is able to overcome drought stress following planting, and to establish successfully in the area.

In literature, various reports have been cited on adaptability circumstances of *Juniperus* genus in arid zone and semi-arid zone of the world. According to Bjugstad and Ardell (1984), survival rate of *J. phoenicea* seedling at the end of the 2nd year was 32%. In the report of Khosrojerdi et al. (2009, on *J. polycarpos*) and Ghasemi (1996, on *J. excelsa*) survival rate of seedlings was 88.61% and 96.3%, respectively. This is while that Khademi et al. (2005) observed the full mortality of *J. Virginiana* plantation after 10 years. The findings on *Pinus* genus were different, too. In this respect, it can be paid to some researches including Oliet et al. (2002), who showed that 88.5% of seedlings of *P. halepensis* grown in a semi-arid area survived at the end of the 1st growing season. They observed that *P. halepensis* seedling is successfully able to overcome drought stress after planting and to establish in this area.

Generally, native drought-resistant and compatible species with high ecological elasticity are able to a large extent recompense the damages induced by strict environmental variations and water deficiency. Particularly, *J. excelsa* and *J. ashei* that with penetration of their roots in soil depths and stone layers access soil moisture and overcome drought stress (Weaver and Jurena, 2008). As a result, plantation with these species in arid and semi-arid areas is of high success.

In this investigation, in different years, with enhanced irrigation in the dry seasons, height growth and stem collar base growth of *J. excelsa* were increased. Because, the drought stress causes moisture stress in plant and threats its growth, consequently moisture required of plant decreases the drought stress and increases the growth (Matice, 1982; Lantz et al., 1988). The findings of the current investigation are in line with Brisette and Chamber (1992) on *Pinus echinata*, Kowsar (1995) on *Cupressus arizonica*, Antonio (2001) on *Pinus halepensis*, and Svistula and Tarasenko (1985) on *Juniperus* genus.

5. General conclusion

From the results of the present research it can be concluded that at the end of third year the growth characteristics of rainfed planted seedlings did not much differ with those in watering treatments, particularly in watering applied with interval of 40 days. Likewise, establishment of seedlings responded well (about 47%) to rainfed status. As a whole, although under rainfed status survival rate of the *J. excelsa* seedlings was satisfactory; however, for caution and help in higher assurance of establishment and growth it is better that for plantation development of this species in this arid area and the same ecological regions some irrigations to be applied in the primary years.

6. Acknowledgment

Many thanks go to the Tarbiat Modares University and the Municipality of Mashhad (Khorasan Razavi Province) for the financial support provided. We also would like to acknowledge Dr. E. Khosrojerdi and Eng. H. Daroudi for their kind assistance rendered in this research. Our sincere appreciations also go to all anonymous experts in the Research Center of Agriculture and Natural Resources of Khorasan Razavi Province for the help provided during the course of the field work.

7. References

Alrababah, M.A.; Bani-Hani, M.G., Alhamad M.N. & Bataineh, M.M. (2008). Boosting seedling survival and growth under semi-arid Mediterranean conditions: Selecting appropriate species under rainfed and wastewater irrigation. *Arid Environments*, 72, 1606- 1612

Antonio, R.; Luis, G. & Jose, A. (2001). Effect of water stress conditioning on morphology, physiology and field performance of *Pinus halepensis* Mill. seedlings. *New Forests*, 21, 127–140

Baquedano F.J.; & Castillo, F. (2006). Comparative ecophysiological affects of drought on seedlings of the Mediterranean water-saver *Pinus halepensis* and water-spenders *Quercus coccifera* and *Quercus ilex*. *Trees*, 20, 689–700

Bjugstad, A.J.; & Ardell. (1984). Shrub and tree establishment on coal spoils in northern high plains, USA. *Environmental Geochemistry and Health*, 6, 3, 127- 130

Boers, TM. (1994). Rain water harvesting in arid and semi-arid zones. International Institute for Land Reclamation and Improvement, Wageningen, The Netherlands, 133 pp.

Brisette, J.C.; & Chamber, J.L. (1992). Leaf water status and root system water flux of short leaf pine (Pinus echinata Mill.) seedling in relation to new growth, after transplanting. Tree physiology, 11, 3, 289-303

Castro, J.; Zamora, Jose, R.A. & Jose, M. 2005. Alleviation of summer drought boosts establishment success of Pinus sylvestris in a Mediterranean mountain: an experimental approach. Plant Ecology, 181, 191–202

Garcia, D. (2001). Effects of seed dispersal on Juniperus communis recruitment on a Mediterranean mountain. Journal Vegetation Sciences, 12, 839–848

Ghasemi, R. (1996). Determining best transfer age of Juniperus polycarpos seedling from nursery to plantation site. Iranian Forest and Rangeland Institute, 76p.

Hampe, A. & Petit, R.J. (2010). Cryptic forest refugia on the 'Roof of the World'. New Phytologist, 185, 5–7

Khademi, A.; Adeli. E., Babaei, S. & Mattaji, A. 2005. Study of afforestation (Khojin Forest Park & Hiroabad) in Khalkhal area and present adaptable species. Journal Agricultural Sciences, Islamic Azad University, 11, 4, 60-68

Khosrojerdi, E.; Daroudi, H. & Namedoust, T. (2008). Effect of nursery plant and slope exposure on survival and growth of Juniperus excelsa seedlings in Hezarmasjed forests. Iranian Biology Journal, 21 (5): 760-768.

Koroori, S and Khoshnevis, M. (2000). Ecology of Juniperus sp. habitats in Iran. Research Institute of Forest and Rangeland, 229, 208p (In Persian)

Kowsar, A. 1995. Application of tar in rainfed plantation and effect of the occurred run-off on success and growth of Robinia pseudoacacia, Cupressus arizonica and Fraxinus rotundifolia. Iranian Forest and Rangeland Institute, 43, 79p.

Kozlowski, T. (1987). Water Deficits and Plant Growth, Vols. I and II. Academic Press, New York.

Lantz, C.W., Baldwin, B.L. & Barnett, J.P. 1988. Plant them deep and keep those roots straight U.S.A Dept. Agric. Forest Service, Management Bulleti, RG-MB, 27, 2 p.

Lichter, J. (2000). Colonization constraints during primary succession on coastal Lake Michigan sand dunes. Journal Ecology, 88, 825–839

Maria, J.; Benayas, R. Lopez, A., Garcla, C., Camara, N., Strasser, R. & Gomez, A. (2002). Early establishment of planted Retama sphaerocarpa seedlings under different levels of light, water and weed competition. Plant Ecology, 159, 201–209

Matice, C.R. (1982). Comparative performance of paper pot and bare root trees in experiments established in northern Ontario from 1977-1980. No. 585044. Matcam Forestry consultants, Inc. Update Rrp, 147 p.

Oliet, J.; Planelles R., Lopez, M. & Artero, F. (2002). Soil water content and water relations in planted and naturally regenerated Pinus halepensis Mill. Seedlings during the first year in semiarid conditions. New Forests, 23, 31– 44

Sanchez-Coronado, ME.; Coates, R., Castro-Colina, L., Buen, AG., Paez-Valencia, J., Barradas VL., Huante P. & Orozco-Segovia, A. (2007). Improving seed germination and seedling growth of Omphalea oleifera (Euphorbiaceae) for restoration projects in tropical rain forests. Forest Ecology and Management, 243,1, 144-155

Svistula, G.E. and Tarasenko, I.M. (1985). Increasing the ecological capacity of the lower Dnieper sands. Lesovod Stvo.I .Agrolesomel Ioratsiya, 10, 13-16

Shirzad, M.A. (2009). Methods of green space development in south of Mashhad by Juniperus excelsa. M.Sc. thesis, Faculty of Natural Resources and Marine Sciences, Tarbiat Modares University, Noor, Iran, 131p. (In Persian)

Shirzad, M.A. & Tabari, M. (2010). Effect of some environmental factors on woody plant species in Juniperus excelsa habitat of Hezarmasjed mountains. Journal of Technology and Environment Sciences. (Accepted, In Persian)

Tabari. M.; Shirzad, M.A., Khosrojerdi, E. & Daroodi, H. (2011). Effect of seedling transfer age and soil bed on growth and early establishment of Juniperus excelsa M. Bieb. seedlings in southern hills of Mashhad. Iranian Forest and Poplar Researches, 19, 1,119-127 (In Persian)

Weaver, J.A. & Jurena, P.N. (2008). Response of newly established *Juniperus ashei* and *Carex planostachys* plants to barrier-induced water restriction in surface soil. *Journal of Arid Environments*, 73, 267–272

Zar, J.H. (1999). Biostatistical analysis. Prentice Hall International, Inc 66 pp.

Zare, H. (2001). Native and exotic coniferous in Iran. *Iranian Forest and Rangeland Institute, 498p.*

Surface Infiltration on Tropical Plinthosols in Maranhão, Brazil

Alba Leonor da Silva Martins[1], Aline Pacobahyba de Oliveira1[1],
Emanoel Gomes de Moura[2] and Jesús Hernan Camacho-Tamayo[3]
[1]Embrapa Solos – National Center of Soil Research - CNPS,
[2]Maranhão State University,
[3] Colômbia National University, Bogotá
[1,2]Brazil
[3]Colombia

1. Introduction

Plinthosols are soils that have inherent limitations such as the formation of highly condensed layers, surface crust formation and restricted water percolation, thereby hampering their management for agricultural production. According to Embrapa (2006), this kind of soil when subjected to the continual process of wetting and drying encourages the development of a plinthic horizon, which means that the soil remains saturated and is subject to water table level fluctuations.

In order to reduce the impact of the limitations of these soils, proper conservative management systems have to be applied. These systems shall have to take into account the soil quality, improving water retention, reducing soil's lost, increasing soil's biological activity, improving exchanging of heat and gases between soil and atmosphere as well as the availability of air, water and nutrients to plants. Thus, understanding the relationship between infiltration and physical attributes in this soil is important for ensuring adequate soil and crop management.

Water infiltration into the soil is one of the basic factors for estimating irrigation intensity according to the plants' requirements; with the aim of avoiding problems of run-off and surface degradation. Ascertaining the infiltration rate and accumulated water quantity is of great importance due to their application in the agriculture and environment, allowing for estimating surface run-off, suspended particle transport, sediment availability, and aquifers capacity to be recharged, defining irrigation systems and studying the effects of different soil management practices (Machiwal et al., 2006; Strudley et al., 2008).

This chapter presents a characterization of that soil and of a conservative management system in a research that evaluated the variability of water infiltration and its relation to soil's physical attributes.

2. Description of Plinthosols

Plinthosols are soils with plinthite, petroplinthite or pisoliths. Plinthite is a Fe-rich, in some cases also Mn-rich, humus-poor mixture of kaolinitic clay (and other products of strong weathering such as gibbsite) with quartz and other constituents that change irreversibly to a layer with hard nodules, a hardpan or irregular aggregates on exposure to repeated wetting and drying. Petroplinthite is a continuous, fractured or broken sheet of connected, strongly cemented to indurated nodules or mottles. Pisoliths are discrete strongly cemented to indurated nodules. Both petroplinthite and pisoliths develop from plinthite by hardening (FAO, 2006).

Formation of Plinthite involves the following processes: (1) accumulation of sesquioxides, through: (a) relative accumulation as consequence of the removal of silica and bases by ferralitization, and/or (b) absolute accumulation through enrichment with sesquioxides from outside (vertical or lateral, see Figure 1). (2) segregation of iron mottles, caused by alternating reduction and oxidation. In times of water saturation, much of the iron is in the ferrous form, has high mobility and is easily redistributed. When the water table falls, this iron precipitates as ferric oxides that will not, or only partially, redissolve in the next wet season (Driessen & Dudal, 1989).

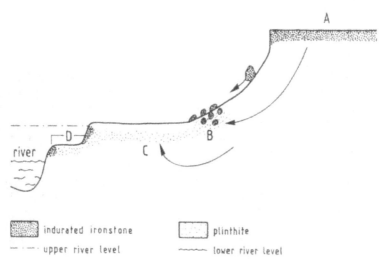

A: Indurated ironstone (massive iron pan or gravel) capping an old erosion surface.
B: Plitnthite and ironstone (gravel and boulders) in a colluvial footslope (subject to iron-rich water seepage).
C: Plinthite in soils of a low level plain (river terrace) with periods of high groundwater.
D: Along the banks of rivers where plinthite becomes exposed and hardens to ironstone. Source: Driessen & Dudal (1989).

Fig. 1. Genesis of Plinthosols. Four physiographically distinct landscape positions where plinthite and ironstone occur.

Global extent of Plinthosols is estimated at some 60 million ha. Soft plinthite is most common in the wet tropics, notably in the Eastern Amazon basin, in the Central Congo basin, and parts of Southeast Asia. Extensive areas with pisoliths and petroplinthite occur in the Sudano-Sahelia zone, where petroplinthite forms hard caps on top of uplifted/exposed landscape elements. Similar soils occur in the Southern African savannah, in the Indian subcontinent, and in drier parts of Southeast Asia and Northern Australia (FAO, 2006).

Many of these soils are known as: Groundwater Laterite Soils, Perched Water Laterite Soils and Plintossolos (Brazil); Sois gris latéritiques (France); and Plinthaquox, Plinthaqualfs, Plinthoxeralfs, Plinthustalfs, Plinthaquults, Plinthohumults, Plinthudults and Plinthustults (United States of America).

In Keys to Soil Taxonomy (Soil Survey Staff, 1992) these soils are included in the Alfisol, Inceptisol, Oxisol and Ultisol orders. In that classification system the occurrence of plinthite is considered to be an additional differentiating characteristic, not a diagnostic horizon, and the definition of plinthic subgroups is not complete.

In Brazilian Soil Classification System (Embrapa, 2006) the Plintosols classes were defined as a soil consisting of mineral material, with the plinthic horizon or concretionary/ petroplinthic solodic horizon, or in one of the following conditions:

(1) Starting within 40 cm from the surface, or

(2) Starting within 200 cm from the surface when preceded by gley horizon, or immediately below the horizon A or E, or other horizon that shows pale colors, variegated or mottled in abundance (> 20 % by volume), a matrix of reddish or yellowish hue 5 Y or that have shades 7.5 YR, 10 YR or 2.5 Y, with chroma less than or equal to 4.

In this definition, it is important to consider the depth, what is the plinthic or lithoplintic/petroplinthic horizon and the amount of plinthite and, or concretions.

Soils with plinthic horizons occur in floodplains, areas with flat relief or gently sloping ondulating and less frequently in geomorphic depression areas. They also occur in shoulders or floodplain areas, under presence and oscillation of the water table, either from periodic flooding or waterlogging, restrictict effect on water percolation or runoff. Concretionary horizon soils have better drainage and higher positions in relation to soils with plinthic horizon, and are most frequent in the Amazon region of Brazil (Embrapa, 2006).

Soils with plinthite and, or ironstone (petroplinthite) are common in Northeastern Brazil (Anjos et al., 1995, 2007). According to Embrapa (1986), the main areas in Brazil with plinthite or petroplinthite occur in the Amazon region (upper Amazon), Amapá, Marajo Island, Maranhão lowland, Northern of Piauí, Southwest of Tocantins, Northern Goias, Pantanal and Bananal Island (Figure 2).

A reconnaissance Soil Survey of the State of Maranhão, Brazil (Embrapa, 1986) showed that associations between soils in which Plinthosols are the main component constitute approximately 44,000 km² of the whole state.

In humid tropics in the Northern center of Maranhão, Brazil, Plinthosols from Itapecuru Formation or colluvial sediments are formed from ferruginous sandstone (Anjos et al., 1995), where conditions of high rainfall alternates with prolonged period of high decrease in rainfall and present various problems such as the formation of condensed layers, surface crust and, consequently, poor drainage, limiting the harvest conditions of the farmers.

Plinthosols presents serious management problems. Waterlogging and low natural fertility are their main limitations. If the plinthite layer hardens, for example, because of deep drainage or erosion, ironstone will limit the possibilities for root growth and lower the water storage capacity of the soil. Most petroferric soils are unsuitable for arable farming; they are used for extensive grazing and firewood production (Driessen & Dudal, 1989).

Legenda
■Plinthosols

Fig. 2. Occurrence of in Brazil. Source: Embrapa, 2011.

3. Family farming in the humid tropics of Maranhão

Family farmers in the humid tropics in the Northern center of Maranhão, Brazil, practice itinerant agriculture, or slash and burn, over the Plinthosol prevalent in the region. The smallholder farms occupy areas ranging from 1 to 5 hectare, mainly cultivating rice (*Oryza sativa*), corn (*Zea mays L.*), beans (*Phaseolus vulgaris*) and cassava (*Manihot utilissima*).

This shifting cultivation system of the soil, predominantly in the humid tropics, has been practiced for centuries and occupies near 30 % of arable land, holding between 300 and 500 million people among the poorest in the world (Von Uexkul and Mutert, 1990).

After burning, the crops established in the first year produce relatively well, reaching levels of field production higher than those obtained in areas without burning and without fertilizer (Van Reuler and Jansenn, 1993). However, from the second year of cultivation begins a phase of decline in productivity that can be attributed to the chemical fertility reduction, more pronounced in soils of low fertility, and increased weed infestation, more damaging in soils with high natural fertility (Sanchez, 1982 adapted by Ferraz Junior (2004).

The emphasis of this intensification should not be intensive soil tillage and external inputs use. It should promote ways to incorporate management of biological cycles and environmental interactions that determine the sustainable productivity of the agroecosystem (Harwood, 1996). Therefore, this production model has been studied in search of more sustainable alternative proposals.

4. The alley cropping system

The alley cropping system is one of the most simple agroforestry systems, which combines in the same area tree species, preferably legumes and perennial or annual crops of economic interest. Legumes are planted in single or double rows, spaced 2 to 6 m. The branches of the plants are periodically cut at a height from 0.1 to 0.5 m, and lines of crops of economic interest are added, serving as cover plants and green manure (Kang et al, 1990, Szott et al, 1991; Cooper et al 1996).

This practice, traditionally used in tropical regions of Africa and Asia, has led to improvement in soil chemical attributes (organic carbon and nutrients) in the superficial layer, compared to other practices such as monoculture. The improvement has been attributed to more effective recycling of nutrients, pruning or litter biomass. In addition, the forest species show beneficial effects for their deeper roots that reduce leaching losses and higher soil cover that provides protection against erosion (Mafra et al, 1998).

This agricultural system is based on the following principles: soil fertility regeneration, mulch formation, weeds elimination, nitrogen fixation, nutrient recycling, which consequently provides improved soil quality (Kang 1997).

5. Surface infiltration in Plinthosols

Local research was conducted in 2005 and aimed to determine the spatial variability of surface infiltration in an Argiluvic Plinthosol (Embrapa, 2006) or Plinthic Kandiudult (Soil Survey Staff, 1992). This is presented in detail in the following sections.

5.1 Study area

The study was carried out in Miranda del Norte (Maranhão, State, Brazil), located at 3^0 36' South latitude and 44^0 34' West longitude (Figure 3) on a 1 hectare plot planted with corn (Zea mays L.) using agro-forest system, in an area where family agriculture predominates. The predominant soil is an Argiluvic Plinthosol, with medium texture and over flat relief.

According to Köppen's climatic classification, the region is located in an Aw' – humid tropical climate, having an average temperature of 27°and 1,600 to 2,000 mm year[-1] rainfall, concentrated in December and June, with a drier climate predominating in the other months.

5.2 Sampling and laboratory tests

A 113-point mesh (Figure 4) was designed for sampling, having a regular 10 m distance between points in the experimental area, considering the 44 rows of legumes, along the corn rows. Disturbed and undisturbed soil samples were taken at each point at 0 - 0.20 m depth. The disturbed samples were air-dried until they reached an equilibrium point; once being clod, they were passed through a 2 mm mesh for granulometric analysis. Sand, silt and clay content were thus established by means of a 10 g sample subjected to slow agitation for 16 h, in 100 ml solution with NaOH (0.1 mol L $^{-1}$); clay content was obtained by means of an aliquot drawn out with a pipette (Gee and Bauder, 1986). The undisturbed samples were taken from 100 cm^3 volumetric rings to determine bulk density, macroporosity, microporosity and total porosity using a tension table, following the methodology described by Kiehl (1979).

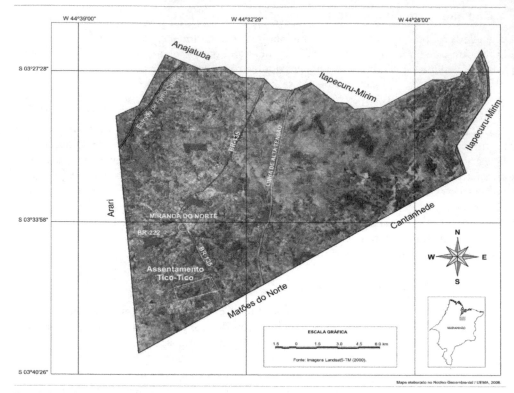

Fig. 3. Localization of the experimental area.

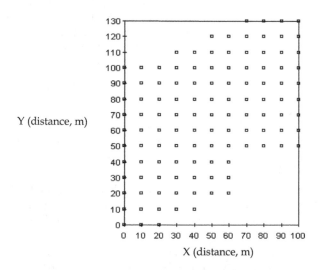

Fig. 4. Regular points mesh in experimental area of Argiluvic Plinthosol.

Infiltration rate was determined during a dry period, at each sampling point, by means of a 15 cm diameter ring, using Guelph permeameter modified by the Agronomical Institute of Campinas (IAC), thus guaranteeing constant water load. Infiltration rate was calculated using Reynolds and Elrick's equation (1985), modified by Vieira (1988):

$$I = 60\left(\frac{D_p}{D_a}\right)^2 . Q \qquad (1)$$

Where I is the infiltration rate of water in saturated soil (mm h^{-1}), Dp is the diameter of the Guelph permeameter (9 cm), Da is the ring diameter (15 cm) and Q is the estimated water flow in saturated soil (mm min^{-1}).

5.3 Statistical analysis

Descriptive statistics were initially used to analyzing the data, using SPSS 16.0 software for calculating the mean, median, coefficient of variation (CV), minimum, maximum, skewness, kurtosis and Kolmogorov-Smirnov test. Normality was thus verified, a pattern which is not indispensible when related to geoestatistical techniques but does give better predictions when data are fit to a normal distribution (Diggle and Ribeiro, 2000). The classification proposed by Warrick and Nielsen (1980) was used for CV analysis: it indicates low CV variability for values below 12 %, average CV variability between 12 % and 60 % and high CV variability when greater than 60 %. Pearson linear correlation was also carried out to identify the correlation between infiltration and the soil's different physical properties.

The physical properties and infiltration's spatial dependence were analyzed by using semivariograms. The theory of regionalized variables was taken into consideration for the fit of theoretical experimental semivariogram models; it has different methods for analyzing spatial variation, one of them being semivariogram (Vieira, 2000) which is estimated by:

$$\gamma(h) = \frac{1}{2N(h)} \sum_{i=1}^{N(h)} \left[Z(x_i) - Z(x_{i+h}) \right]^2 \qquad (2)$$

where $Z(x_i)$ and $Z(x_{i+h})$ are observed values of a variable, separated by a distance h, and $N(h)$ is the number of pairs of experimental observations, separated by distance h. The semivariogram is graphically represented by $\gamma(h)$ versus h. The theoretical model's coefficients are estimated from the fit of the model: nugget effect (C0), sill (C0+C1) and range (A).

The degree of spatial dependence (DSP) was also estimated, based on the ratio between the nugget effect and the sill (C/Co+C), being considered strong for DSP when above 75 %, moderate DSP between 25 % and 75 %, and weak for DSP less than 25 % (Cambardella, 1994). GS+ software (Gamma Design Software, 1998) was used to calculating the semivariograms; adopting the largest determination coefficient value (R2), the lowest value for the sum of the square of the residue (SQR) and crossing a validation coefficient (CVC) close to one as criteria for selecting the model. Once spatial dependence had been determined, infiltration in the nonsampled areas was predicted by kriging, represented on contour map that was drawn up using Surfer software (2000).

5.4 Results and discussions

The average and median values were similar for each one of the studied properties, indicating symmetrical distributions. This was verified by skewness and kurtosis values being close to zero for all properties. They approached normal distribution, except for infiltration (Table 1). These properties' normality pattern has also been reported by Ramírez-López et al. (2008). The Kolmogorov-Smirnov test confirmed normal distribution for most properties, except for infiltration, silt and microporosity.

Property	Mean	Median	CV, %	Minimum	Maximum	Skewness	Kurtosis	K-S
Infiltration, mm h-1	36,03	30,00	64,15	3,00	90	0,73	-0,38	*
Sand, %	25,63	26,00	12,43	18,00	34,00	0,18	-0,11	
Silt, %	49,35	51,50	19,03	24,00	65,00	-0,66	-0,32	*
Clay, %	24,56	24,00	36,42	7,00	45,00	0,45	-0,65	
Bd, g cm-3	14,13	14,10	4,32	12,60	15,30	-0,42	-0,05	
Map, %	11,85	11,73	19,22	7,23	17,78	0,46	-0,06	
Mip, %	42,91	43,47	7,22	34,82	50,19	-0,34	0,09	*
PT, %	56,73	56,92	7,73	47,55	67,05	0,00	-0,12	
PT, %	56,73	56,92	7,73	47,55	67,05	0,00	-0,12	

Table 1. Descriptive measurements of soil infiltration and physical properties. K-S: Kolmogorov-Smirnov test, *: significant the 5 %; Bd: bulk density, Map: macroporosity, Mip: microporosity, PT: total porosity.

According to Warrick and Nielsen's criteria (1980), it was seen that infiltration had high variability (64.15 % CV). Rodríguez-Vásquez et al. (2008) also found high variability for this property in a high silt content Andisol . Bulk density was the property showing the lowest CV (indicating low variability) together with microporosity and total porosity. Low variability for bulk density and total porosity has also been reported by Ramírez-López et al. (2008). The other properties showed average variability. Clay was the soil particle showing the greatest variability.

Total porosity was the only property which did not fit a theoretical semivariogram model, indicating lack of spatial autocorrelation, showing a pure nugget effect (Table 2). The spherical model was the predominant model among the properties analyzed; it fits infiltration, silt, clay and microporosity data. Sand, bulk density and macroporosity data fit in exponencial models.

Models fit adequately, given the coefficient determination of values which were always above 0.75. Sand and clay were the properties having the best fit for the models. The coefficient of crossed validation was close to 1 for most properties, whereas bulk density had the lowest value (0.81). Infiltration showed the greatest range and was the only property having a range greater than 100 m.

Property	Model	Co	Co+C	A (range,m)	C/Co+C	R2	CVC
Infiltration	Spherical	207,70	415,50	118,90	0,50	0,79	0,96
Sand	Exponential	0,93	11,62	43,80	0,92	0,94	1,05
Silt	Spherical	27,30	84,69	50,80	0,68	0,80	0,98
Clay	Spherical	24,70	85,05	55,80	0,71	0,92	1,00
Bd	Exponential	4,2E-4	38,6E-4	35,10	0,89	0,86	0,81
Map	Exponential	0,28	3,28	29,10	0,91	0,77	0,97
Mip	Spherical	1,10	6,09	24,90	0,82	0,84	0,96
PT	NE	13,69	13,69				

Table 2. Fit in theoretical semivariogram parameters for soil infiltration and physical properties. NE: nugget effect; CVC: coefficient of cross validation, Bd: bulk density, Map: macroporosity, Mip: microporosity, TP: total porosity.

Infiltration did not have a significant correlation with the physical properties (Table 3). It can be observed that infiltration was directly related to silt content and inversely to clay content. The same pattern has been reported by Rodríguez-Vásquez et al. (2008). This relationship between infiltration and silt and clay content is verified on the contour maps (Figure 5) where areas having a greater infiltration rate correspond to areas of greater silt content and lower clay content.

Macroporosity and microporostity were the properties having the least range. Regarding the degree of spatial dependence (DSD), infiltration showed the lowest value, having moderate spatial dependence. Clay and silt also showed a moderate DSD. The other properties showed high DSD, having values greater than 0.80.

A significant and inverse correlation between sand, clay and microporosity was oberserved. Silt and clay content showed a strong inverse correlation, as can be seen in the contour maps (figure 5) where areas of greater silt content correspond to areas of lower clay content and vice versa.

	Infiltration	Sand	Silt	Clay	Dd	Map	Mip	PT
Infiltration	1							
Sand	0.058	1						
Silt	0.125	-0.127	1					
Clay	-0.198	-0.163*	-0.912**	1				
Dd	-0.098	-0.187	-0.082	0.165	1			
Map	-0.129	-0.211	-0.194	0.223	-0.13	1		
Mip	-0.057	-0.137*	-0.067	0.147	0.144*	0.255	1	
PT	-0.021	-0.197	-0.156	0.226	0.064*	0.698**	0.654**	1

*Correlation was significant at the 0,05 level ; **Correlation was significant at 0,01 level, Bd:bulk density, Map: macroporosity, Mip: microporosity, PT: total porosity.

Table 3. Pearson correlation between soil infiltration rate and physical properties.

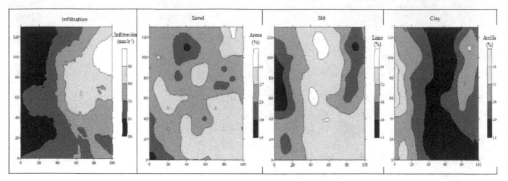

Fig. 5. Contour maps of infiltration and sand, silt and clay contents.

Bulk density also had a positive and significant correlation with microporosity, leading to moderately similar contour maps (Figure 6). Total porosity (a property in which spatial dependence was not found) had an inverse correlation with sand and silt content and a direct correlation with clay content, as well as with bulk density, macroporosity and microporosity.

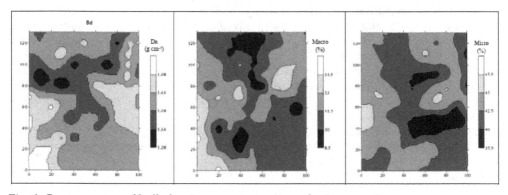

Fig. 6. Contour maps of bulk density, macroporosity and microporosity.

The contour maps confirmed the relationship between the variables analyzed in this study (figure 5 and 6) as well as the spatial variability existing in the soil, indicating the convenience of carrying out localized soil management practices according to the conditions of the area and crop requirements. This could potentially lower production costs and soil degradation which results from conventional management.

6. Conclusions

Surface infiltration and soil properties analyzed by means of descriptive statistics and geostatistics revealed the soil's high variability in its infiltration and physical property patterns, this being an important tool for decision-making since it leads to establishing the soil quality parameters which are directly related to agricultural production.

The linear correlation estimated for the properties led to identifying which ones were most closed related; such relationships were spatially confirmed by means of contour maps obtained by kriging when the properties fit the theoretical semivariogram models.

7. Acknowledgments

We would like to thank the Maranhão State University and the researchers Humberto Gonçalves dos Santos, José Francisco Lumbreras and Gustavo de Mattos Vasques of the National Center of Soil Research (Embrapa Solos) for providing support for the first author.

8. References

Anjos, L.H.C.; Franzmeier, D.P. & Schulze , D.G. (1995). Formation of soils with plinthite on a toposequence in Maranhão state, Brazil. *Geoderma*. v.64, pp.257-279.

Anjos, L.H.C.; Pereira, M.G., Pérez, D.V. & Ramos , D.P. (2007). Caracterização e Classificação de Plintossolos no município de Pinheiro - MA. *Revista Brasileira de Ciência do Solo*. v. 31, p.1035-1044.

Cambardella, C.A.; Moorman, T.B.; Novak, J.M.; Parkin, T.B.; Karlen, D.L.; Turco, R.F. & Konopka, A.E. (1994). Field-scale variability of soil of soil properties in central Iowa soils., *Soil Science Society America Journal*, v.58, n.5, pp.1501-1511.

Cooper, P.J.M.; Leakey, R.R.B.; Rao, M.R. & Reynolds, L. (1996).*Agroforestry and the mitigation of land degradation in the humid and sub-humid tropics of Africa. Experimental Agriculture*. Cambridge, v.21, pp.235-290.

Diggle, P. J. & Ribeiro, J.R. (2000). *Model Based Geostatistics*, 1.ed. São Paulo: Associação Brasileira de Estatística, 129p.

Driessen, P.M. & Dudal, R. (1989). *Lecture Notes on the Major Soils of the World*. Geography, Formation, Properties and use of the. Agricultural University Wageningen. 296p.

EMBRAPA. (1986). *Levantamento exploratório de solos do Estado do Maranhão*. Vols I e II. Boletim de Pesquisa 35. Empresa Brasileira de Pesquisa Agropecuária. Serviço Nacional de Levantamento e Conservação de Solos, SUDENE/DRN, Rio de Janeiro, Brazil.

EMBRAPA – Centro Nacional de Pesquisa de Solos. (2006). *Sistema Brasileiro de Classificação de Solos*. Rio de Janeiro: Embrapa Solos. 2.ed. 306p.

EMBRAPA – Centro Nacional de Pesquisa de Solos. (2011). *Solos do Brasil : Calendário*. Rio de Janeiro: Embrapa Solos.

FAO. (2006). Food and Agriculture Organization. IUSS Working Group. *WRB - World reference base for soil resources. A framework for international classification, correlation and communication*. 2.ed. World Soil Resources Reports. N.103. Rome: FAO.

Ferraz Junior, A.S.L. (2004). O cultivo em aléias como alternativa para produção de alimentos na agricutura familiar do trópico úmido. In: MOURA, E.G.(org.) *Agroambientes de transição entre o trópico úmido e o semi-árido do Brasil: atributos, alternativas; uso na produção familiar*. São Luís: UEMA.

Gamma Design Software. (1998). Geostatistics for the environmental sciences (version 5.1 for windows). In: *http://www.famat.ufu.br/ednaldo/ednaldo.htm*.

Gee, G. W., Bauder, J. W. (1986) Particle-size analysis. In: KLUTE, A. *Methods of soil analysis*. 2a. Ed. Madison: American Society of Agronomy, part 1, pp. 383-411.

Harwood, R.R. (1996). Development pathways towards sustainable systems following slash and burn. *Agriculture, Ecosystems and Environment*, Amsterdam, v.58, pp.75-86.

Kang, B.T. (1997). Alley cropping – soil productivity and nutrient recycling. *Forest Ecology and Management*, v.91, pp.75-82.

Kang, B.T.; Reynolds, L. & Atta-Krah, A.N. (1990). Alley farming. *Advances in Agronomy*, New York, v.43, pp.315-359.

Kiehl, E. J. (1979). *Manual de edafologia, relações solo* – planta. São Paulo, Ceres. 264p.

Machiwal, D.; Jha, M.K. & Mal, B.C. (2006). Modelling Infiltration and quantifying Spatial Soil Variability in a Wasteland of Kharagpur, India. *Biosystems Engineering*, United Kingdon, v.95, n.4, pp.569– 582.

Mafra, A.L. et al. (1998). Produção de fitomassa e atributos químicos do solo sob cultivo em aléias e sob vegetação nativa de cerrado. *Revista brasileira de Ciência do Solo*, Campinas, v.22, pp.43-48.

Ramírez-López, L., Reina-Sánchez , A., Camacho-Tamayo, J. H. (2008). Variabilidad espacial de atributos físicos de un Typic Haplustox de los llanos orientales de Colombia. *Engenharia Agrícola*, Vol.28, No. 1, pp.55-63.

Rodríguez-Vásquez, A. F., Aristizábal-Castillo, A. M., Camacho-Tamayo, J. H. (2008). Variabilidad espacial de los modelos de infiltración de Philip y Kostiakov en un suelo ándico. *Engenharia Agrícola*, Vol. 28, No. 1, pp. 64-75.

Soil Survey Staff. (1992). *Keys to Soil Taxonomy*. 5th ed. U.S. Dept. of Agriculture, SMSS Tech.Monogr.19. Pocohantas Press, Blacksburg V.A.

Surfer For Windows. Realese 8.0. (2000). *Contouring and 3D surface mapping for scientist´s engineers*. User´s Guide. New York: Golden software, Inc.

Strudley, M.W., Green, T.R., Ascough II, J.C.(2008).*Tillage effects on soil hydraulic properties in space and time: State of the science.*, Soil and Tillage Research, Vol., No. 1,p. 4-48.

Szott, L.T.; Palm, C.A.; Sanchez, P.A. (1991). Agroforestry in acid soils of humid tropics. *Advances in Agronomy*, Washington, v.45, pp.275-301.

Vieira, S.R. (1988). Permeâmentro: novo aliado na avaliação de manejo do solo. *O agronômico*, v. 47/50, pp.32-33.

Vieira, S.R. (2000). Geoestatística em estudos de variabilidade espacial. In: *Tópicos em Ciência do Solo*. Sociedade Brasileira de Ciência do Solo,Viçosa, v.1, 352p.

Von Reuler, H.; Janssen, B.H. (1993). Nutrient fluxes in shifting cultivation system of southwest Cotê d'Invore. I Dry matter production, nutrient system content and nutrient release after slash and burn for two vegetation. *Plant and Soil*. Amsterdan, v.154, pp.169-177.

Von Uexkull, H.R.; Muterte, E. (1990). Conversion of tropical rain forest into plantation and arable land with due attention to the ecological and economics aspects. *Plant Reserarch and Development*, Tubingen, v.32, pp.71-85.

Warrick, A.W.; Nielsen, D.R. (1980). *Spatial variability of soil physical properties in the field.Applications of soil physics*, New York: Academic Press, Hillel, D. (ed.), p.319-344.

Part 3

Examples of Irrigation Systems

10

Irrigation of Field Crops in the Boreal Region

Pirjo Mäkelä, Jouko Kleemola and Paavo Kuisma
Department of Agricultural Sciences,
University of Helsinki & Potato Research Institute,
Finland

1. Introduction

The Boreal region is indicated as the green areas in Figure 1, covering northern, central and southern boreal zones according to the definition of Hämet-Ahti (1981).

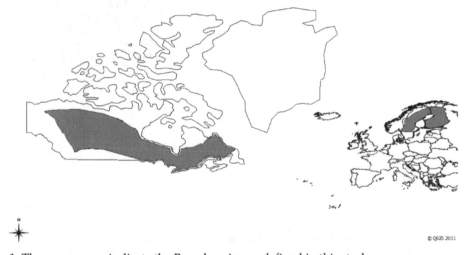

Fig. 1. The green areas indicate the Boreal region as defined in this study.

Nearly the entire agricultural area in the Boreal regions shown in Figure 1 lies in Canada (total of 68 million hectares; STATCAN, 2009). Sweden has about 2.6 million hectares (Jordbruksverket, 2011) and Finland 2.3 million hectares (TIKE, 2010). Most of Sweden's agricultural area is, however, located in the southern part of the country, outside the Boreal region.

2. Specific climate features of the Boreal region

The length of the growing season is limited by late spring frosts, early autumn frosts and solar availability in the Boreal region (Mela, 1996; Olesen & Bindi, 2002). The length of the growing season is typically 180 days in the southern areas whereas in the northernmost areas, where agricultural production occurs in the Boreal region, it is only 120 days (Mukula & Rantanen, 1987; ATLAS, 2009).

A part of the annual precipitation falls as snow in the Boreal region, a significant part of which can be lost from agricultural fields as surface runoff when the snow melts. Moreover, a part of the precipitation may be lost as evaporation or sublimation directly from snow (Perlman, 2011). Evaporation losses are considered to be small, however, at least according to a study conducted in Finland in the early 1970s (ref. Kuusisto 2010).

When estimating how much water is available for crop growth in the Boreal region, it is reasonable to divide annual precipitation into two components, as shown in the schematic water balance figure (Figure 2).

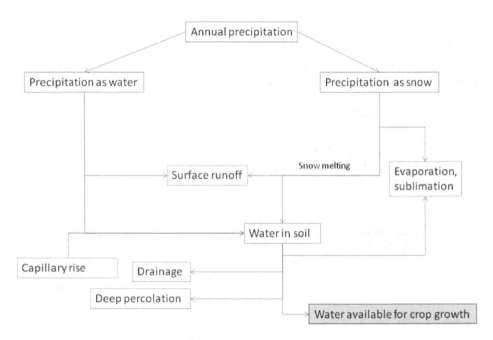

Fig. 2. A schematic representation of water balance in agricultural fields in the Boreal region.

The total volume of water available for crop growth is determined by the water-holding capacity of the soil, contributed by precipitation, minus evaporation, runoff and drainage losses during the growing season. Depending on the soil type and structure, a small volume of water can be lost by deep percolation and some can be gained by capillary rise of water from deeper soil layers. Groundwater can be a significant water resource in soils with a shallow water table (Muellera et al., 2005).

2.1 Precipitation and evapotranspiration

Annual precipitation exceeds potential evapotranspiration (measured often with Class A evaporation pan) in Scandinavia and eastern and northern parts of continental Canada (Figure 3). Thus, annual water deficit is negative in large areas within the Boreal region (ATLAS, 2007; Wallen, 1966; Ilmatieteenlaitos, 2011; Järvinen, 2007).

It has to be noted that data from Canada did not include an estimate of evaporation during the period outside the growing season. This value can be about 10 % of the annual potential evapotranspiration (Wallen, 1966). Thus, adding it into our data would result in only minor changes in Figure 3.

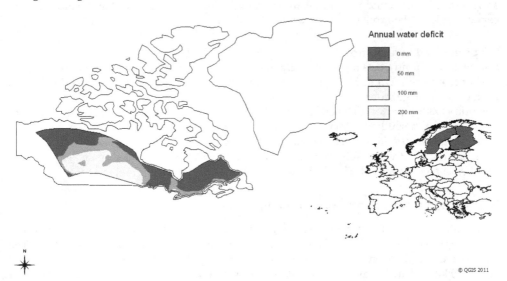

Fig. 3. Annual water deficit (potential evapotranspiration - precipitation) in the Boreal region.

Despite the annual deficit being negative in the Boreal region, it can be positive during the first half of the growing season, for example, in the period from May to July (Pajula & Triipponen, 2003): the average deficit is 100 – 250 mm in Finland during this period (Figure 4).

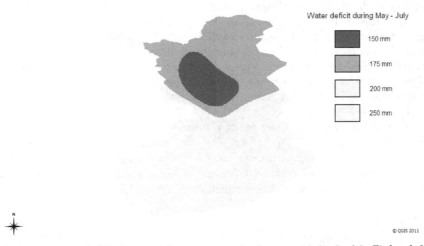

Fig. 4. Average water deficit (potential evapotranspiration - precipitation) in Finland during May - July.

The Atlas of Canada indicates that British Columbia and southern Alberta contain areas of severe moisture deficit at some times during most summers (ATLAS, 2007). Such seasonal drought affects agricultural production. For example, Hooli (1971) and Peltonen-Sainio et al. (2011) reported that early summer drought decreases yields of field crops in Finland. Spring sown crops in particular may suffer from drought during this period if soil water-holding capacity is low when root systems are not fully developed.

2.2 Spring and autumn frosts

Spring and autumn frosts represent a climatic risk for field crops grown in the Boreal region. Spring crops are sown in May and harvested in August – September, and frost during this period can damage crops.

Solantie (1980) reported detailed data for frosts in Finland. It is common to have 3 night frosts in June in southern Finland, in areas away from lakes and the Baltic Sea. Frosts are even more common in northern Finland where there can be 4 night frosts in June. It is not common to have frost in July, but in August the risk of frost increases again. The average number of night frosts is 1.5 in southern Finland and 2.5 in northern Finland.

3. Irrigation methods and possibilities

Canada has most of the irrigated field area in the Boreal region, about one million hectares, which represents about 1.5 % of the entire agricultural area in Canada (ATLAS, 2007). The irrigated area in Sweden is about 55 000 hectares (about 2 % of all agricultural land, Brundell et al., 2008) and in Finland it is 88 000 hectares (4.4 % of all agricultural land, Pajula & Triipponen, 2003).

It is not common practice to irrigate field crops (cereals, oilseed crops, potatoes, sugar beet, forage) in the Boreal region. For example, for field crops other than potato (*Solanum tuberosum* L.) and sugar beet (*Beta vulgaris* L. var. *altissima*), irrigated crops represented only 1% of the total crop area in Sweden in 2006. In contrast, the entire area of potato for processing was irrigated in 2006. Also table potatoes (59 %) and sugar beet (16 %) were irrigated fairly often in Sweden (Brundell et al., 2008). Finnish reports indicate the irrigated area to be divided between crops such as cereals and sugar beet in a similar way as in Sweden, but exact data are difficult to locate (Pajula & Triipponen, 2003). Nevertheless, the irrigated area of potato is markedly larger in Sweden than in Finland.

The situation in the Nordic region is opposite to that in Canada, where field crops comprise 53 % of the irrigated area. Feed crops are the second most common crop to be irrigated (33 % of the irrigated area). It is less common to irrigate vegetables and fruits in Canada (STATCAN, 2009).

Regardless of the application, a key set of resources determines the maximum return on investment (Holzapfel et al., 2009). At farm level, this means using available data on climate, water availability, soil, crop, economic and social circumstances, and evaluating the potential distribution system (Pláyan & Mateos 2006). Irrigation types include surface or flood irrigation, overhead, drip and sub-irrigation.

3.1 Surface irrigation

In surface irrigation, water moves over and across the land by simple gravity flow. The most widely techniques on surface irrigation are furrow irrigation and flood irrigation. These systems, however, are not common on field crops in Boreal region.

3.2 Overhead irrigation

Overhead irrigation includes different types of sprinkler and centre-pivot irrigation systems that are used to supply water in a similar way to natural rainfall (Brouwer et al., 1988) and can be used for most crops. This does not set strict constraints on the slope or soil type of the field as long as the application rate is adjusted to avoid ponding and runoff.

Sprinkler systems are used for 60 % of the irrigated area in Finland (Pajula & Triipponen, 2003) and sprinklers and pivots are used for virtually all of the irrigated field and feed crops in Canada, whereas micro-irrigation is the method most commonly used for fruits and vegetables (STATCAN, 2009).

3.3 Drip irrigation

Drip or trickle irrigation uses slow flow of low pressure, well filtered water in a (permanent) network of small diameter polythene tubes that open near the root zone (Shock, 2006). According to the location of tubes in the soil, drip irrigation can further be divided as a surface drip and subsurface drip (Sammis, 1980). According to Shock (2006) drip irrigation can be used whether the water demand is high or low.

Drip irrigation is considered to use water more efficiently than overhead systems (Shock, 2006). In an experiment on potatoes in Sweden, there was no significant difference in water use between these methods, but water use efficiency was better in the drip system, producing 11 % higher total yield and 28 % more marketable ware potato yield (Wiklund & Ekelöf,2006).

Drip irrigation is adaptable to irregularly shaped fields, it is not affected by wind, it can be designed so that the wheel traffic rows are dry to allow tractor operations at any time, and it can easily be automated (Shock, 2006). Precise application of nutrient is also possible using drip irrigation. Its disadvantages include high establishment, labour and consumables costs, as the fine tubes require inspection, maintenance and generally annual replacement (Forsman, 2006).

3.4 Subirrigation

Subirrigation is a method in which water is added below the soil surface into a controlled drainage system (Broughton, 1995), but is not as common as sprinkler irrigation. Subirrigation requires a flat field with a slope of less than 1-2% (Pajula & Triipponen, 2003) and sandy soil in which the capillary rise is high and soil permeability is sufficiently fast and uniform within a field (Geohring et al., 1995; Muellera et al., 2005). Optimum water table depth is related to soil type and crop grown on the field (Muellera et al., 2005).

Gilliam & Skaggs (1986) listed five environmental benefits that derive from using subirrigation: 1) more complete use of fertilizers, 2) reduced nutrient losses, 3) reduced leaching of pesticides and herbicides, 4) stabilized crop production and 5) more food grown on flat land, thereby removing erosion-prone land from cultivation and increasing land area for forests, grasslands, wildlife habitats and recreational uses. Moreover, farmers may benefit from increased yields due to subirrigation. Among others, the studies of Drury et al. (1996) and Madramootoo et al. (2001) support the environmental issues listed by Gilliam & Skaggs (1986). Efficient control of nutrient losses requires a dual-level subirrigation system where there are separate pipes for drainage and irrigation water. Otherwise, there may be significant loss of water and nutrients when the system is switched from irrigation to drainage mode (Melvin & Kanvar, 1995).

Subirrigation may promote substantial yield increases if the field is suited to this irrigation method. Yield increases of 20 to 40 % have been reported for maize (*Zea mays* L.) and soybean [*Glycine max* (L.) Merr.] in humid areas (Broughton, 1995).

Subirrigation and controlled drainage could be used more in the Boreal region (Pajula & Triipponen, 2003). For example, it was estimated by Puustinen et al. (1994) that 770 000 hectares in Finland could be equipped with controlled drainage systems instead of the current 34 300 (Pajula & Triipponen, 2003). This last figure overestimates the current value of subirrigation, because although the drainage systems installed under most Finnish arable fields can fairly easily be used for water supply, this seldom happens in practice. The main barriers to using subirrigation are steepness of slope, soil types that do not transport water fast enough, and high construction costs (Pajula & Triipponen, 2003).

3.5 Availability of irrigation water

The Boreal region is characterized by areas in which there is no risk of water scarcity. Alcamo et al. (2000) determined this based on calculating the ratio of water withdrawals and water availability in different river basins. Water withdrawal includes all water that is taken from rivers/lakes, evaporation and deep percolation. Water availability is determined as a sum of annual runoff plus shallow groundwater recharge. The Boreal region has low ratio values, in the 0 – 0.1 range, which means there is enough water for domestic, industrial and agricultural use according to Alcamo et al. (2000).

Annual precipitation usually exceeds potential evapotranspiration in the Boreal region (Figure 3), which means that surface water is usually available for irrigation. This assumption is supported by a survey made in Sweden that reported surface water being the most common source (53 %) for irrigation (Brundell et al., 2008). Groundwater was used by 31 % of the farms that irrigated, and stored water was used by 16 % of the farms. However, another Swedish study reported the percentage of farms using surface water to be much higher, 80 % (Anonymous, 2007). Surface water was the most common source (37 % of the farms that irrigated) of irrigation water also in Canada (STATCAN, 2009). However, it was almost as common to use off-farm water (34 %). Underground water or well water was a fairly common source for irrigation (21 %).

Local conditions markedly affect the availability of irrigation water in a specific field. Irrigation costs increase rapidly when the distance of the field from the irrigation water source increases. The difference in altitude between field and water source has even more effect on the costs (e.g., Dumler et al., 2007). Therefore, the availability of irrigation water that is accessible at reasonable cost has to be estimated separately for each field.

4. Effect of short drought periods on crop development, growth and yield

The drought periods are usually short in the Boreal region, lasting for two to four weeks. However, in rainfed areas, even such a short period without rain can result in yield losses related to water deficit. The southern areas in Finland, especially the south and west coasts, are prone to spring droughts. The drought periods can sometimes occur after mid-summer, during the grain filling period. Boyer (1982) stated that drought is one of the most yield-limiting factors worldwide. Peltonen-Sainio et al. (2011) estimated that on average short drought periods have caused yield losses of 7-17 % in Finland within the past 30 years.

4.1 Early season drought and irrigation

Water deficit early during the season, in the spring, affects the emergence of spring-sown crops. Seeds either do not germinate or the germination processes are interrupted if there is not enough water in the surrounding soil. This results in formation of uneven plant stands (Figure 5) and causes both reduction in yield and quality. In sparse plant stands weeds have more space for growth, and for example in the case of cereals, if additional tillers are formed the grain size is reduced in them in comparison with in the main culm, and the tillers reach maturity later than those of the main culm. At later stages, during vegetative growth of cereals, water deficit affects the solar energy capture capacity of cereals, mainly decreasing the fraction of photosynthetically active radiation intercepted, but also radiation use efficiency (Olesen et al., 2000). In general, the amount of vegetative biomass at the time of anthesis is related closely to the final number of grains (reviewed in Passioura & Angus, 2010). For spring cereals, long coleoptiles and vigorous seedling emergence are important traits in Boreal conditions, where there is uneven precipitation and a short growing season (reviewed in Mäkelä et al., 2008).

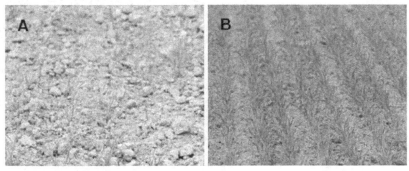

Fig. 5. Seeds germinate unevenly or germination processes are interrupted in dry soil, resulting in uneven and sparse plant stands (A). Uniform and vigorous plant stands develop when soil moisture, among other factors, is optimal for germination (B).

In spring barley (*Hordeum vulgare* L.), early season water deficit reduced both the sink and source capacity of plants through decrease in photosynthesis and thereby dry weight accumulation and number of grains (Rajala et al., 2011). According to Jamieson et al. (1995), a water deficit related decrease in dry weight accumulation of barley during the early phases of growth is mainly due to reduced radiation use efficiency, which does not recover completely after stress relief. In oat (*Avena sativa* L.), vegetative growth was limited most due to early occurrence of a water deficit, which was related to decrease in gas exchange and increase in osmolyte synthesis, e.g. proline (Peltonen-Sainio & Mäkelä, 1995).

Water deficit occurring at an early stage can, however, have more wide-ranging effects. In wheat (*Triticum aestivum* L.), pre-anthesis water deficit significantly increased floret abortion and formation of sterile florets (Rajala et al., 2009). Also, depending on cultivar, pollen fertility can decrease by over 50 % when wheat is exposed to a water deficit (Briggs et al., 1999). Saini & Aspinall (1981) reported that water deficit induced formation of shrivelled anthers and non-viable pollen. Even a short duration of water deficit at the young microspore stage induces sterility in wheat, whereas ovaries are not affected (Ji et al., 2010). Supply of assimilates seemed to be closely connected to floret abortion and formation of sterile florets (Rajala et al., 2009). In drought-stressed wheat, the sugar delivery to the reproductive organs, for example pollen grains, is limited, resulting in absence of starch (Saini & Westgate, 2000). However, Lalonde et al. (1997) stated that this was merely a consequence of earlier changes such as suppression of invertase activity, triggering sterility and delayed or premature degeneration of male cells. In small-grain cereals, determination of grain size begins already before anthesis, during sporophyte development, whereas grain number is controlled throughout spike development, at the earliest stage of male gametophyte formation (Ji et al., 2010).

Spring oilseed rape [*Brassica napus* L. ssp. *oleifera* (DC.) Metzg.] responded to early season water deficit with decreased leaf area, which was later also observed as halved dry matter by the time of flowering (Mogensen et al., 1996). The early season stress affected also the seed quality of oilseed rape, seed protein content and glucosinolate concentration increasing markedly in stressed plants (Jensen et al., 1996). The increase in glucosinolate concentration, from approximately 12 to 20 µmol g^{-1}, was in relation to plant water status, whereas the protein content probably increased after active uptake of nitrogen followed by transport into seed after growth retardation (Jensen et al., 1996). Winter turnip rape [*Brassica rapa* L. ssp. *oleifera* (DC.) Metzg.] is most sensitive to water deficit during flowering. Water deficit at the flowering stage caused over 100 growing degree days later maturity, 15 % lower seed oil content, and 64 % lower yield than in a well-watered control (Tesfamariam et al., 2010).

Cool-season grain legumes grown in the Boreal region are rather drought sensitive; faba bean (*Vicia faba* L.) being the most sensitive (F.L. Stoddard, personal communication). Furthermore, the European faba bean genotypes are more drought sensitive than genotypes from North Africa and Latin America (reviewed in Stoddard et al., 2006). The differences between genotypes in drought tolerance are marked. Some have high water use efficiency, indicating good capacity for photosynthetic maintenance and transpiration efficiency, some are able to adjust the leaf temperature, and thus stomatal conductance, more efficiently according to prevailing conditions (reviewed in Khan et al., 2010). Unlike cereals and oilseed crops, legumes have *Rhizobium* in their root nodules, which fix nitrogen. Drought limits the number of nodules in faba bean roots (reviewed in Serraj et al., 1999) and thus decreases further the nitrogen availability for the crop. All the steps of establishment of *Rhizobium*-legume symbiosis are sensitive to water deficit. This can be related to limited growth and movement of *Rhizobium* in soil during drought, thus restricting the infection process needed for nodule formation. The root hair infection and formation of infection threads is also sensitive to water deficit (reviewed in Serraj et al., 1999). Moreover, the supply of carbohydrates from the plant to the *Rhizobium* decreases due to decreased photosynthetic activity of the legume (Gálvez et al., 2005).

White clover (*Trifolium album* L.) is one of the most drought sensitive components of pastures (Skinner et al., 2004). In pastures, irrigation increases the number of tillers of grasses and the longevity of clover and its dry mass yield, the increment being higher the older the clover pasture. The main reason for the success of clover is increased stolon formation induced by irrigation (García et al., 2010).

Potato is sensitive to water deficit and soil water availability is considered as a major limiting factor in its production and quality (Phene & Sanders, 1976). Water deficit easily leads to significant yield and quality losses. In trials at Potato Research Institute in Finland in 1989-1995 and 1999-2002, irrigation increased tuber yield, starch yield and marketable ware potato yields by 9.1 %, 9.2 % and 22.2 %, respectively (Potato Research Institute, unpublished data). On the other hand, over-irrigation promotes diseases as well as leaching of nutrients and pesticide residues towards groundwater (Pereira & Shock, 2006). In particular, late irrigation at the senescence may increase risks of tuber diseases (Wikman et al., 1996).

Water has a definitive influence on tuber bulking (Wikman et al., 1996). Especially during mid-bulking, three to six weeks after tuber initiation, potato needs an even supply of water (Scherer et al., 1999). Water deficit during tuber bulking usually affects total tuber yield more than quality (King & Stark, 1997).

On average, stored soil and spring moisture supplies are adequate from planting to tuber initiation. Lack of water before tuber initiation reduces tuber set per stem (MacKerron & Jeffries, 1986). Dryness at the tuber initiation phase decreases tuber number (King & Stark, 1997; Scherer et al., 1999). In Finnish conditions, the most decisive phase on tuber formation is about 40 days after planting. Effects on the final tuber yield, however, are seldom clear in these phases (Wikman et al., 1996).

Early season drought is also known to shorten the developmental phases of cereals (Figure 6). In triticale (*Triticosecale*), the phase from emergence to anthesis and to maturity was reduced both in calendar days and in thermal time. The acceleration of development was probably related to a water-deficit induced increase in canopy temperature of 1.3 °C (Estrada-Campuzano et al., 2008). Water deficit results in stomatal closure, and thus decreased gas exchange and transpiration, which decreases the cooling inside leaf cells and leads to increased temperature in leaf tissues. Water deficit affects not only the growth and development of aerial plant parts but also the roots. The seminal roots are shorter and the root volume is lower in cereals, such as barley (Sahnoune et al., 2004), wheat (Guedira et al., 1997), and oat (Larsson & Górny, 1988) when grown under water deficit. In barley, root growth was restricted mostly in deeper (> 30 cm) soil layers (Sahnoune et al., 2004).

Fig. 6. Water deficit hastens the development of cereals. Non-irrigated wheat reached anthesis stage (A, B), whereas irrigated wheat (C, D) was not yet at heading stage on 2 July, 2011.

A common problem in the Boreal region, connected to early season drought, is inhibition of early tillering, which can result in formation of late tillers after rainfall at anthesis (Mäkelä & Muurinen, 2011). These tillers produce spikes which, however, reach maturity several weeks later than the majority of the spikes (Figure 6), causing delays in harvesting, increases in drying costs, and quality losses due to sprouting or immature grains (Kivisaari & Elonen, 1974). According to Kivisaari & Elonen (1974), irrigation of spring barley at an early growth stage (twice 30 mm) significantly decreased moisture content at harvest, number of immature grains and late tiller formation. Even though the number of spikes per plant decreased, single grain weight, number of grains per spike and thus, grain yield, increased significantly. The final increase in grain yield was nearly 70 %.

Fig. 7. Non-irrigated winter wheat cv. Olivin (A) stand includes late-formed tillers, which are still green, whereas the main crop has already reached maturity. Irrigated crop (B) has reached maturity evenly.

Irrigation of spring wheat (35 mm) at the vegetative stage increased the number of spike-bearing tillers and grain yield (Table 1) and decreased the formation of tillers at later growth stages (Figure 7). However, single grain weight, harvest index and number of grains per spike were not significantly different between irrigated and non-irrigated wheat. In the spring, the duration from seeding to anthesis can be lengthened with irrigation by five to seven days (Cutforth et al., 1990), allowing more time for accumulation of vegetative mass and formation of floret structures.

Location Cultivar	Grain yield g m^{-2}	Single grain weight mg	Grains per spike	Spikes per m^2	Harvest index
Location 1, Amaretto					
Irrigated	809	38.4	29	729	0.45
Non-irrigated	406	33.9	27	445	0.42
SEM (df 5)	64.7	2.1	0.7	49.6	0.035
Location 2, Amaretto					
Irrigated	603	41.5	20	738	0.53
Non-irrigated	359	39.5	15	622	0.52
SEM (df 5)	33.5	0.62	1.5	48.5	0.009

Table 1. Grain yield, number of grains per spike, single grain weight, number of spikes per m^2 and harvest index for irrigated and non-irrigated spring wheat cv. Amaretto in 2011 at two different locations (Location 1, Laukka 60 15' 00'' N, 23 05' 00'' E, alt. 14.0 m and Location 2 Tamminiemi, 60 24' 00'' N, 25 40' 00'' E, alt. 23.5 m). Plant stands were irrigated (30-35 mm) once at tillering stage.

It also should be pointed out that in coastal areas of Finland, irrigation is sometimes needed to enhance the germination of autumn-sown winter cereals.

4.2 Late season drought and irrigation

Late season drought affects the grain-filling period of both spring and winter cereals. After anthesis, water deficit mainly caused decreases in grain yield and fewer grains in barley (Rajala et al., 2011). At that stage, the limited availability of water and assimilates affects the translocation processes, from sources to sinks, filling grains. In oat, however, water deficit at post-anthesis mainly decreased the number of grains (Peltonen-Sainio & Mäkelä, 1995), similarly to wheat (Rajala et al., 2011), probably due to limited source capacity. Fábián et al. (2011) showed that water deficit occurring after pollination, during early grain development stage, decreases the embryo size and increases the degradation of cell layers surrounding the ovule. It also affects the distribution of starch granules A and B, favouring the A-type. Similarly as for drought occurring early in the season, late season drought also shortens the grain-filling period (Fábián et al. 2011), thus decreasing the grain yield by limiting the length of assimilate transport to grains.

A wet winter may increase the drought vulnerability of winter cereals. As a response to waterlogging, aerenchyma-forming nodal roots located in the top 20 cm develop at the cost of deeper seminal roots. During water deficit in the summer, the nodal roots cannot extract water from the deep soil layers, which multiplies the drought effect and grain yield losses (Dickin & Wright, 2008). In general, root growth is reduced under water deficit, but the reduction is more severe and permanent after anthesis than before terminal spikelet formation. This effect is cumulative at the grain-filling stage, since water uptake rates of roots decrease in relation to crop life (Asseng et al., 1998).

Stem reserves stored prior to anthesis have been considered as an important source of assimilates to maintain grain-filling processes, especially in environments with terminal stress, where the contribution of stem reserves can be as high as 90 % of the final yield (Asseng & Herwaarden, 2003). Foulkes et al. (2007) reported that stem reserves correlated positively with grain yield also in Britain, where rainfall is unpredictable, as it is in the Boreal region.

In spring oilseed rape, water deficit occurring during the seed filling stage had no effect on total dry matter but decreased the number and size of seeds. Stressed plants also had fewer siliquee in comparison with well-watered plants (Jensen et al., 1996). The explanation was offered by Richards & Thurling (1978), who found that in oilseed rape dry weight accumulation before flowering was the major factor affecting seed yield when water deficit occurred after flowering, whereas in turnip rape dry weight accumulation after flowering was more important. However, Tesfamariam et al. (2010) found that water deficit during the seed-filling period hastened maturity over 100 growing degree days. Flower dropping caused by water deficit limits the grain yield in grain legumes. It decreases the number of sinks and may cause feedback inhibition of photosynthesis. On the other hand, poor carbohydrate translocation to the reproductive sinks may limit both seed size and number (Tesfaye et al., 2006).

Water deficit limits sugar beet root and sugar yield and quality throughout Europe (Hoffman et al., 2009), even though it has been commonly considered to be a drought tolerant crop (Vamerali et al., 2009). Water deficit decreases the processing quality of roots since the potassium and α-amino nitrogen concentrations increase due to osmotic adjustment and nitrogen metabolism (Bloch & Hoffmann, 2005). It also affects the decay of sugar beet fibrous roots, which form a dense and deep system, even though their fraction of biomass is 3 - 10 % of the total biomass (Brown et al., 1987). Under water deficit the longevity of fibrous roots increases. Life span of roots is longest at a depth where there is least water available. This decreases the cost of carbon for renewal of roots (Vamerali et al., 2009).

In sugar beet, water deficit decreased the leaf lifespan by 20 days and net photosynthesis of leaves. Moreover, the photosynthesis of mature leaves could not fully recover from water deficit (Monti et al., 2007), decreasing the ability of the crop to capture solar energy, and thus leading to decreased root and sugar yield. According to Ober et al. (2005) the sugar beet ideotype for environments with limited water availability should be able to maintain a green and succulent canopy late into the season and thus have low leaf senescence and specific leaf weight as well as high transpiration rate and ability to use water from deep soil layers. That will ensure good resource capture ability of the crop and good yield. For sugar beet, late season irrigation induces formation of deep roots (>1 m) and reduction of penetration resistance in deep soil due to irrigation may further induce root growth (Vamerali et al, 2009). Camposeo & Rubino (2003) stated that low interval drip irrigation of sugar beet can increase the root:shoot ratio by 35 % and thus result in yield increases. In Sweden, irrigation throughout the growing season increased the root yield almost 14 t/ha and sugar yield by 2.6 t/ha. Late season irrigation increased root yield by 9.6 t/ha and sugar yield by 1.8 t/ha. Irrigation did not affect markedly the root quality (Persson, 1993). In Finland, late season irrigation increased root yield by 18 t/ha and sugar yield over 2.5 t/ha (Erjala, 2002).

5. Effect of frost on crop development, growth and yield

Spring and autumn frost can occur every now and then in the Boreal region, causing yield losses. To avoid this, farmers have to delay sowing spring crops and use early maturing cultivars. Thus, the growing season is limited and leads to indirect yield losses. Frosts can damage spring crops in particular. Frost damage occurs in the spring most typically, but sometimes also in the autumn. This can be explained by the farmers' experience and knowledge-based selection of cultivars with adequate growing time.

5.1 Spring frosts

The early vegetative stage of spring crops is vulnerable to frosts (Andrews, 1987). Most severe damage caused by spring frosts occurs usually in sugar beet and oilseed brassica crops, when the entire canopy can be killed (Figure 9). In these species the apical bud is located over the insulating soil surface and is thus sensitive to damage. In cereals such as oat, wheat, barley and maize, the apical bud is located below the insulating soil surface and is not easily damaged. Frost damage mainly slows down biomass accumulation of cereals due to decreased photosynthesis following membrane leakage and reduced mesophyll conductance, decreased photosynthetic leaf area, and thus causes yield decrease (Andrews, 1987). According to Gusta & O'Connor (1987), frost also slows down the development rate of oilseed rape and mustard (*Sinapis alba* L.) by approximately nine days and spring cereals by up to ten days. On the other hand, McKenzie et al. (1982) reported a reduction in number of days to heading in barley after frost damage and concluded that it was due to response to prevailing photoperiod, temperature and soil moisture.

Fig. 9. Spring frost damage in A) sugar beet and B) barley. Some of the sugar beet plantlets have been completely destroyed by frost. The apical bud of sugar beet is the uppermost part of the plant and thus sensitive to damage. Barley survives the frost even though the leaves are damaged because the apical bud is sheltered below the soil surface during the vegetative stage.

Gusta & O'Connor (1987) found that there were no cultivar differences in frost tolerance for mustard and oilseed rape, but the developmental stage did affect the tolerance. Plantlets having 4 – 6 leaves were 1 – 2 °C more frost tolerant than plantlets at the cotyledon stage. Frost tolerance of turnip rape is significantly improved already after a two-day hardening in the field, in terms of death of unhardened plantlets after a night frost and survival without damage of hardened plantlets (K. Pahkala, personal communication). However, hardening resulted in reduced plant height and leaf area. The cultivar differences in growth response to hardening as well as in frost (-5 °C) tolerance were significant. During hardening the hexadecatrienoic acid (16:3) concentration of leaves increased, whereas there was no change in α-linolenic acid (18:3) concentration (Pahkala et al., 1991).

Among spring cereal species at the two-leaf stage, oat is most frost sensitive and wheat most frost tolerant (Gusta & O'Connor, 1987). However, at anthesis, barley is most frost tolerant, probably as it is pollinated in the boot (Passioura & Angus, 2010). McKenzie et al. (1982) found significant differences among spring barley cultivars in frost tolerance when plants were subjected to -5.6 °C at the two-leaf stage. In wheat, spring frost at the two-leaf stage increased the number of spikes and delayed maturity for one to ten days. The delay in maturity was less obvious in barley and wheat. Frost treatment increased wheat yield but tended to decrease the barley and oat yield (Gusta & O'Connor, 1987). In Finland, barley has been considered to be more frost tolerant than wheat and usually frost has induced formation of spike-bearing tillers in barley. According to McKenzie et al. (1982), it seems that early maturing cultivars are more sensitive to frost than late maturing cultivars. Cool-season grain legumes are tolerant to frost at the vegetative stage, but pollen and flower structures are sensitive to low temperatures (Stoddard et al., 2006; Link et al., 2010; F. L. Stoddard, personal communication).

In the Boreal region, winter rye can sometimes be damaged by spring frost at the time of pollination, causing incomplete fertilization of florets and leading to reduced grain yield. A low temperature of -4 °C caused disruption of membrane structures in wheat spikes at the heading stage (Marcellos & Single, 1984). That resulted in destruction of entire spikes on some occasions when temperature decreased 1 °C below a cultivar-dependent threshold temperature. On the other occasions, the spike consisted of both fertile and sterile spikelets. The occurrence of sterile spikelets did not depend on their position within the spike, whereas the higher order florets within the spikelet remained fertile in many cases

(Marcellos & Single, 1984). Marcellos & Single (1984) explained this phenomenon by discontinuous extension of ice nucleation within the spike. An ice front can move from a single nucleating point on a leaf at approximately 120 cm/min, but it is delayed by nodes in the stem, rachis and spikelets (Andrews, 1987). However, in wheat, cold stress similarly to drought, induces irreversible damage in pollen development at the young microspore stage and thus formation of infertile pollen grains (Ji et al., 2011). This damage is probably similar to that in pollen development in rye.

5.2 Autumn frosts

Autumn frosts can occur during the grain filling stage of cereals. The risk increases with the late maturing cultivars. Frost affects the biological processes of grain filling and thus the grain weight remains low, the grain is of poor quality and in some cases only empty hulls remain. Occurrence of frost has resulted in formation of shrivelled and shrunken wheat grains and in less severe cases blistering of the grain surface and thus, low grain weight during the grain filling stage (Cromey et al., 1998). Cromey et al. (1998) suggested that the damage was due to death of outer tissues of grains, most damage occurring in the testa and pericarp and to some extent in the endosperm. The pericarp and testa were crushed and the dead cells were in a loose arrangement in frosted grains. Starch accumulation was reduced severely, probably since the number of living endosperm cells able to take up carbohydrates was decreased. However, the surviving tissues could compensate to some degree for the damage. Cromey et al. (1998) concluded that the original aleurone layer could have been reformed from the outermost endosperm cells, and therefore the germination ability of the grains was maintained. When frost occurred at the early dough stage, the germination of wheat decreased from 98 % to 62 % and barley from 82 % to 40 %, and oat completely failed to germinate, but at late dough stage the germination was not affected in any of the species studied (Gusta & O'Connor, 1987). However, at the grain filling stage wheat seems to be the most frost tolerant cereal species in general (Gusta & O'Connor, 1987). In brassicas, the seed yield, and especially the oil quality, is most affected by frost (Gusta & O'Connor, 1987). Pahkala et al. (1991) reported a decrease in α-linolenic acid and an increase in oleic acid concentration in ripening seeds of turnip rape after frost. Cool-season grain legumes are tolerant to frost at later stages of pod filling (Stoddard et al., 2006; Link et al., 2010; F. L. Stoddard, personal communication).

Frost also severely affects the baking quality of the grains and in many cases results in rejection of the yield by the milling industry. Dexter et al. (1985) reported decreased visual quality, grain weight and flour yield and increased ash content and grain hardness of frosted wheat grains. Falling number was only slightly lower in frosted grain in comparison with unfrosted grains. Flour was darker the more severe the frost damage and protein content of the grains was similar to that of unfrosted grains. The increased hardness and maltose content of grains, as well as loss of gluten functionality, resulted in increased flour starch damage. However, no sprout damage was observed, even though the falling number decreased and maltose content increased (Dexter et al., 1985). The estimate of yield loss in small-grain cereals caused by autumn frost varied from 13 to 33 %, although it could have been even more due to loss of shrivelled grain during harvest (Cromey et al., 1998).

5.3 Frost protection

Irrigation is successfully used to prevent damage due to night frost. In direct protection, the temperature of the plant is kept above the freezing point by using latent heat fusion of water. The crop is irrigated for the whole time that temperature is at or below zero, forming a film of ice on the plant which releases energy, preventing the tissue from freezing. Irrigation is started when temperature decreases to +1-0 °C, and it is stopped when the ice on the plant starts to melt (van der Gulik & Williams, 1988; Ingvarsson, 1992; Svensson, 2003). In this way, irrigation can protect the crop until -7 – -8 °C (Ingvarsson, 1992).

In the indirect method, the soil is irrigated on the morning of the day when frost is anticipated. Wet soil traps plenty of warmth during a day and releases it during the night, preventing frost damage up to -2 – - 3 °C. The shelter effect is highest about 30 cm above the soil surface, so indirect frost protection is suitable only for low growing crops (Ingvarsson, 1992).

Irrigation as a direct protection against freezing temperature is mostly used in the production of berries and fruits where inputs are high (van der Gulik & Williams, 1988). In Boreal region, especially in Nordic countries, first-early potatoes are so valuable that irrigation generally is used to prevent frost damage (Wikman et al., 1996).

6. Future challenges

Alcamo et al. (2000) suggested that industrial and economic growth in regions under severe water stress (high water withdrawal compared with water availability) may be limited in the near future. Climate change may decrease water availability and thus increase water stress in those areas in the longer term. This means that agriculture will have to compete harder with domestic and industrial water use. The Boreal region will, however, remain a region without water stress in the longer term (Alcamo et al., 2000).

Climate change will probably increase the frequency of drought spells also in the European Boreal region (Olesen et al., 2011). The changes are expected to be, however, smaller than in the southern parts of Europe. The increase in drought frequency would increase the need for irrigation also in northern areas. The risk of late spring frosts and early autumn frosts is likely to decrease due to climate change (Olesen et al., 2011). There is still an avenue for plant breeding in improving frost tolerance at the grain filling stage in cereals since genetic variation for the trait is known to exist (Marcellos & Single, 1984), as well as at early growth stages. In dry areas, especially in a Mediterranean climate, selection of drought tolerant crop species and cultivars has been considered to be one of the most important means of alleviating agricultural problems (Ashraf et al., 1992). Drought has not commonly been considered a serious problem in the Boreal region. However, in the future, more attention should be paid to methods and technologies for alleviating short-term drought problems.

7. References

Anonymous, 2007. Vattenuttag och vattenanvändning i Sverige 2005. *Statistics Sweden* 2007. Available from
http://www.scb.se/Pages/PublishingCalendarViewInfo_259923.aspx?PublObjId=6598

Alcamo, J.; Henrichs, T. & Rösch, T. (2000). World water in 2025. Kassel World Water Series, Report No 2. Available from
http://www.usf.uni-kassel.de/ftp/dokumente/kwws/kwws.2.pdf

Andrews, C.J. (1987). Low-temperature stress in field and forage crop production – an overview. *Canadian Journal of Plant Science*, 67, 1121-1133, ISSN 0008-4220

Ashraf, M.; Bokhari, H. & Cristiti, S.N. (1992). Variation in osmotic adjustment of lentil (*Lens culinaris* Medic) in response to drought. *Acta Botanica Neerlandica*, 41, 51-62, ISSN 0044-5983

Asseng, S.; Ritchie, J.T.; Smucker, A.J.M. & Robertson, M.J. (1998). Root growth and water uptake during water deficit and recovering in wheat. *Plant and Soil*, 201, 265-273, ISSN 0032-079X

Asseng, S. & van Herwaarden, A.F. (2003). Analysis of the benefits to wheat yield from assimilates stored prior to grain filling in a range of environments. *Plant and Soil*, 256, 217-229, ISSN 0032-079X

ATLAS. (2007). The Atlas of Canada. Available from http://atlas.nrcan.gc.ca

Bloch, D. & Hoffmann, C. (2005). Seasonal development of genotypic differences in sugar beet (*Beta vulgaris* L.) and their interaction with water supply. *Journal of Agronomy and Crop Science*, 191, 263-272, ISSN 0931-2250

Boyer, J.S. (1982). Plant productivity and environment. *Science*, 218, 443-448, ISSN 0036-8075

Briggs, K.G.; Kiplagat, O.K. & Johnson-Flanagan, A.M. (1999). Effects of pre-anthesis moisture stress on floret sterility in some semi-dwarf conventional and conventional height spring wheat cultivars. *Canadian Journal of Plant Science*, 79, 515-520, ISSN 0008-4220

Broughton, B.S. (1995). Economic, production and environmental impacts of subirrigation and controlled drainage, In: *Subirrigation and drainage*, H.W. Belcher & F.M. D'Itri, (Eds.), 183.191, CRC Press, ISBN 1-56670-139-2, Boca Raton, Florida, USA.

Brown, D.M. & Blackburn, W.J. (1987). Impacts of freezing temperatures on crop production in Canada. *Canadian Journal of Plant Science*, 67, 1167-1180, ISSN 0008-4220

Brundell, P.; Kanlen, F. & Westöö, A-K. (2008). *WaterUuse for Irrigation*. Report on Grant Agreement No 71301.2006.002-2006.470. Statistiska Centralbyrån, Statistics Sweden. Available from http://circa.europa.eu/Public/irc/dsis/envirmeet/library?l=/meetings_2008_arc hive/statistics_09-101008/background_documents/sweden_irrigationpdf/_EN_1.0_&a=d

Brouwer, C.; Prins, K.; Kay, M. & Heibloem, K. (1988). *Irrigation Water Management: Irrigation Methods*. Training manual No. 5. Available from http://www.fao.org/docrep/S8684E/S8684E00.htm

Brown, K.F.; Messem, A.B.; Dunham, R.J. & Biscoe, P.V. (1987). Effect of drought on growth and water use of sugar beet. *Journal of Agricultural Science*, 109, 421-435, ISSN 0021-8596

Camposeo, S. & Rubino, P. (2003). Effect of irrigation frequency on root water uptake in sugar beet. *Plant and Soil*, 253, 301-309, ISSN 0032-079X

Cromey, M.G.; Wright, D.S.C. & Boddington, H.J. (1998). Effects of frost during grain filling of wheat yield and grain structure. *New Zealand Journal of Crop and Horticultural Science*, 26, 279-290, ISSN 0114-0671

Cutforth, H.W.; Campbell, C.A.; Brandt, S.A.; Hunter, J.; Judiesch, D.; DePauw, R.M. & Clarke, J.M. (1990). Development and yield of Canadian western red spring and Canada prairie spring wheats as affected by delayed seeding in the brown and dark brown soil zones of Saskatchewan. *Canadian Journal of Plant Science*, 70, 639-660, ISSN 0008-4220

Dexter, J.E.; Martin, D.G.; Preston, K.R.; Tipples, K.H. & MacGregor, A.W. (1985). The effect of frost damage on the milling and baking quality of red spring wheat. *Cereal Chemistry*, 62, 75-80, ISSN 0009-0352

Dickin, E. & Wright, D. (2008). The effects of winter waterlogging and summer drought on the growth and yield of winter wheat (*Triticum aestivum* L.). *European Journal of Agronomy* 28, 234-244, ISSN 1161-0301

Drury, C. F., Tan, C. S., Gaynor, J. D., Oloya, T. O. & Welacky, T. W. (1996). Influence of controlled drainage-subirrigation on surface and tile drainage nitrate loss. *Journal of Environmental Quality*, 25, 317-324, ISSN 0047-2425

Dumler, T.J.; O'Brien, D.M. & Rogers, D.H. (2007). Irrigation Capital Requirements and Energy Costs, 24.8.2011, Available from
http://www.ksre.ksu.edu/library/agec2/mf836.pdf

Erjala, M. (2002). Kastelukokeen tulokset. *Juurikassarka*, 1, 16-17, ISSN 0789-2667

Estrada-Campuzano, G.; Miralles, D.J. & Slafer, G.A. (2008). Genotypic variability and response to water stress of pre- and post-anthesis phases in triticale. *European Journal of Agronomy*, 28, 171-177, ISSN 1161-0301

Fábián, A.; Jäger, K.; Rakszegi, M. & Barnabás, B. (2011). Embryo and endosperm development in wheat (*Triticum aestivum* L.) kernels subjected to drought stress. *Plant Cell Reports*, 30, 551-563, ISSN 0721-7714

Forsman, K. (2006). Perunan kastelumenetelmät vertailussa. *Tuottava Peruna*, 33, 16-17, ISSN 0787-670X

Foulkes, M.J.; Sylvester-Bradley, R.; Weightman, R. & Snape, J.W. (2007). Identifying physiological resistance in winter wheat. *Field Crops Research*, 103, 11-24, ISSN 0378-4290

Geohring, L.D; van Es, H.M. & Buscaglia, H.J. (1995). Soil water and forage response to controlled drainage, In: *Subirrigation and drainage*, H.W. Belcher & F.M. D'Itri, (Eds.), 183-191, CRC Press, ISBN 1-56670-139-2, Boca Raton, Florida, USA

Gilliam, J.W & Skaggs, R.W. (1986). Controlled agricultural drainage to maintain water quality. *ASCE Journal of Irrigation and Drainage Engineering*, 112, 254-263, ISSN 0733-9437

Gálvez, L.; González, E.M. & Arrese-Igor, C. (2005). Evidence for carbon flux shortage and strong carbon/nitrogen interactions in pea nodules at early stages of water stress. *Journal of Experimental Botany* 56, 2551-2561, ISSN 0022-0957

García, J.A.; Piñeiro, G.; Arana, S. & Santiñaque, F.H. (2010). Moisture deficit and defoliation effects on white clover yield and demography. *Crop Science*, 50, 2009-2020, ISSN 0011-183X

Guedira, M.; Shroyer, J.P; Kirkham, M.B. & Paulsen, G.M. (1997). Wheat coleoptiles and root growth and seedling survival after dehydration and rehydration. *Agronomy Journal*, 89, 822-826, ISSN 0002-1962

van der Gulik, T.W. & Williams, R.J. (1988). *B.C. Frost Protection Guide*, Irrigation Industry Association of British Columbia, ISBN 0-7726-0705-2, Vernon, B.C., Canada

Gusta, L.V. & O'Connor, B.J. (1987). Frost tolerance of wheat, oats, barley, canola and mustard and the role of ice-nucleating bacteria. *Canadian Journal of Plant Science*, 67, 1155-1165, ISSN 0008-4220

Hooli, J. (1971). Säätekijöiden vaikutuksesta viljelykasvien satoihin ja vesitalouteen. English summary: Effect of weather on water economy and crop yields. *Tieteellisiä julkaisuja*, Helsingin teknillinen korkeakoulu, No. 35. 244 p. ISSN 0355-774X

Hoffmann, C.M; Huijbregts, T.; van Swaaij, N. & Jansen, R. (2009). Impact of different environments in Europe on yield and quality of sugar beet genotypes. *European Journal of Agronomy*, 30, 17-26, ISSN 1161-0301

Holzapfel, E.A.; Pannunzio, A.; Lorite, I.; Silva de Oliveira, A.S. & Farkas, I. (2009). Design and management of irrigation systems. Review. *Chilean Journal of Agricultural Research*, 69, 17-25 ISSN 0718-5820

Hämet-Ahti, L. (1981). The boreal zone and its biotic subdivision. *Fennia*, 159, 69-75, ISSN 0015-0010

Ilmatieteenlaitos. (2011). Suomen nykyilmastoa kuvaavat lämpötilan, sateen ja lumensyvyyden keskiarvot. Available from http://ilmatieteenlaitos.fi/pitkan-ajan-tilastot

Ingvarsson, A. (1992). Bevattning. Ekologisk trädgårdsodling. Från teori till praktik. Jordbruksverket (SJV). Available from http://www.vaxteko.nu/html/sll/sjv/utan_serietitel_sjv/UST92-3/UST92-3N.HTM

Jamieson, P.D.; Martin, R.J.; Francis, G.S. & Wilson, D.R. (1995). Drought effects on biomass production and radiation-use efficiency in barley. *Field Crops Research*, 43, 77-86, ISSN 0378-4290

Jensen, C.R.; Mogensen, V.O.; Mortensen, G.; Fieldsend, J.K.; Milford, G.F.J.; Andersen, M.N. & Thage, J.H. (1996). Seed glucosinolate, oil, and protein contents of field-grown rape (*Brassica napus* L.) affected by soil drying and evaporative demand. *Field Crops Research*, 47, 93-105, ISSN 0378-4290

Ji, X.; Shiran, B.; Wan, J.; Lewis, D.C.; Jenkins, C.L.D.; Condon, A.G.; Richards, R.A. & Dolferus, R. (2010). Importance of pre-anthesis anther sink strength for maintenance of grain number during reproductive stage water stress in wheat. *Plant, Cell and Environment* 33, 926-942, ISSN 0140-7791

Ji, X.; Dong, B.; Shiran, B.; Talbot, M.J.; Edlington, J.E.; Hughes, T.; White, R.G.; Gubler, F. & Dolferus, R. (2011). Control of ABA catabolism and ABA homeostasis is important for reproductive stage stress tolerance in cereals. *Plant Physiology*, 156, 647-662, ISSN 0032-0889

Jordbruksverket. (2011). Available from http://www.sjv.se

Järvinen, J. (2007). Haihdunta. In: *Hydrologinen vuosikirja 2001-2005, Suomen ympäristö*, No. 44, J. Korhonen (Ed.), pp. 169-175. ISBN 978-952-11-2930-8

Khan, H.R.; Paull, J.G.; Siddique, K.H.M. & Stoddard, F.L. (2010). Faba bean breeding for drought-affected environments: A physiological and agronomic perspective. *Field Crops Research*, 115, 279-286, ISSN 0378-4290

King, B.A. & Stark, J.C. (1997). Potato Irrigation Management. University of Idaho, Cooperative Extension System, Bulletin 789, Available from http://www.cols.uidaho.edu/edcomm/pdf/BUL/BUL0789.pdf

Kivisaari, S. & Elonen, P. (1974). Irrigation as a method preventing detrimental late tillering of barley. *Journal of the Scientific Agricultural Society in Finland* 46, 194-207, ISSN 0024-8835

Kuusisto, E. (2010). Väärinymmärretty lumi, 14.12.2010, Available from http://www.ymparisto.fi/default.asp?contentid=68383

Lalonde S.; Beebe, D.U. & Saini, H.S. (1997). Early signs of disruption of wheat anther development associated with the induction of male sterility by meiotic-stage water deficit. *Sexual Plant Reproduction*, 10, 40-48, ISSN 0934-0882

Larsson, S. & Górny, A.G. (1988). Grain yield and drought resistance indices in oat cultivars in field rain shelter and laboratory experiments. *Journal of Agronomy and Crop Science*, 161, 227-286, ISSN 0931-2250

Link, W.; Balko, C. & Stoddard, F.L. (2010). Winter hardiness in faba bean: Physiology and breeding. *Field Crops Research*, 115, 287-286, ISSN 0378-4290

Madramootoo, C.A.; Helwig, T.G. & Dodds, G.T. (2001). Managing water tables to improve drainage water quality in Quebec, Canada. *Transactions of the ASAE*, 44, 1511-1519, ISSN 0001-2351

Marcellos, H. & Single, W.V. (1984). Frost injury in wheat ears after emergence. *Australian Journal of Plant Physiology*, 11, 7-15, ISSN 1445-4408

MacKerron, D.K.L. & Jefferies, R.A. (1986). The influence of early soil moisture stress on tuber numbers in potato. *Potato Research*, 29, 349-359, ISSN 0014-3065

McKenzie, J.S.; Faris, D.G. & De Pauw, R.M. (1982). Influence of simulated spring frost on growth and yield of three barley cultivars. *Canadian Journal of Plant Science*, 62, 81-88, ISSN 0008-4220

Mela, T.N.J. (1996). Northern agriculture: constraints and responses to global climate change. *Agricultural and Food Science in Finland*, 5, 229-234, ISSN 1459-6067

Melvin, S.W. & Kanvar, R.S. (1995). Environmental and economic impacts of a recycling subirrigation-drainage system. , In: *Subirrigation and drainage*, H.W. Belcher & F.M. D'Itri, (Eds.), 183-191, CRC Press, ISBN 1-56670-139-2, Boca Raton, Florida, USA

Mogensen, V.O.; Jensen, C.R.; Mortensen, G.; Thage, J.H.; Koribidis, J. & Ahmed, A. (1996). Spectral reflectance index as an indicator of drought of field grown oilseed rape (*Brassica napus* L.). *European Journal of Agronomy*, 5, 125-135, ISSN 1161-0301

Monti, A; Barbanti, L. & Venturi, G. (2007). Photosynthesis on individual leaves of sugar beet (*Beta vulgaris*) during the ontogeny at variable water regimes. *Annals of Applied Biology*, 151, 155-165, ISSN 0003-4746

Muellera, L.; Behrendtb A.; Schalitzb, G. & Schindler, U. (2005). Above ground biomass and water use efficiency of crops at shallow water tables in a temperate climate. *Agricultural Water Management*, 75, 117-136, ISSN 0378-3774

Mukula, J. & Rantanen, O. (1987). Climatic risks to the yield and quality of field crops in Finland. I. Basic facts about Finnish field crops production. *Annales Agriculturae Fenniae*, 26, 1-18, ISSN 1459-6067

Mäkelä, P. & Muurinen, S. (2011). Uniculm and conventional tillering barley accessions under northern growing conditions. *Journal of Agricultural Science*, in press, ISSN 0021-8596

Mäkelä, P.; Muurinen, S. & Peltonen-Sainio, P. (2008). Spring cereals: from dynamic ideotypes to cultivars in northern latitudes. *Agricultural and Food Science*, 17, 289-306, ISSN 1459-6067

Olesen, J.E. & Bindi, M. (2002). Consequences of climate change for European agricultural productivity, land use and policy. *European Journal of Agronomy*, 16, 239-262, ISSN 1161-0301

Olesen, J.E.; Jørgensen, L.N. & Mortensen, J.V. (2000). Irrigation strategy, nitrogen application and fungicide control in winter wheat on a sandy soil. II. Radiation interception and conversion. *Journal of Agricultural Science*, 134, 13-23, ISSN 0021-8596

Olesen, J.E.; Trnka, M.; Kersebaum, K.C.; Skjelvåg, K.; Seguin, B.; Peltonen-Sainio, P.; Rossi, F.; Kozyra, P. & Micale, F. (2011). Impacts and adaptation of European crop production systems to climate change. *European Journal of Agronomy*, 34, 96-112, ISSN 1161-0301

Ober, E.S.; Le Bloa, M.; Clark, C.J.A.; Royal, A.; Jaggard, K.W. & Pidgeon, J.D. (2005). Evaluation of physiological traits as indirect selection criteria for drought tolerance in sugar beet. *Field Crops Research*, 91, 231-249, ISSN 0378-4290

Pahkala, K.; Laakso, I. & Hovinen, S. (1991). The effect of frost treatment on turnip rape seedlings, ripening seeds and their fatty acid composition, *Proceedings of GCIRC Eight International Rapeseed Congress*, D.I. McGregor, (Ed.), pp. 1749-1753, Saskatoon, Saskatchewan, Canada, July 9 -11, 1991

Pajula, H. & Triipponen, J-P. (2003). Selvitys Suomen kastelutilanteesta, esimerkkialueena Varsinais-Suomi. *Suomen ympäristö*, No. 629. ISBN 952-11-1416-9

Passioura, J.B. & Angus, J.F. (2010). Improving Productivity of Crops in Water-Limited Environments, In: *Advances in Agronomy*, D.L. Sparks, (Ed.), Vol. 106, pp. 37-75, Academic Press, ISBN 978-0-12-381035-9, Burlington, MA, USA

Peltonen-Sainio, P.; Jauhiainen, L. & Hakala, K. (2011). Crop responses to temperature and precipitation according to long-term multi-location trails at high-latitude conditions. *Journal of Agricultural Science* 149, 49-62, ISSN 0021-8596

Peltonen-Sainio, P. & Mäkelä, P. (1995). Comparison of physiological methods to assess drought tolerance in oats. *Acta Agriculturae Scandinavica, Section B – Plant Soil Science*, 45, 32-38, ISSN 0906-4710

Pereira, A.B. & Shock, C.C. (2006). Development of irrigation best management practices for potato from a research perspective in the United States. *Sakia.org e-publish*, 1, 1-20

Perlman, H. (2011). The Water Cycle: Sublimation. Available from http://ga.water.usgs.gov/edu/watercyclesublimation.html)

Persson, R. (1993). Bevattningstidpunkter i sockerbetor. Available from http://www.vaxteko.nu/html/sll/slu/meddelande_sjfd/MSJ40/MSJ40W.HTM

Phene, C.J. & Sanders, D.C. (1976). High-frequency trickle irrigation and row spacing effects on yield and quality of potatoes. *Agronomy Journal*, 68, 602-607, ISSN 0002-1962

Playán, E. & Mateos, L. (2006). Modernization and optimization of irrigation systems to increase water productivity. *Agricultural Water Management*, 80, 100-116, ISSN 0378-3774

Puustinen, M.; Merilä, E.; Palko, J.& Seuna, P. (1994). Kuivatustila, viljelykäytäntö ja vesistökuormitukseen vaikuttavat ominaisuudet Suomen pelloilla. *Vesi- ja ympäristöhallinnon julkaisuja - sarja A*. ISBN 951-47-9883-X, ISSN 0786-9592.

Rajala, A.; Hakala, K.; Mäkelä, P.; Muurinen, S. & Peltonen-Sainio, P. (2009). Spring wheat response to timing of water deficit through sink and grain filling capacity. *Field Crops Research*, 114, 263-271, ISSN 0378-4290

Rajala, A.; Hakala, K.; Mäkelä, P. & Peltonen-Sainio, P. (2011). Drought effect on grain number and grain weight at spike and spikelet level in six-row spring barley. *Journal of Agronomy and Crop Science*, 197, 103-112, ISSN 0931-2250

Richards, R.A. & Thurling, N. (1978). Variation between and within species of rapeseed (Brassica campestris and B. napus) in response to drought stress. II. growth and development under natural drought stresses. *Australian Journal of Agricultural Research*, 29, 479-490, ISSN 0004-9409

Sahnoune, M.; Adda, A.; Soualem, S.; Harch, M.K. & Merah, O. (2004). Early water-deficit effects on seminal roots morphology in barley. *Comptes Rendus Biologies*, 327, 389-398, ISSN 1631-0691

Saini, H.S. & Aspinall, D. (1981). Effect of water deficit on sporogenesis in wheat (*Triticum aestivum* L.). *Annals of Botany*, 48, 623-633, ISSN 0305-7364

Saini, H.S. & Westgate, M.E. (2000). Reproductive Development in Grain Crops During Drought, In: *Advances in Agronomy*, D.L. Sparks, (Ed.), Vol. 68, pp. 59-96, Academic Press, ISBN 978-0-12-00768-4, Burlington, MA, USA

Sammis, T.W. (1980). Comparison of sprinkler, trickle, subsurface and furrow irrigation methods for row crops. *Agronomy Journal*, 72, 701-704, ISSN 0002-1962

Scherer, T.F.; Franzen, D.; Lorenzen, J.; Lamey, A.; Aakre, D. & Preston, D.A. (1999). Growing Irrigated Potatoes. AE-1040 Available from http://www.ag.ndsu.edu/pubs/plantsci/rowcrops/ae1040w.htm

Serraj, R.; Sinclair, T.R. & Purcell, L.C. (1999). Symbiotic N2 fixation response to drought. *Journal of Experimental Botany* 50, 143-155, ISSN 0022-0957

Shock, C. (2006). Drip Irrigation: An Introduction. Sustainable Agriculture Techniques EM 8782-E. Available from http://www.extension.oregonstate.edu/catalog/pdf/em/em8782-e.pdf)

Skinner, R.H.; Gustine, D.L. & Sanderson, M.A. (2004). Growth, water relations, and nutritive value of pasture species mixtures under moisture stress. *Crop Science*, 44, 1361-1369, ISSN 0011-183X

Solantie, R. (1980). Kesän yölämpötilojen ja hallojen alueellisuudesta Suomessa. *Maataloushallituksen Aikakauskirja*, 4, 18-24, ISSN 0355-0486

STATCAN. (2009). Statistics Canada. Available from http://www.statcan.gc.ca/pub/16-001-m/2009008/userdef-defusager1-eng.htm

Stoddard, F.L.; Balko, C.; Erskine, W.; Khan, H.R.; Link, W. & Sarker, A. (2006). Screening techniques and sources of resistance to abiotic stresses in cool-season food legumes. *Euphytica* 147, 167-186, ISSN 0014-2336

Svensson, S.-E. (2003). Bevattning i grönsaksodling. Ekologisk odling av grönsaker på friland. Jordbruksverket 2003. Available from http://www2.sjv.se/webdav/files/SJV/trycksaker/Pdf_ovrigt/p7_12.pdf

Tesfamariam, E.H; Annandale, J.G & Steyn, J.M. (2010). Water stress effects on winter canola growth and yield. *Agronomy Journal*, 102, 658-666, ISSN 0002-1962

Tesfaye, K.; Walker, S. & Tsubo, M. (2006). Radiation interception and radiation use efficiency of three grain legumes under water deficit conditions in a semi-arid environment. *European Journal of Agronomy*, 25, 60-70, ISSN 1161-0301

TIKE. (2010). Maataloustilastot. Available from http://www.maataloustilastot.fi

Vamerali, T.; Guarise, M.; Ganis, A. & Mosca, G. (2009). Effects of water and nitrogen management on fibrous root distribution and turnover in sugar beet. *European Journal of Agronomy*, 31, 69-76, ISSN 1161-0301

Wallen, C. C. (1966). Global solar radiation and potential evapotranspiration in Sweden. *Tellus*, 18, 786-800, ISSN 2153-3490

Wiklund, A. & Ekelöf, J. (2006). Droppbevattningens inverkan på kvalitet och kvantitet hos *Solanum tuberosum* L, i jämförelse med konventionell spridarbevattning. Bachelor (Kandidatuppsats) inom Hortonomprogrammet DSH. SLU Alnarp, Institution för landskaps- och trädgårdsteknik. *Rapport*, 3, 1-33, ISSN 1652-1552

Wikman, U.; Torttila, A.; Virtanen, A. & Kuisma, P. (1996) Perunan Vesitalous ja Sadetus. *Perunantutkimuslaitoksen julkaisu* 3, 1-31, ISSN 0787-7323

11

Land Flooding Irrigation Treatment System for Water Purification in Taiwan

Yu-Kang Yuan

National Yunlin University of Science and Technology, Taiwan, R.O.C.

1. Introduction

The primarily focus of the agricultural development is to create an appropriate and suitable environment for growing plants. To do so, supplying plants sufficient growing factors would be necessary, including water, nutrition, sunlight, adequate temperature and air. For water, a construction of irrigation system is one of the vital methods of supplying sufficient water.

Trace back to Taiwan's history, irrigating system began really early. However, due to the features of the natural geology and people migration, the accumulation and application of water resource is not average. From the era of Japanese occupation (1895-1945), Irrigation Waterway Popularization was being promoted (Taiwan Governor-General Office, 1901), which started by integrating private irrigation waterways to public ones. Afterwards, here comes the establishment of farmland cultivating and water conservation system, first began in the north Taiwan and then to the middle, south, finally from west to east.

In 1949, while the R.O.C. government retreated to Taiwan, the managing organization still remained as it was during Japanese occupation, that is, the Irrigation Association as to keep the public legal entity. Therefore, the agricultural irrigation system could keep operating appropriately.

However, with Taiwan's miraculous economy took-off, needs from various kinds of livelihood, industry, and infrastructures gradually occupied the farmland. Furthermore, because of inferior drainage facilities that utilized original irrigation waterways as sewage drainages, polluted crops incidents emerged in endlessly.

Due to the 21st century global changes issue, Taiwan's water resource now faces a more devastating situation caused by climate abnormality. According to the data resourced from Taiwan Water Resource Bureau (2011); currently, there are 69 dams among all now have severely siltation problem that leads to the deficiency of water retention capacity. Under the circumstance, reducing water expenditure becomes one of the solutions to the future water usage; moreover, sewage recycling and reuse also can be other conditions which we should take into consideration.

Facing the dilemma, the application of Land Treatment System (LTS) plays an important role inside; especially the offspring from the trend that can cope with it, flood irrigation. It allows the sewage which is about to flow into the irrigation waterway to be purified by the interaction of earth, plants, and microorganism. For countries or regions with limited water resource, the design, construction, and maintenance of LTS are quit cost-efficient and its degree of technique is low. Therefore, in Taiwan, it is a possible method to deal with the sewage recycling issue for future climate changes. Later on, this chapter would take the last sections of LTS as examples and also; from a long-term observation, it would analyze and explain its improvements to the water quality.

2. History of irrigation system in Taiwan

2.1 Features of topography and environment in Taiwan

Two-thirds of the total area in Taiwan is covered by forested mountains and the remaining area consists of hilly country, platforms and highlands, coastal plains and basins. The longest main river is 186 km, and the shortest one is merely 41 km; also, most of the river basins are among precipitous mountains and most of this rainfall rushes directly to the sea, is hard to keep. (Picture 1)

Picture 1. Satellite image of Taiwan

In terms of rainfall, since Taiwan is located in the sub-tropical area, at typhoon path, most rainfall concentrates in this abundant rainy season. (April to October, about 78%) (Picture 2) (Northern Region Water Resources Office, Water Resources Agency, Ministry of Economic Affairs, 2009). Therefore, the amount of runoff is rich with sufficient water resource; on the contrast, during the dry season from October to next May, the amount of runoff occupies only one tenth of the abundant rainy season resulted by the dificiency of rainfall. As the consequence, the problems of extremely and seasonal lack of water resource have been quite a challenge to Taiwan.

In Taiwan, the annual average rainfall is 2,510 millimeters, the volume of it is 90.5 billion cubic meters. Most of rainfall comes from typhoon season, which is 2.6 times the globally averaged annual precipitation. However, bacause of the spatial and rainfall period limitation, only 15% of the amount of rainfall is available to use; moreover, since the population density in Taiwan is quite high, the per capita rainfall is 4,250 cubic meters (in the context of 21 million population), which is one sixed of globally average. As a result, Taiwan ranked as the 18th water shortage country worldwide. (Li, 2003) Therefore, in the constraint of the topography, rainfall issue, and under the circumstance as an isolated sea island, water resource is precious to Taiwan.

Precipitation

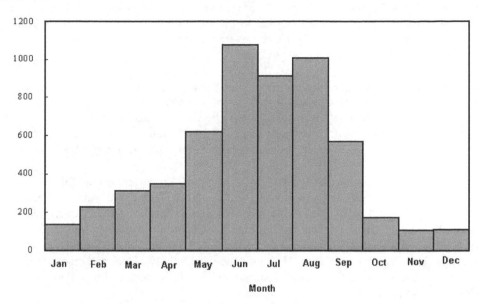

Picture 2. Taiwan's Monthly average rainfall bar chart, 2009

In addition, according to the research done by The Asian Development Bank, in Asia, the yearly per capita rainfall is around 26,007 tons, but the volume in Taiwan is only 4,250 tons, which is far behind Asia's average value. In terms of the volume of water, the yearly per capita rainfall in Taiwan is merely higher than Pakistan (1,800 tons), China (2,000 tons), India (2,200 tons), Sri Lanka (2,400 tons), Afhanistan (2,500 tons), Thailand (3,400 tons), and other six countries. It can be seen from that, though there is plenty of rainfall in Taiwan, resulting by the factors of geology and population, Taiwan still is a "water in shortage" country.

According to the statistic from Water Resource Agency, Minister of Economy, the annual amount of water usage is about 18 billion tons, occupies 20% of the amount of annual rainfall, the rest of it returns to the nature either by evaporation or flow into the sea. As for the 18 billion water supply, only 15.7 billion of them fits into safe water supply capacity. Among them, 11.7 billion tons is from surface water, 4 billion tons from groundwater, and the rest insufficient part would be made up by over pumping groundwater; which means at least 2.3 billion tons of groundwater ever year. As a result, it leads to the problems such like groundwater drawdown, land subsidence, salt water intrusion, and water deterioration.

Besides, the amount of annual rainfall is not the only difference, but also regional divergence (north, middle, south, and east) in terms of the runoff during dry and wet seasons. As for the rigional ratial comtrast, it's 1:3.5 at north, 1:7 at middle, 1:10 at south, and 1:4 at east.The tremendous difference of runoff ratial during dry and wet seasons is one of the main causes which lead to the seasonal water shortage. Among them, the contrast and water shortage possibility during dry season is much higher in the south of Taiwan; furthermore, affected by regional water resource issue and the deficiency of pipeline construction, water distribution remains a hard task and can easily influence the industry producing capability and livelihood water supply. (Water Resource Agency, Ministry of Economy, 2001). To this day, if there is any other severe dry season coming or globally climate changes that cause the abnormal to the rainfall, water shortage can happen even during wet season that bring the sufficient water resource, let along dry season with limited water supply, it will only getting worse.

2.2 Water conservation constructions in Japanese occupation

During Japanese occupation, the Taiwan Governor-General Office published "The Rules of Public Irrigation Waterway", which integrated all private owned ditches or waterways to a public irrigation system; meanwhile, various kinds of water conservation planning and facilities were being promoted to the expectation of the land settlement policy, as to make Taiwan an agriculture base.

The hydraulic engineering carried out by Taiwan Governor-General Office included irrigation project, land improvement project, drainage engieering, old waterways renovaing project and brought in the new water conservation techniques; among them, Tao-Yuan Canal and Chiayi-Tainan Canal are the biggest irrigation waterway projects. Because of these water conservation facilities and engineering projects, they efficiently increase the available irrigation ground area and rise up the productivity of rice and sugar. Since then, Taiwan's agricultural development foundation had been setted up, the influence was profound even now.

In Febuary, 1908, Taiwan Governor-General Office issued new law and began to operate the construction of official irrigation waterway. Afterwards, in 1910, water conservation combination rules of public irrigation waterways was published as to organize the management organization. During this period, because the water conservation environment in Tao-Yuan tableland accorded with official irrigation water subsidy conditions, this region was listed as one of the official irrigation system planning area. In 1913, a severe drought damage occurred in Taiwan and alerted Taiwan Governor-General Office; thus, the engineering in Tao-Yuan irrigation canal was accelerated. In terms of the canal project, it contains two part, one is the official irrigation waterway, which started construction in 1922 and finished in 1924 and began irrigating that year; another is the combination project of water conservation, which consisted of constructing, renovating reservoirs and separated channels, starting from 1916 and finished in 1928.

Because of the permission of building Tao-Yuan Canal, it simulated the engineering project of Chiayi-Tainan Canal, which started constructing in September, 1920 and finished in May, 1930. The predicting irrigating area was 145,500 hectares, but after operating for one year, the actual irrigating area was about 131,920 hectares. Nevertheless, contrasting to Tao-Yuan Canal, with only 22,310 hectares irrigating area, the benefit brought by Chiayi-Tainan Canal was hard to count.

Resulted by irrigating system, dry lands gradually tranformed to paddy fields. In 1928, after full operating of Tao-Yuan Canal, area of paddy lands rocketed up obviously; especially for coastal regions, some even increased over 40% (picture 3). By using the the altitude gap and interceptive return flow from rivers, it enabled low-lying coastal areas with underdeveloped agriculture resulted by factors such as barren grounds, sea wind, salt spray and irrigation shortage areas to transform from low productivity dry lands to high productivity paddy lands.

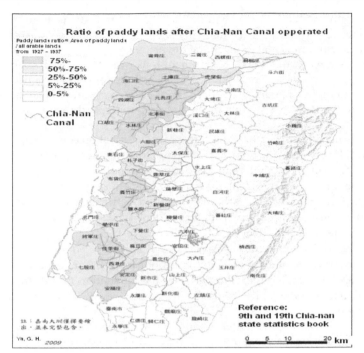

Picture 3. Paddy lands ratio after Chiayi-Tainan Canal operated. (Reference: 9th and 19th Chia-nan state statistics book)

Although Chiayi-Tainan Canal was the biggest water conservation construction during Japanese occupation, the irrigating region was still far bigger than it could supply. Therefore, the water supply couldn't satisfy all farm reservoirs. Under the circumstance, the organization supplyed water by taking turns; meanwhile, farmers were forced to practice rotation of planting crops which are rice, sugarcane and others, and that is the so-called the "Three-year rotation". Though farmers couldn't choose which crop to grow by themselves, after the canal started supplying sufficient irrigation water, dry farmlands around Chiayi-Tainan Canal soon transformed into paddy lands; hence, crop productivity, land value rose up and farmers' lives also improved.

2.3 Development of water conservation after R.O.C. government retreated to Taiwan

During World War II, because of centrol supervision, goods and materials deficiency, the mainteinance of water conservation facilities was difficult. With series of disasters and bombing at that period of time, mutiple water conservation utilities were damaged or couldn't operate appropriately. For instance, the Tseng-wen Stream intake in Chiayi-Tainan Canal was the water abstraction utility that was piled up by water conservation wire barriers; however, later on it collapsed because of rusty iron. At that time, iron and wires are under supervision; therefore, they couldn't but renovate it with concrete. In 1941, a flood crashed it again; thus, renovation of water conservation facilities became one of the post-war tasks.

In addition to the renovation of water conservation facilities, the taking over and recovering of water conservation managing organization were also vital issues during post-war period.

Moreover, after WW II, Taiwan's political and ecnomical status transformaiton were extreme; as a result, the water conservation managing organizaiton had to keep adjusting its structure as to fit environmental changes and development needs.

At the initial stage when the R.O.C. government took over Taiwan, all water conservation combinational orgainizations were adopting systems in Japanese occupation. The only difference was to replace regions to districts. Since a huge amount of Japanese left their job, Taiwanese had the opportunity to fill in the vacant position and then recovered the water conservation's operation.

Since there was no similar organizations like water conservation combinations in Mainland China, the R.O.C. government followed Japanese water conservation system for about a year; afterwards, to in line with the spirit of democracy and autonomy, central officials then revised the rules of electing president, that is, replacing officially appointment with popular vote. On December 26th, 1946, Taiwan Chief Executive Office comanded all districts to elect the first water conservation combination president while the articles of orgainization remained the same.

From 1953, the R.O.C. government successively implemented a five-terms four-year economical construction, aiming to "cultivate industry with agriculture, and develop agriculture with industry." Within this mission, the foundation of industry development was founded but, on the other hand, contrainted agriculture, transferring plenty of resources to industry.

In November, 1956, in order to advocate the "land to the tiller act" policy, the R.O.C. government issued "The Articles of Irrigation Association of Taiwan Province" which consolidated the original 40 water conservation boards to 26 farm irrigation associations. However, from 1960, the value of industry product first surpassed agriculture; therefore, more and more villege labors moved to either cities or industries. As a consequence, labor shortage gradually became common situaion; thus, wages and commodities then got higher and higher. While industry development expanded generously, population and arable lands; however, became smaller and resulted in impoverished finacial problem of farm irrigation association. In order to keep operating, quarrels and disputes occurred between water conservation associations. In 1982, the volatile political situation in Taiwan became stablized and the water conservation orgainizations operated regularly, the R.O.C. government then revised the "Organic Statute of Farm Irrigation Association", which enabled all water conservation associations to operate autonomously and gave all assocaitons public legal entities. The R.O.C. government would select 2 to 3 people as appointment of candidates of the organization president , then members could elect one of them. In June, 2001, "Organic Statute of Farm Irrigation Association" was revised again, which enabled members to elect the boards and the president of the association.

All in all, before 1970, the agricultural irrigation system constructions mainly focused on renovating damaged water conservation facilities; afterwards, in order to increase the crop productivity, "water conservation engineering" then concentrated on improving the original utilities' efficiency. Moreover, the exploitation of water resource then transformed to power generation and supply potable water or other multiple purposes from agricultural irrigation.

Resulted by the shift of national policy, areas of farm lands became much smaller in order to meet the needs of industry and livelihood development. In January, 2000, after the third-reading of "Agricultural Development Act" in the legislative Yuan, the R.O.C. government lifted the embargo on farm lands, arable lands free on trade, relaxing the restrictions of farm segmentation, no limitations of farm segmentation inheritage, and allowed farmers with farmhouse non-ownership status to build farmhouses, officially broke the "farmlands will be used as agricultural purposes" policy. As a result, it withered Taiwan's agriculture; what's more, environmental impacts such as sewage pervading problems are currently grand challenges to Taiwan.

3. Coping strategy to water shortage

People have right to use water resource, and government has to guarantee this right. Due to causes like climate, geological environment and other factors, though Taiwan remains a sea island with the amount of rainfall that is 2.6 times global average volume, the per capita rainfall is only one fifth of global average volume. Therefore, Taiwan is a rainly country but remains a water shortage region; in that case, in terms of coping this problem, there is no shirking the responsibility to the government.

Traditionally, whenever there is a water problem, after traceing it to its causes, there are always two possible reasons, water shortage and deterioration. However, ever since the climate abnormality intensified, there is a significant difference of water resource problem comparing to the past. Which can be indicated from a simple localized problem, slowly expanding to a regional one, even a whole area's evironmental issues; worst of all, becomes a significant restriction to economical development.

In Taiwan, because of government's negligence and inappropriate policy, we will need a new concept while dealing with water resource problems, especially for climate abnormality changes in 21st century.

3.1 Impacts of global warming

During the past one hundred years, the annual mean temperature rose more than one celcius (picture 4), global warming effect is obvious. The main reasons are the global warming effect and regional human exploitation. (Chang, 2011) Generally speaking, in the half of 20th century (before global warming), the temperature rose slowly and the changes of rainfall was rather stable with timely wind and rain. However, in 1950, after the annual mean temperature first exceeded the average value for a century period of time, water conservation and precipitation evolved to a completely different type of problem, included increased level of scale, dry season period became longer, and the rainfall intensity rose up significantly after 1980. Basically, they all had something to do with global warming in some measure. It shows the operation of nature was transforming swiftly, extremely and interactively, which could be as a reaction to the increasing temperature; meanwhile, it affected Taiwan's water resource problems profoundly.

After global warming tendency effect in evidence since 1950, there were some features of the rainfall: (1) Precipitation tended to bipolar distributed; divergence of south and north region gradually expanded; (2) the whold island's rainfall periods decreased; (3) all districts' intensity of rainfall strengthened; worst of all, due to global warming, it decreased the temperature difference which also leaded low humidity to atmosphere, causing frosting and fogging time reduced; finally, shortering rainfall days annually.

On the contrary, the average intensity of rainfall had an opposite effect comparing to the change of rainfall days. Since 1950, the long-term intensity of rainfall indicated an increasing value. (Wang, 2004; Lin, 2007). Especially in wet seasons, as a result of the increasing of precipitation in the north of Taiwan, the increasing ratio of intensity of rainfall was twice times east. In southern west, the record was broke almost every year and became a vital precaution movement in Taiwan.

Average annual temperature

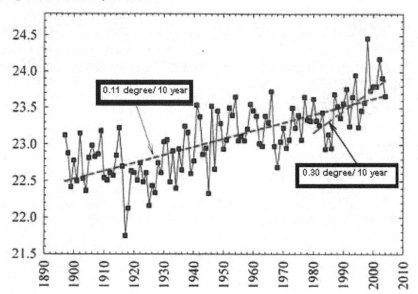

Picture 4. Temperature Increasing in the Past Century (Wang, 2005)

3.2 A predicament of lacking enough sewage pipes

In 2000, total livelihood water consumption was 3,633 megatons (MT), industry water consumption was 1,870 megatons, and agricultural water consumption was 12,318 megatons; which occupied total water consumption of 20.4%, 10.5%, and 69.1%, repectively. In Taiwan, the dominant source of sewage came from household sewages (about 40%-50%). Therefore, in order to improve the household environmental sanitary, constructing public sewage pipes system has become an urgent environmental protection infrastructure. (Environmental Quality Protection Fundation, 2000)

In Taiwan, the popularization rates of the sewage pipe intallation were 0.45% – 46.30% in 2000 and the country's average value was 6.92%, which was quite a divergence between rural and urban areas. (See chart 1) (Construction and Planning Agency Ministry of the Interior, 2000) Untill in April, 2011, the country's popularization rate had risen up to 26.92 %, which was still lower than the presetted 35.8% (Environmental Protection Administration Executive Yuan, 1998). In order to figure out on how to solve the serious sewage pollution, it's the government, academic fields and civil orgainization's joint responsibility.

Area	Taipei	Kaosiung	Taiwan
Pupularization rate（%）	46.30	11.25	0.45

Table 1. Popularization Rate of Sewage Pipes, April, 2000

The tendency of water environmental deterioration makes us concern, especially water pollution.The aggravation of water pollution resulted in multiple contradictions between regions, urban and rural areas such as, how to transfer and release the original agricultural irrigation water in order to put down the disputes among agriculture, residents, and industry.

Also, the severe water pollution demolished the ecological environment profoundly. The emerging of Lifestyles of Health and Sustainability (LOHAS) made the middle class reconsider their health and living environment; for example, purchasing farmlands and builds farmhouses, liberating themselves from busy city lives, but resultes in a more serious pullution. Besides, the farmland segmentation makes ensuing water conservation planning hard to practice.

During the past, building constructions didn't include separating the allocation of rainfall and sewage. Moreover, illegal buildings also unabled the processing of constructing sewage pipes as to centralize the sewages. Illegal farmhouses on farmlands often directely drains the wated water into irrigation waterways. (see picture 5) Thus, solving illegal buildings would be the first stop in order to construct sewage pipelines. Moreover, it's difficult to install the sewage pipelines for legal users even the distribution of pipelines were already setteled because of factors like to conduct construction across private lands.

Picture 5. Community sewage directely drains to irrigation waterways

Eventually, the sewage would flow into dams, streams, rivers, and to the sea. There are 21 principle rivers, 29 secondary rivers and 79 ordinary rivers in Taiwan; in 2001, the total lenth of all rivers is 2,934 kilometers, among all, unpolluted lenth of rivers is 1808.88 kilometers, light polluted rivers for 287.62 kilometers, meduim polluted rivers for 451.3 kilometers, and heavy polluted rivers for 386.2 kilometers; which occupies 61.65%, 9.08%, 15.38% and 13.16%, respectively. During the past decades, the statistic has shown that fewer and fewer unpolluted rivers exised, which means water pollution is getting worse. As a result, the amount of available water keep diminishing.

3.3 Approaches to broaden water resources of income and reduce expenses

We can never put too much emphasis on the importance of water resource since it plays an important role to civilizations on earth. Though water covers 70.9% of the Earth's surface, oceans hold 97% of surface water, plain water holds 3%; in terms of plain water, glaciers and polar ice caps 75%, rivers, dirt, atmosphere, lakes and biological organisms occupy 3%; the rest available water for humankinds is less then 1%. The human civilizations and developments heavily rely on water resource.

Since a long time ago, living near the river makes people easy to utilize water; thus, most human civilizations originated from banks of significant rivers. With the development of civilization, humankinds grow plants, fertilize, clean objects, accumulate garbages and so on, the wastedwater and sewage that penetrates from the wastes often flows into rivers or streams. Furthermore, fertilizing and cleaner chemicals such as phosphor, nitrogen, bisphenol A (BPA), nonylphenol (NP), plasticizer and other heavy metals all causes sever water pollution, diminishing available water resources.

Due to the high density of phosphor and nitrogen under water which can intrique eutrophication easily, plenty of algen grow and cover the water surface, even sunshines. As a consequence, it results in the death of fish, shrimp and other underwater plants. Besides, the decomposition of either animals or plants needs oxygen in the water and form into the anoxic vicious cycle. "Eutrophication" indicates the excessiveness of nutrition materials in the water, which leads to a situation of algae bloom. In terms of its nutrients, there are roughly two main sources: (1) Natural eutrophication: It refers to the sedimentaion and accumulation of nutrients in a lake of natural ecosystem, which causes eutrophication step by step. In other words, it means the lake is getting old and it a normal but must successsion in natrual ecosystem. (2) Artificial eutrophication: artificial eutrophication occurs when human activity introduces increased amounts of these nutrients, which speed up plant growth and eventually choke the lake of all of its animal life. For instance, growing and fertilizing plants and lake recreation activities, pouring amounts of nutrients into ponds, rivers, lakes or dams, which often lead to the propagation of plankton underwater. Finally, the ecosystem underwater changes swiftly and water quality will become deteriorated. Usually the phenomenon can be form up in just a few years. (Wen, 1995)

In Taiwan, the water resources supply mainly come from dams, rivers, and groundwater; to this day, there are 109 dams and dikes. In terms of the functions, it included power generating, irrigation, water supply, travelling and flood protection. Due to natrual causes and over-manmade exploitation, dam siltation has been a serious problem. In the past, there was a misconception, that is to consider Taiwan as a island with rush rivers, centralized rainy seasons, water shortage problem in winter, and the only way to solve it was to construct dams. However, the truth is, because of steep mountains, a huge amounts of dirt, rubbles, and R.O.C.ks would be flooded into dams and caused siltation after rainfall. Take the Kaopinghsi weir which is located in Kaoping River as an example, whenever there is a heavy rain, the weir would act more like a silt arrester rather than a weir. Also, resulted by the high turbidity of dirt and R.O.C.ks that affects the height of water, intaking water gradually becomes a challenge.

Picture 6. Shihman Reservoir. The picture was took after two weeks of Matsa typhoon, the reservoir water looks pure but the water supply reservoir was muddy. It was because of the dirt which came from the upper reaches of the river silted into the reservoir, and after generating power at the power plant, it became muddy and then flew into the lower reaches of the river. So far, water towers had been constructed on the reservoir in order to intake the pure water on the top. (Li, On 20th Auguest, 2005)

For solutions, first of all, is to broad water resources; constructing sustainable water collection gallery on principle riverbeds. (see picture 7)

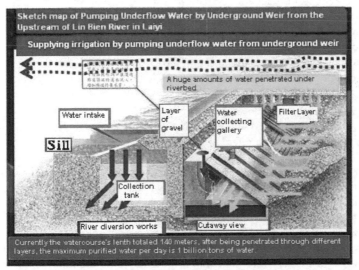

Picture 7. Pingtung: Sketch map of Pumping Underflow Water by Underground Weir from Upstream of Lin Bien River in Laiyi (Lin, 2011)

This type of water collection gallery originated from the underflow water that was about 2 meters deep under the riverbed; in other words, while the surface water flew (surface runoff), it would also penetrate down through the layer of gravel to the bottom of riverbed (underground runoff) and formed into underflow water. In July, 1996, typhoon Herb brought a huge amounts of rainfall to Taiwan. According to the statistic, in just one day, there was 0.6 billion tons of water being flushed into the sea (almost equal to the total amounts of water usage in Kaosiung City yearly); meanwhile, the amounts of underwater flow was counted under 10 meters of the riverbed was also about 0.6 billion tons, this indicated the abundance of the underwater flow under riverbeds. In addition, by R.O.C.ks and rubbles' filtration, the water quality was quite superior. In other words, the gravel on riverbeds is the natural water supply and purification plant.

The second approach is to reduce water expenses. Generally speaking, after irrigation water flew through fields and it would become wasted water; finally flushed into irrigation waterways with household sewage. Therefore, the irrigation pollution possibility would go up significantly; after flowing into rivers, it often causes unavailable water in low reaches of rivers. Besides, residents in Taiwan often are lack of proper water usage habbits because of cheap water rates. So, even there is any excellent water conservation construction planning, it still can't satify to make up the load, as a matter of bad water usage habbits. To this day, the cost of broadening new water resources has grown higher and it takes more time to finish a project. What's more, the amounts of safe water supply are currently severely insufficient; thus, it's urgent to publish relevant water reducing policy.

According relevant researches (Li, 2003), here are some available coping strategies as followed: (1) rise up the ratio of recycling industrial wastewater to 65%; (2) promote the system of separating potable and available water in daily lives; (3) advocate the construction of dual water supply system and sewage pipelines in order to rise up the reusing rate of household water; (4) government should build up a precipitation recycle and reuse demonstration; also, issue relevant regulations for residential and industrial areas; (5) for users who tranferred agricultural water to residential water should make efforts to increase the waterways' water delivery efficiency; (6) advocate the installation of water-saving equipments during dry seasons as to decrease the amounts of water supply; (7) regularly examine water supply pipelines in order to drop water leakage rate; (8) enhance water usage management so that the water reducing efficiency will go up.

3.4 Value of wastewater recycle and reuse

According to Water Resources Agency, Ministry of Economy, the definition of "water reclamation", the coverage water refers to wastewater from industries, homes, rainfall storage, seawater and agricultural irrigation water, excluding "before-waterways-water" defined by Water-Conservancy Act. Therefore, the function of water reclamation is to recycle all water in which from rainfall, household, and enterprise wastewater in line with the drainage area regulation and then purify them until to meet all specific water quality standards, so as to achieve the goal. The Water Resources Agency classfied the resources of water to runoff, domestic sewage (meaning same as "livelihood wastewater"), industrial wastwater, agricultural wastwater, aquacultural wastwater, livestock wastwater reuse and so on.

Domestic sewage means all wastewater that originates from humankinds' daily lives, which includes cooking, cleaning, pouring, drinking and so on. According to the statistic shown by the Water Resources Agency; in 2003, the amounts of livelihood water supply reached 3.55

billion tons, which held 20.20% of total amount. The per capita water usage was about 200-250 liters. Since there are various kinds of water using purposes, the deterioration levels are different. In terms of the sewages, excepting wastewater that needs to be purified by septic tanks, others like wastewater from cooking, cleaning, bathing; all were mixed together and then was drained into the sewage pipelines. Speaking of the pollutants in sewage, they are mainly organic materials, suspended solids, nitrogen and phosphorus nutrients, colon bacillus and so on.

In the light of statistics from the Water Resources Agency; in 2003, the amounts of agricultural irrigation water was 12.43 billion tons, occupied 70.70% of tatal agricultural water usage. In Taiwan, the irrigation water primarily comes from Irrigation Association's waterway system. From the survey of the amounts of agricultural water usage shown by the Council of Agriculture, it reported that the irrigation water originated from rivers, pumping surface warter (85.2%), dams and reservoirs (9.1%), and groundwater (4.4%).

So far, the R.O.C. government only started the counseling of the recycling of agricultural irrigation water without practicing the recycle of irrigation water pragmatically. In addition, the amounts of agricultural water usage declined after Taiwan joined the World Trade Organization (WTO); thus, transferring part of the available water would rise the water resource operating efficiency. Furthermore, although overpumping groundwater is the dominant cause of subsidence, still, pumping groundwater within limits is one effective method to make up the lack of water resource.

For irrigation water reuse, livelihood irrigation assocations usually utilize the irrigating water return, espacially for dry type water usage of 1.40/ sec/ ha using mostly by water conservation associations where locates in the west of Taiwan. During the dry season, in order to solve the water shortage problem, the irrigation efficiency can be rose up by utilizing water ruturn as to supply other waterways. In terms of the amounts of water and its quality, measurements and statistical analysis were not practiced but only setted monitors on irrigation waterways in order to test the water conductivity. While reusing the water, in addition to concern the water resource, local terrains and irrigation facilities are also factors which could affect the benefits of reusing water return. Although some water recirculation area had the potetial of water return, the water resource would be deteriorated and unable to use because of sharing irrigation and drainage waterways, let along the pollution caused by community wastewater which also made irrigation water unavailable to use anymore. (Water Industry Information Network, 2011)

If we plus the amounts of agriculture irrigation water from rivers and groundwater which totalled 10.59 billion tons to 3.55 billion tons of domestic sewage, the tatal amounts is 14.14 billion tons. After deducting the number of diminished water effect such as penetration, evaporation and plants absorbing, in the end we'll roughly have 7.07 billion tons of drainage water. According to a report (Irrigation Association, 2003), the amounts of industry water usage was 1.61 billion tons, which held 9.1% of total water usage. If the agriculture water return could be utilized for reusing the water, the industry water needs still could be satisfied even during dry seasons.

4. The combination and application of irrigation system and sewage reclamation

Owing to Taiwan's unique terrains, climate, and history, considering current situation of water shortage issues, processing the planning and application of Land Treatment System will be a decent choice in terms of its low cost, low tech, enviornmental education, and easier maintenance. Comparing to industrial treatment (such as Wastewater Treatment Plant), the Land Treatment System is a more cost-efficient sewage treatment facility in prospect.

4.1 Operation and structure of land treatment system

The Land Treatment System can be classified into four types, which includes: (1) The Slow Rate (SR) method, (2) Rapid Infiltration (RI), (3) Overland Flow (OF) and (4) Wetland System; their operating effects and availabilities are as followed:

Classification.	Yearly load. (m/a).	Occupancy area. (hm2/ km3).	Gradient ;. soil.	Groundwater depth . (m).	Water directions.	Water quality variation.	Purposes.
The Slow Rate Method (SR).	0.3~1.53	24.2~119.	With crops, less than 20%, if none, less than 40% ; average soil.	0.61~2.44.	Evaporation, groundwater, few runoff.	BOD, SS and Nutrient are eliminated; increasing TDS	Produce as many plant products as possible.
Rapid Infiltration (RI).	3.35~153	0.21~10.7.	Either ; coarse sand, sandy loam.	3.05.	Partly by evaporating, groundwater, no runoff.	Diminishing BOD、SS ; TDS uninfluenced.	Reinjecting or filtrating water; fewer plant products
Overland Flow (OF).	1.53~7.6	4.85~24.2.	2%-8 % ; clay, silty soil.	Not regulated	Partly by evaporating or groundwater, surface runoff.	Diminishing BOD,SS and Nutrient ; increasing TDS.	Plant growing
Wetland System	18.25.	1.64.	1%-8 %; soil at upper layer with gravels at bottom	Not regulated	Partly by evaporating or groundwater, surface runoff.	Diminishing BOD,SS and Nutrient ; increasing TDS.	Plant growing

(Wang, 1990)

Table 2. Comparison of these four types of Land Treatment Techniques

The system operation includes physical sedimentation and filtration, chemical adsorption, ion exchange, oxidation-reduction reaction, decomposition of biological metabolism and so on. In addition, plant absorption also contributes to the system. Explanations are as followed:

Sedimentation: When sewage flows into the system, in response to flow retardation, suspended pollutants then settles down. As a matter of fact, this process is similar to sediment basins in wastewater treatment plant.

Filtration: Sewage filtration is process in the basement soil by blocking or retarding sewage particles; which is similar to sand filter bed in water treatment plant.

Adsorption: Particles at basement soil or sediment are capable of absorbing organisms or odor materials. Besides, other plants such as Pistia stratiotes L., Eichhornia crassipe, and Commelina communis L. are possessed of absorbing heavy metals.

Ion exchange: Particles in basement soil often possess various charged ions; thus, ion exchanges occur between basement soil particles with ionic contaminants (heavy metals and other salt chemicals) or nutrients (ammonium ion, nitrate, and phosphate), finally eliminate them from sewage.

Oxidation and reduction: Due to the difference of oxygen supply and oxygen consumption rate, aerobic and anaerobic environments are formed on system's surface layer, roots of water plants or wetland surfaces, resulting in the divergence of oxidation reduction potential. In an aerobic environment, high oxidation reduction potential stimulates chemical oxidative reaction to sewage pollutants; on the contrary, oxygen reduction occurs when in anaerobic situations. Oxidation reduction reaction changes pollutants' chemical property, making them either harmless materials or easier treatment processes.

Metabolic degradation: A huge amounts of microorganism exist in the system; thus, metabolic degradation takes place in terms of organisms and nutrients in sewage and then turns them into either harmless chemicals or simple metabolic waste. Speaking of microorganisms, bacteria are the dominant decomposers, followed by fungus. As for protozoan, they balance the number of bacteria as predators. Algae in microorganisms are primary producers in ecological environment; for instance, they conduct photosynthesis and then release oxygen, absorbing inorganic nutrients during metabolism, resulting in water purification.

Plants functions: Generally speaking, there are some aquatic plants and semi-aquatic plants existing in the system, like Water Hyacinth, Commelina Communis L., Pennisetum Purpureum, Cattail, Bulrush and so on; just like algae, plants conduct photosynthesis, in addition to release oxygen and absorb inorganic salt, their roots provide mediums for bacterial growth. (Yang, 1996)

Structurally, the Land Treatment System includes oxidation ponds, water storage ponds, irrigation system, and drainage system; for the types of construction, it is mainly conducted by ecological engineering. Concerning factors are as followed:

1. Stepwise aeration: Taking advantage of the altitude difference, stacking gravels and rocks to form a simple water barrier as to increase the area of water's in tough with the air, finally rise up dissolved oxygen value.

2. Oxidation pond: Also named stabilization ponds. It's an ancient sewage treatment technique; advantages are as followed: using the terrain thoroughly and carries out reclamation of sewage (low energy consumption). Furthermore, there are several creatures living inside, which includes bacteria, fungus, algae, protozoan, metazoan and aquatic plants. Due to green plants' photosynthesis (especially from algae), they supply sufficient oxygen for other organisms under water, and enable them to conduct sewage purifications, including dilution effect, sedimentation, flocculation, aerobic and anaerobic metabolic decomposition, functions from aquatic vascular plants and planktons. (Li, 2002)

3. Water storage pond: After sewage flowing through oxidation pond, the next stop is the water storage pond. In this process, the flow slows down, triggering sedimentation.

4. Land treatment techniques: To this day, developed sewage treatment types can be classified as: (1) The Slow Rate Method (SR): Also named slow rate infiltration. It refers

to apply wastewater on a land with vegetation at a slow rate to avoid wastewater runoff and it is treated both by plants and microorganisms present in the soil. (2) Rapid Infiltration (RI): The RI process uses the soil matrix for physical, chemical, and biological treatment, which is similar to simulate the soil matrix as a chromatography column carrier (3) Overland Flow (OF): It is applied to the top portion of a sloping land grown over with grass and flows down the terrace to a runoff collection channel at the bottom of the slope. Some wastewater evaporate while others then flow into collection pipes. (4) Wetland System: This method covers part of the Land Treatment System and Aquatic biological treatment. To make short of this matter, filling in standard packages (e.g. gravel) at a settled gradient on a low laying land with settled length breadth ratio and slope at bottom. Then, growing good processing performance, high survival rate, long growth cycle, beautiful, and cost-efficient plants (such as Phragmites communis) in order to form a ecological Wetland System. (Wang, 1990,2000; Yuan et al, 2001)

4.2 Theory and application principles of ecological engineering

In 1938, Germany, Seifert first addressed the concept of near-natural river and stream control; in 1962, H.T. Odum brought the phrase "Ecological Engineering". Trace back the history of humankind's development, it repeatedly revealed that how mankind ignore and disrespect Mother Nature. Therefore, many scholars had addressed their perceptions and researches until the book Ecological Engineering was published , Mitsch, Jorgensn, 1989, the concept of combining ecology and engineer officially spread to people. (Lin and Qiu, 2003; Mitsch, 1998) Moreover, in 1993, as a result of a series of significant activities and campaigns such as holding seminars in Washington, publishing special topics on ecological engineering in important scientific journals like Environmental Science and Technology, and the establishment of International Ecological Engineering Society in Utrecht, Holland, all affected and founded the foundation of ecological engineering.

In 1989, Mitsch and Jorgensen systematically sorted out other scholars' comments, defining the essence and connotation of ecological engineering which included: (1) Self-design; (2) eco-system conservation; (3) based on solar energy; (4) it's a part of, not apart from nature. Thus, based on those principles, while constructing ecological engineering projects, considerations need to be made as followed: (1) Safety issue; (2) biotope conservation near by constructions; (3) reducing impacts on ecology system during constructing; (4) the must follow-up treatments to ecological environment. (Lin & Qui, 2003) In short, following standard principles and rules when starting a ecological engineering, making appropriate designs according to local resources and situations in order not to process it in a mess or being reproached.

In the past, firm, durable, once and for all structures were engineering purposes; therefore, reinforced concrete works became the best choice. From village gutters, irrigation waterways, retaining walls to drainage channels (picture 8), all adopted this method. Meanwhile, in order to maintain headcounts saving, low time-consuming and construction convenience, engineering heavily depended on amounts of machines. As a consequence, finished constructions were always extraordinary huge or became standard template structures, building by one-size-fit-all molds. For instance, common seawalls, riverbank constructions, and wave energy dissipating concrete blocks belong to the category. Those undue size and occasionally impractical reinforced concrete works not only damaged the nature, but also harmed creatures' habitats (Shie, 2001); most importantly, making them the victims when disasters happen.

Picture 8. On the left side: Irrigation and drainage waterways constructed with reinforced concrete; on the right side: Covering slope collapse with cement concrete structure then grew plants on the steep hillside.

Settings of Land Treatment System structurally belong to artificial facilities; however, it could evolve with natural ecology by creature's sustainable self-purification ability. Land Treatment System is capable of various scales; also, it's adjustable in regard of to local terrace or spaces. Therefore, basing on those basic principles, successful results can be achieved.

For rivers and streams, relevant materials can be collected locally in order to build revetments, consolidation works, weirs, fish ladders and so on. Because of this, not only the construction costs can be low down, but also benefits local environments, visual landscape, ecological restoration, green planting and the outcome of engineering. Even the construction was damaged by flood such as being crashed, covered, and submerged of sandy soil, driftwood or trash, the renovation cost would be much cheaper. Most importantly, without those huge steel and concrete wastes that were difficult to clean, the maintenance such as clean away rubbish and weeding would be easier.

Regarding to the water consumption in Taiwan, agriculture water use ranks the highest (more than 70%); nevertheless, if with the industrial transformation and the adjustments of water resource use as well as the reuse of irrigation water resource from these enormous water conservation facilities, the agriculture water consumption would decrease effectively. At the same time, groundwater pumping could also declined; in that way, the problems of Chia-Nan area subsidence would be improved; thus, future water resource can be stored. Furthermore, irrigation quality precautionary system could be installed if we constructed an irrigation area on slopes near by waterway entrance; by doing this, industrial and domestic sewage, relevant agriculture loss could be avoided, safeguarding to farmers and consumers. Moreover, structuring the Land Treatment System at back end of the whole irrigation system (or Aquatic Plant System) will make agricultural drainage more eco-friendly. Since fertilizer and agrochemical utilizations are unavoidable during agricultural activities in Taiwan, preventions of underground water contamination could be made by setting up an artificial treatment system before drainage gets into ordinary rivers or streams. Especially for troublesome chemicals like nitrate-nitrogen can be effectively decreased after this water treatment system. Also, during fallow durations, the system would prevent nitrate-nitrogen from draining into aquifers in order not to pollute groundwater. (Washington State Department of Ecology, 2004)

5. Application example of land treatment system

5.1 Land treatment system at national yunlin university of science and technology

National Yunlin University of Science and Technology was founded in 1991. Prior to the founding of the school, a huge sugarcane field lied on the region with a principle irrigation waterway across the campus from east to west; after detailed planning and construction, it became a beautiful connection of rivers and lakes. In 2000, in light of achieving the mission of sustainable campus, those widened irrigation waterways construction of present lakes were renovated to a Land Treatment System, as to cope with sewage that flows into the school. This land system of ecological engineering can be divided into the following parts: (1) Step aeration; (2) oxidation pond; (3) water storage pond and (4) irrigated area (picture 9). With the widely distributed system, the principle of natural design for existing plants were adopted; that is, with the changes of seasons and temperatures, plant succession enabled the formation of new species approximately every two months, especially in irrigation area.

(a)   (b)

(c)   (d)

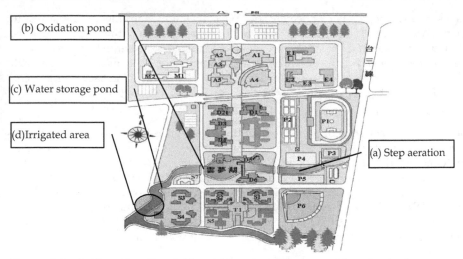

(b) Oxidation pond

(c) Water storage pond

(d)Irrigated area

(a) Step aeration

Picture 9. Main Parts that Consist Land Treatment System and location in the campus map: (a) Step aeration, (b) oxidation pond, (c) water storage pond and (d) irrigated area.

5.2 Functions of water purification

Generally speaking, common river pollution resources included industrial wastewater, livestock wastewater, irrigation runoff, and domestic sewage. For the waterways' external pollution sources, those many came from upper course of agricultural drainage and livelihood wastewater; therefore, nitrogen and phosphorus contaminations are the dominance. However, the nutrients transformation in the irrigation area covered physical, chemical and biological activities (picture 10); including soil filter retention, physisorptions, chemisorption, biological decomposition and phytoextraction.

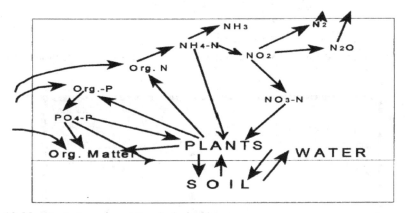

Picture 10. Nutrients transformation in irrigation area

5.2.1 The removal of nitrogen

The element nitrogen (has the symbol N) in sewage is classified to four types when flowing into the soil: organic nitrogen, ammonian, nitrite and nitrate. Organic nitrogen mainly presents as a suspended state, affecting by filter retention and then mineralizes to ammonian (Wang, 1990). Ammonian can be eliminated after Atmospheric volatile, organification, and plants absorption; moreover, with an aerobic environment, because of oxidation, nitrogen would transform from nitrosomonas to nitrite; next, oxidized with nitrobacteria to nitrate, finally being eliminated in result of denitrification bacteria returning to nitrogen. Referring to the monitoring data, explanation is as followed:

1. The removal of nitrite

Even though the concentration of nitrate in livelihood and agricultural wastewater is lower, a few amounts of nitrate (100PPb) will cause great damage to human bodies. According to research, for eliminating nitrate in irrigation area, the effectiveness is superior, In light of the difference of nitrate concentrations, the flow concentrations of section 1 and section 2 (picture 11 and 12) are between 9-270, 16-94 ppb, respectively; moreover, the eliminating rate reaches 47-97%, 82-97%, respectively. In terms of the inferior eliminating rate in section 1, the reason might be from its bad plant growing since there are not too many dominant plants existed; therefore, eliminating rate in section 1 is not stabilized.

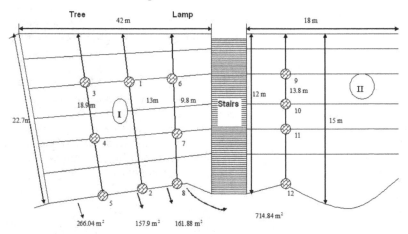

Picture 11. Sampling distribution of irrigation area (Stairs in the middle, section 1 at left side, section 2 at right)

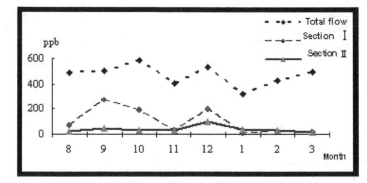

Picture 12. Nitrate concentration of flow from section 1and 2 in irrigation area

2. The removal of ammonian

Picture 13 refers to the ammonian concentration of effluent water of section 1 and 2 in irrigation area; from this, it indicates the tendency similarity with picture 4. According to scholars Sz Gung, Tzeng and Jr Cheng, Jang, 1996, by conducting clayey soil treatment, approximately 16% ammonian in sewage would be absorbed by plants, 14% became organic ingredients after organification, 24% were removed either by denitrification or volatilization. Therefore, plants play an important role when removing water ammonian; thus, comparing the plants growing situation on both sections, the results were: In December, plants growing situation in section 2 was better than section 1, so the ammonian removing rate was higher in section 2 (92%); however, in March, plants growing situation got flourished in section 1 (mainly were Commelina paludosa), so ammonian removing rate in section 2 worsened (81%) than that in section 1 (84%).

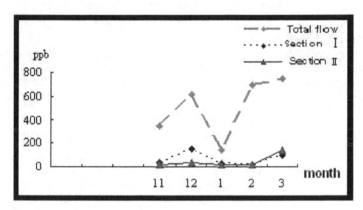

Picture 13. The Concentration of Ammonian in Effluent Water of Section I in Irrigation Area

3. The removal of nitrate

In light of the removal of nitrate in Land Treatment System, usually it's nothing more than either plants absorption or biological denitrification and there are two environmental properties that would stimulate denitrification. (1) Anoxic sediments. (Oxidation reduction potential less than 300mV) (2) Carbon fuel supply from plants growing (Baker, 1998). In light of picture 14, it reveals the instable removal rate of nitrate, ranging from 5%-60%; nevertheless, from the perspective of plants absorption, nitrate is the main nutrient to plants, usually by root absorption. While comparing plants with dominant species of each month, Ageratum was the main dominant plant species in January, differentiating to other dominant plant species; also, the climate got warmer after January, which could be another factor that caused plants absorption. In terms of microbial denitrification, influence factors included pH, carbon nitrogen ratio and oxidation reduction potential (ORP), there were no specific findings during comparison. However, with higher plants density, there will be more microorganisms being absorbed around the roots. In addition to denitrification under anoxic situations, other microorganisms such as mycorrizae (Tsai, 1994) could accelerate the speed of plants absorption, and increase the removal rate of nitrogenous compounds in nitrate.

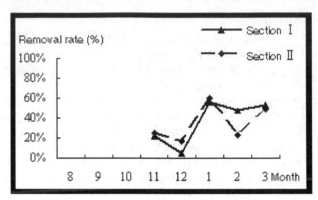

Picture 14. Nitrate Removal Rate Comparison of Effluent Water in Section I

5.2.2 The removal of phosphorus

In terms of removing phosphorus in land treatment, soil and plants absorptions are the dominant approaches. (Tzeng, Jang, 1996) Total phosphorus includes orthophosphate, polyphosphate and organophosphorus; almost all inorganic phosphates exist in phosphate types. Such as orthophosphate (PO_43-,HPO_42-,H_2PO_4-) or polyphosphate(P_2O_74-,P_3O_105-,HP_3O_92-,CaP_2O_72-).

From the aspect of removing orthophosphates, research shows that the removal rate is higher during spring and summer (about 35%-80%); whereas in autumn and winter, the rate is much lower (less than 57%); sometimes release phenomenon even occurs. From other documents (Wu, 2001; Tanner etal. ,1993), similar situations also were mentioned; possible reasons were because of higher temperatures during spring and summer so that lead to prosperous plants and microorganisms activities, resulting in the strengthen of absorption and metabolism. However, the removal of total phosphates ranged from 23%-73%.

5.2.3 Removal of other materials

According to researches, Commelina communis L. had the strongest absorption and resistant abilities to iron; for cadmium absorption and resistant ability, Patchouli thistle and Bidens had better performance; for absorbing and resisting magnesium, all plants affected well, especially Mikania cordata. As for copper and zinc absorption and resistant ability, Patchouli thistle and Commelina communis L. were superior to the former, whereas Ageratum to the latter; Micrantha, Bidens and Humulus functioned averagely. (Jeng, 2003)

In addition to functions of plants above, soil microorganisms surely could also absorb the heavy metal element: manganese; the best absorption time is 96 hours and its adsorbent concentration could rise to 1000 ppm (mg/L) whereas the maximum limitation of resisting manganese is 2000 ppm. (Yuang, Fang, Chang, 2003)

Besides, relevant researches about microorganisms' absorption and resistant ability of copper, zinc and cadmium showed that soil microorganisms which existed near by Commelina communis roots could exposure to maximum limitation ranges- copper (2500 – 3000 ppm), zinc (16800 – 19200 ppm) and cadmium (2400 – 3000 ppm); this indicated how tough the microorganism could take while resisting heavy medals, and was in line with the reaction of the plant (Commelina communis). (Yuang, Guo, 2003)

Furthermore, irrigation area not only had a clearly processing function to heavy medals, but also superior removal effects on the weed killer- glyphosate. Basic on the research done by

Yuan, Chang and Fan in 2004, after applying glyphosate in irrigation area, the dominant plant which absorbed the most glyphosate was Mikania Cordata, the absorption amounts was 2986.53 mg glyp/kg (plant net weight); next came Humulus and the worst absorption ability belonged to Bidens. In terms of those three plants' metabolic condition of glyphosate, Mikania Cordata was capable of extremely high metabolic effect, thus could eliminate about 90.1% of glyphosate in plants, next came Humulus japonicus, the least was Bidens.

As a consequence, plants can play the role of decontamination only with microorganisms' assistance, if not, it would not be functional. Therefore, the basic irrigation system based on plants-microorganism-soil can actually process water purification thoroughly, and all water quality parameters are in line with water quality standard.

5.3 Impacts of global changes

The flora composition has an obviously water purification function but is influenced by temperature changes. Therefore, observing the flora composition changes, especially the succession of dominant plants with high level of affection, can be an useful water quality standard indicator.

Humulus japonicus was the most stable and dominant plant before 2006, though not thoroughly in irrigation area, still contributed to water purification in certain level (Jen, 2004; Huang, 2011). However, because of global changes, during May 2009 to March 20011, Humulus japonicus only dominated 30% of all regions.

According to Huang's research (2011), the removal affection of nitrate nitrogen in Land Treatment System would rise up as the temperature got higher; the effect would decline if the average temperature decreased. This result was in accordance with relevant documents, showing the removal rate of ammonian in artifitial wetland system was about 0.3-1.1g/m2/d, and would change by the changes of temperature. It was because of plants grow slower during winter, thus it decreases the nutrients usage; as a result, the nitrogen removal rate would be affected by temperature obviously, with higher temperature comes better removal rate (Sikora et al., 1995 ; Van Oostrom, 1995). As for orthophosphate, the correlation was not something obvious.

Moreover, with adequate climate, the flowering period of Mikania cordata starts from October to December annually, after this period, the orthophosphate removal rate will decrease. However, due to La Nina occurred during 2009-2010, at this period, rainfall and temperature changed rapidly which lead to either earlier or delayed flowering period of Mikania cordata.

5.4 Regular harvesting in irrigation area

Using plants as a water purification approach has been proved in terms of the long term observation in irrigation area. Yet, with the growing of plants, the problem of leaching out nitrogen and phosphorus is the current issue. According to research, after harvesting plants in irrigation area, the nitrate removal rate would drop to the lowest on the third day, gradually rise to 10%; for orthophosphate removal rate, it would maintain stable on the third day and increase to positive value on the thirtieth day. What peculiarly is that in the harvesting area, the average removal rate of chemical oxygen demand (COD) was 41.41%, comparing to the figure 26.09% and 22.63% in contrast area, the former one was much higher obviously (Huang, 2011). Since the harvesting season is in summer with higher temperature, plants grew much faster, thus it was fit for harvesting.

Therefore, conduct regularly harvesting benefits to refresh the plant system and also a wonderful maintenance approach of water purification.

5.5 Sustainable development

With the rapid development of urbanization in Taiwan, farmlands decrease and polluted irrigation water are undeniable facts. Thus, how to apply current facilities and resources to improve deteriorating environment and ecology will be an urgent issue for Taiwan even other developing countries.

If the government can scheme a new plan for the west of Taiwan with irrigation waterways, making the best of water conservation areas with the concept of community empowerment to conduct sewage ecological treatment by designing superior hardware facilities and simple management and operational guidebook, any unprofessional community citizen can play the role of management and maintenance. Considering the advantages of low cost of settings and simple maintenance work, the outcome in National Yunlin University of Science and Technology can be a good reference.

In 21st century, the influence of global changes gradually becomes serious; meanwhile, drought and flood disasters might be common in the future; therefore, how to cope with this issue has been an inescapable task to face of all countries on earth. Surely Taiwan is no exception; therefore, starting from management by reallocating water resource in order to transfer redundant agricultural irrigation water to livelihood and industry. For polluted irrigation water, streams and rivers, conducting remediation with the concept of Land Treatment System such as process with ecological engineering on slopes of irrigation waterways in water conservation area would be a useful way of finding new sources and reducing water expense. Facing this inescapable challenge, Land Treatment System would be on of the solutions to develop sustainably.

6. Conclusion

The design of Land Treatment System was made to meet the trends of irrigation system. Making use of current irrigation waterways by implementing ecological engineering as to foster mother nature to heal and develop herself and to purify polluted water, thus to increase available water resources. Functioned by the combination of "earth-microorganism-plants" effect, it can be applied not only to livelihood wastewater usage but also irrigation; most importantly, the outcome is satisfying.

7. References

Chen, H. T. (2009). *History of Water Conservancy in Taiwan*, Wu-Nan Book Incorporation, ISBN 978-957-11-5772-6, Taipei City, Taiwan

Environmental Protection Administration, Executive Yuan, R.O.C. (1998). National Environmental Protection Plan(NAPP): Sewer Development in a Long-term Goal, Environmental Protection Administration, Executive Yuan, Taipei City, Taiwan

Environmental Quality Protection Fundation (2000). *The comparison of water poverty index (WPI) in Taiwan*, 02.06.2011, available from
http://www.envi.org.tw/common/download/Papers/Paper2.pdf

Huang, J.T. (2011). The study of the Temperature Effect on Succession of Flora and Water Quality Improvement in Land Treatment System, Master Thesis, Graduate School of Safety Health and Environment Engineering, National Yunlin University of Science and Technology, Taiwan

Jang, Y., Lu, Y. and Shie, B. R. (Translator)(2011). *War of Water Resource: The Fight to Stop Corporate Theft of the World's Water*, pp.26-31, Global Group Holdings LTD., ISBN 978-986-185-561-5, Taipei City.(Translated from Barlow, M. and Clarke, T.. (2002).

Blue Gold: The Fight to Stop Corporate Theft of the World's Water, Global Group Holdings, Ltd., ISBN 978-986-185-561-5, Taipei City.)

Jen, R.H.(2004). To Study of The Relationship Between The Diversity of Plants in a Land Treatment System and Water treatment efficiency, Master Thesis, Graduate School of Safety Health and Environment Engineering, National Yunlin University of Science and Technology, Taiwan

Jeng, Y. L., Jian, S. F. and Huang, J. W. (2003). Study of Native Plants on Absorbing and Accumulating Heavy Medals, Master Thesis, Graduate School of Safety Health and Environment Engineering, National Yunlin University of Science and Technology, Taiwan

Li, J., Yang, S. and Peng, Y. J. (2002). Microorganisms and Water Engineering Treatment, Chemical Industry Publisher, ISBN 7-5025-3847-X, Beijing, China

Li, J. L. (2003). Study of "Post-Dam Era" and Water Resources Policy in Taiwan, National Policy Fundation, S.C. (R.)092-012, (October 2003), 11.06.2011, available from http://old.npf.org.tw/PUBLICATION/SD/092/SD-R-092-012.htm

Lin, Z.Y. and Qiu, Y.W.(2003). *Introduction to Ecological Engineering Methods*, National Taipei University of Technology/Water Environment Research Center, ISBN 957-703-103-X, Taipei, Taiwan

Lin, R. Y. (2007). Warning of Devastating Collapse—Ecological Crisis in Taiwan, *Rhythms Monthly*, vol.107, (June 2007), pp., ISSN 10298371

Lin, S. Y. (2011). Water Resources Changing Policy, *Formosa Weekly*, No.101, 13.06.2011, available from http://www.formosamedia.com.tw/weekly/post_2485.html

Mitsch, W.J.(1998). *Ecological engineering*—the 7-year itch，Ecological Engineering, Vol.10, pp.119-130

Mitsch, W.J. and Jørgensen, S.E.(1989). Ecological Engineering：An Introduction To Ecotechnology, Wiley, New York

Northern Region Water Resources Office, Water Resources Agency, Ministry of Economic Affairs (2009). Water Conservation Classroom, 12.06.2011, available from http://www.wranb.gov.tw/ct.asp?xItem=3528&ctNode=578&mp=2

Odum, H.T.(1962). Man in the ecosystem, In proceedings Lockwood Conference on the Suburban Forest and Ecology, Bull. Conn. Agr. Station 652, Storrs, CT, pp.57-75

Shie, H. L., Chen, Y. S. and Chen, J. P. (2001). Saving Coastal Wetlands- The Application of Ecological Engineering, *Science Monthly*, Vol.32, No.7, (July 2001), pp.594-599, ISSN 0250-331X

Sikora, F.J., Tong, Z., Behrends, L.L., Steinberg, S.L. and Coonrod, H.S.(1995). Ammonium removal in constructed wetlands with recirculating subsurface flow: Removal rates and mechanisms, *Wat. Sci. Tech.*, Vol.32, No.3, pp.193-202, ISSN 0273-1223

Tanner C. C., Clayton J. S. and Upsdell M. P. (1993). Effect of loading rate and planting in constructed wetlands-Removal of nitrogen and phosphorus, Wat. Res., Vol.29, No.1, pp.27-34

Taiwan Governor-General Office (1901). *Rules of Public Irrigation Waterway*, Taiwan Governor-General Office Files, Taiwan Historica, 25.05.2011, available from https://dbln.th.gov.tw/sotokufu/

Tsai, Y. F. (1994). Management of Cymbidium Fertilizer From the Aspect of Science of Plant Nutrition, Agriculture Promoting News in Taichung, Vol. 139, 22.06.2011, available from http://www.tdais.gov.tw/search/book3/131-140/139.htm

Tzeng, S. G. and Jang, J. C. (1996). Polluted Soil Treatment, Vol.9, No.1, pp.52-66

Van Oostrom, A.J.(1995). Nitrogen removal in constructed wetlands treating nitrified meat processing effluent, Wat. Res. Tech., Vol.32, No.3, pp.137-147, ISSN 0273-1223

Wang, J. H. (2005). Global Warming- Key Issues of Sustainable Development in Taiwan, *Agricultural World*, vol.262, (June 2005),pp.48-50, ISBN 0255-5808

Wang, J. H. (2005). Long-term Changes and Impacts of Rainfall in Taiwan, *Agricultural World*, vol.264, (August 2005), pp.62-70, ISBN 0255-5808

Wang, J. S., Huang, S. H. and Cheng, G. P. (1990). Chemical and Water Pollution, Zhongshan University Press, Guangdong, China

Wang, H. S. (1990). *Ecological Basis of Pollution*, Yunnan University Press, ISBN 9787810250177, Yunnan province, China

Wang, H. S. (2000). *Pollution Ecology*, Advanced Education Press, ISBN 9787040079807, Beijing, Chian

Washington State Deptartment of Ecology(2004). Guidance on Land Treatment of Nutrients in Wastewater, with Emphasis on Nitrogen, Publication #04-10-081

Water Industry Information Network (2011). Introduction to Water Conservation Industry, Current Industry Analysis 18.06.2011, available form http://km.wpeiic.ncku.edu.tw/5_industrial/reuse.aspx   [Water   Industry Information Network (2011).]

Water Resources Agency, Ministry of Economy (2001). Program of Water Resources Development in Taiwan, 15.06.2011, available from http://hysearch.wra.gov.tw/wra_ext/WaterInfo/wrproj/main/main.htm

Water Resources Agency, Ministry of Economy (2011). *Introduction to Main Dams*, 25.05.2011, available from http://fhy.wra.gov.tw/ReservoirPage_2011/StorageCapacity.aspx

Water Resources Department, R.O.C. (1997). *Water Conservation in Taiwan*, Water Resources Department, Taichung, Taiwan

Wen, C. G. (1995). *Impacts on Dam Eutrophication of Adding Nutrients During Initial Impoundment*, National Science Council, NSC 83-0410-E-006-017, Taipei City, Taiwan

Wu, J. B., Chen, H. R., He, F., Cheng, S. P., Fu, G. P., Jin, J. M., Chiou, D. R. and Ren, M. S. (2001). Sewage Phosphors Purification of Artificial Wetland System, Acta Hydrobiologica Sinica, Vol.25, No.1, pp.28-35

Yang, L. (1996). Concept of Conducting Wetland Wastewater Treatment and Water Conservation in Wetlands, *Environmental Impacts Assessment Technology Workshop-Conservation and Development of Coastal Areas*, pp.237~244, Kaohsiung City, Taiwan

Yuan, Y. K., Guo, S. Y. and Shen, J. F. (2003). Study of Soil Microorganisms' Absorption and Resistant Ability to Heavy Medals: copper, zinc and cadmium, The 28th Wastewater Treatment Seminar

Yuan, Y. K., Fan, H. T., Chang, J. G., Jang, Y. T., Guo, S. Y. and Jeng, R. H. (2003). Study of Manganese Treatment of Soil Microorganisms in Land Treatment System, The 28th Wastewater Treatment Seminar

Yuan, Y. K., Jeng, R. H., Shr, T. A. and Huang, J. M. (2003). *自Research of Water Purification and Diversity of Plants Through Land Treatment System Under Natural Succession*, The 28th Wastewater Treatment Seminar, 28-29.11.2003, Taichung, Taiwan

Yuan, Y. K., Jang, Y. T. and Fan, H. T. (2004). Application and Influence of Glyphosate to Plants and Water Quality in Flood Area of Land Treatment System, The 29th Wastewater Treatment Seminar, 29-27.11.2004, Tainan, Taiwan

Yuan, Y. K., Jang, Y. T. and Fan, H. T. (2004). Feasibility of Removing Nutrients in Sewage by Using Land Treatment System, *Journal of Science and Technology*, Vol. 13, No. 3, pp. 203-209, ISSN 1023-4500

Yuan, Y.K., Lian, W.F. and Li, Y.Z.(2002). *The Application of Land Treatment System for Wastewater in Conservation of Water Resource*, 12th Conference of The Sewer & Water Environment Regeneration, pp. 177-186, Taipei, Taiwan, August 30, 2002

12

A Review of Subsurface Drip Irrigation and Its Management

Leonor Rodríguez Sinobas and María Gil Rodríguez
Research Group "Hydraulic of Irrigation" Technical University of Madrid,
Spain

1. Introduction

According to the ASAE Standards (2005), subsurface drip irrigation SDI is "the application of water below the soil surface though emitters* with discharge rates generally in the same range as drip irrigation." Thus, aside from the specific details pointed out in this chapter, an SDI unit is simply a drip irrigation network buried at a certain depth.

There is a wide variety of plants irrigated with SDI all over the world such as herbaceous crops (lettuce, celery, asparagus and garlic), woody crops (citrus, apple trees and olive trees) and others such as alfalfa, corn, cotton, grass, pepper, broccoli, melon, onion, potato, tomato, etc.

Over a quarter of a billion hectares of the planet are irrigated and entire countries depend on irrigation for their survival and existence. Growing pressure on the world's available water resources has led to an increase in the efficiency and productivity of water-use of irrigation systems as well as the efficiency in their management and operation. The efficiency of subsurface drip irrigation SDI could be similar to drip irrigation but it uses less water. It could save up to 25% - 50% of water regarding to surface irrigation. Throughout this chapter, the specific characteristics of SDI will be presented and some criteria for its design and management will be highlighted.

1.1 A bit of history

The first known reference of a subsurface irrigation comes from China more than 2000 years ago (Bainbridge, 2001) where clay vessels were buried in the soil and filled with water. The water moved slowly across the soil wetting the plants' roots. SDI, as we know it nowadays, developed around 1959 in the US (Vaziri and Gibson, 1972), especially in California and Hawaii, as a drip irrigation variation. In the 60s, SDI laterals consisted of polyethylene PE or polyvinyl chloride PVC plastic pipes with punched holes or with punched emitters. These systems used to work at low pressures, depending on water quality and filtration systems.

By the 70s, this method extended to crops as citrus, sugar cane, pineapple, cotton, fruit trees, corn, potato, grass and avocado. Its main disadvantages were: the difficult maintenance, the low water application uniformity and the emitters' clogging due to, mainly, oxid and soil particles. Alternatively, the equipment for SDI installation in the field developed (Lanting, 1975), and the fertilizers' injection with the irrigation systems started in Israel (Goldberg and

---

* Note: In SDI, the word emitter is used instead of dripper since the last has a lack of meaning as water does not drip as it does in drip irrigation.

Shmueli, 1970). Further, the quality of commercial emitters and laterals improved and SDI gained respect over other irrigation methods mainly due to the decrease of clogging.

At the beginning of the 80s, factors such as: the reduction of the costs in pipes and emitters, the improvement of fertilizers application and the maintenance of field units for several years promoted the interest in SDI. This has been extended since then.

Fig. 1. Lateral of subsurface drip irrigation.

1.2 Subsurface drip Irrigation in the world

There is a general agreement about the spread of SDI however it is difficult to obtain data to confirm this trend since the surface irrigated with SDI is counted as a drip irrigation in most surveys. In the US, the USDA Farm and Ranch Irrigation Survey (USDA-NASS, 2009) indicates that SDI comprises only about 27% of the land area devoted to drip and subsurface drip irrigation. Nonetheless, this percentage is continuously increasing over drip irrigation as farmers substitute their conventional drip irrigation methods by SDI systems. If this framework continues in the future, SDI would still have a potential increase.

In underdeveloped countries, there are many references about the so called low-cost SDI with rudimentary laterals and low emitter pressures that highlight even more, the SDI potential increase. Likewise, this is enhanced by the savings on water and energy.

1.3 Advantages and disadvantages

Experimental evidences of SDI advantages over other irrigation methods, specifically drip irrigation, are vast. Some advantages and drawbacks of this method, compiled by Lamm (2002) and Payero (2005), are shown below.

Advantages:

- The efficiency of water use is high since soil evaporation, surface runoff, and deep percolation are greatly reduced or eliminated. In addition, the risk of aquifer contamination is decreased since the movement of fertilizers and other chemical compounds by deep percolation is reduced.
- The use of degraded water. Subsurface wastewater application can reduce pathogen drift and reduce human and animal contact with such waters.
- The efficiency in water application is improved since fertilizers and pesticides can be applied with accuracy. In widely spaced crops, a smaller fraction of the soil volume can be wetted, thus further reducing unnecessary irrigation water losses. Reductions in weed germination and weed growth often occur in drier regions.
- Hand laborers benefit from drier soils by having reduced manual exertion and injuries. Likewise, double cropping opportunities are improved. Crop timing may be enhanced since the system need not be removed at harvesting nor reinstalled prior to planting the second crop. On the other hand, laterals and submains can experience less damage and the potential for vandalism is also reduced.
- Operating pressures are often less than in drip irrigation. Thus, reducing energy costs.

Drawbacks:

- Water applications may be largely unseen, and it is more difficult to evaluate system operation and water application uniformity. System mismanagement can lead to under irrigation, less crop yield quality reductions, and over irrigation. The last may result in poor soil aeration and deep percolation problems.
- If emitter discharge exceeds soil infiltration, a soil overpressure develops around emitter outlet, enhancing surfacing and causing undesirable wet spots in the field.
- Timely and consistent maintenance and repairs are a requirement. Leaks caused by rodents can be more difficult to locate and repair, particularly for deeper SDI systems. The drip lines must be monitored for root intrusion, and system operational and design procedures must employ safeguards to limit or prevent further intrusion. Roots from some perennial crops may pinch drip lines, eliminating or reducing flows. Periodically, the drip lines need to be flushed to remove accumulations of silt and other precipitates that may occur in the laterals. Likewise, operation and management requires more consistent oversight than some alternative irrigation systems. There is the possibility of soil ingestion at system shutdown if a vacuum occurs, so air relief/vacuum breaker devices must be present and operating correctly. Compression of laterals by soil overburden can occur in some soils and at some depths, causing adverse effects on the flow.

1.4 Components of SDI units

The typical SDI unit, contrarily to drip irrigation units, is a looped network. It is composed by a submain or feeding pipe, and a flushing pipe that connects all lateral ends (Fig. 1). The looped network enhances the hydraulic variability and thus, the flow variability within the unit is also improved. The pipes and laterals are buried at a certain depth, generally from 7 to 40 cm, supplying the water directly in the crop root area.

The flushing pipe eliminates possible sediments or other water suspended elements. Two relief valves may be placed at the head and tail of the feeding and flushing pipes to release the air at the beginning of the irrigation, and to let it pass when irrigation stops. Thus, soil particles are not introduced into the laterals. Alternatively, some flushing valves for the lateral end are also developed for that purpose (Fig. 3).

A manometer and a flow meter are placed at the unit head to measure the inlet unit pressure and the total unit discharge, respectively. Filters are also installed at the unit head and additional containers with fertilizers and other chemical products are located there, as well.

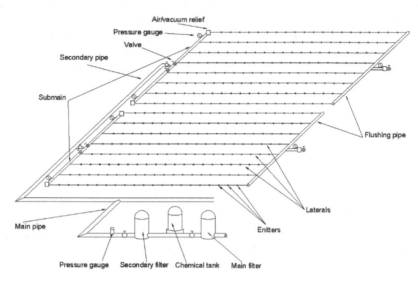

Fig. 2. Scheme of a typical SDI unit.

Emitters in SDI are similar to drip irrigation, but they generally are impregnated with a weed killer to reduce root intrusion at the emitter outlet. In countries where herbicides are forbidden, commercial emitter models are developed with specific geometries to reduce the risk of root intrusion. Likewise, there are some models with elements such as diaphragm and geometries inducing vortical movement, to avoid the entrance and deposition of soil particles (Fig. 3).

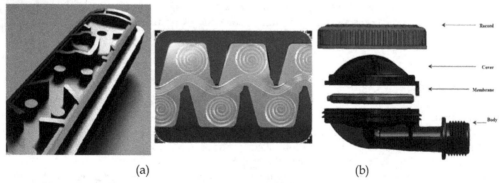

(a)   (b)

Fig. 3. Elements of SDI units: (a) emitter models and (b) flushing bulb. (Courtesy of NaaDanJain Irrigation).

1.5 Field installation of subsurface drip irrigation units

The installation of SDI units in the field shows differences with conventional drip irrigation. In most cases, the soil is chiseled to a depth close to the crop roots length prior the layout of laterals. This will also favor the horizontal water movement. Then, the feeding and flushing pipes are laid on trenches dug following the lateral ends keeping an extra space for the connections between pipes and laterals. Once these are done, the other elements such as: valves, relief valves, flow meters can be installed. Finally, all the subsurface elements are buried.

Among all the installation tasks, the difficult one corresponds to the deployment of laterals. These are introduced into the soil by one or several plows connected to a tractor (Fig. 4). Care must be taking to ensure the laterals are placed following a straight line and at a proper depth, and also that the lateral spacing stays constant.

Fig. 4. Installation of SDI.

2. Effect of soil on emitter discharge

One of the key differences between SDI and surface drip irrigation is that soil properties in fine-pore soils affect emitter discharge. A spherical-shaped saturated region of positive pressure h_s develops at the emitter outlet. Consequently the hydraulic gradient across the emitter decreases reducing emitter flow rate according to the following equation.

$$q = k \cdot (h - h_s)^{\,x} \tag{1}$$

where q is the emitter flow rate, h is the working pressure head, and k and x are the emitter coefficient and exponent, respectively.

In fine-pore soils, the overpressure h_s increases rapidly at the beginning of irrigation until it stabilizes after, approximately, 10 to 15 minutes. Values of h_s from 3 m up to 8 m have been recorded on single emitters in the field within a wide range of emitter discharges (Shani et. al. 1996, Lazarovitch et al. 2005). However for similar flow rates and soils, these values were smaller, within an interval from 1 to 2 m, in laboratory tests since the soil structure increases the soil mechanical resistance to water pressure under field conditions (Gil et al. 2008). Emitter flow reaches a permanent value coinciding with the stabilization of h_s.

Sometimes, the overpressure produces what it is called a chimney effect or surfacing (Fig.2). Overpressures develop locally in the soil displacing the soil components and creating preferential paths. Soil surface is wetted and big puddles are also observed.

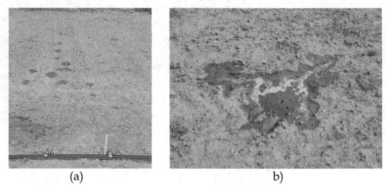

<table>
<tr><td>(a)</td><td>b)</td></tr>
</table>

Fig. 5. Surfacing in SDI units: (a) wetted soil surface above the laterals and (b) detail of the wetted area.

Pressure-compensating and non-compensating emitters exhibit different performance. For non-compensating emitters, discharge reductions from 7 up to 50% have been measured in controlled conditions on single emitters in fine-coarse soils. However, the decrease is less than 4% in sandy soils. For pressure-compensating emitters, flow variation is negligible if h_s stays below the lower limit of the emitter compensation range.

Water movement in SDI can be considered as a buried point source (Fig. 6a). Philip (1992) developed an analytical expression to determine the pressure at the discharge point in a steady-state conditions that was applied by Shani and Or (1995) to relate h_s with the soil hydrophysical properties and the emitter permanent flow rate q as

$$h_s = \left(\frac{2 - \alpha \cdot r_0}{8\pi \cdot K_s \cdot r_0} \right) \cdot q - \frac{1}{\alpha} \tag{2}$$

where r_0 is the spherical cavity radius around emitter outlet, K_s is the hydraulic conductivity of the saturated soil and α is the fitting parameter of Gardner's (1958) of the non-saturated hydraulic conductivity expression.

In moderate flows, h_s is linear and q follows a straight line whose slope depends on r_0, K_s and α. For a given discharge q and radius r_0, the pressure h_s is more sensitive to K_s and less sensitive to α.

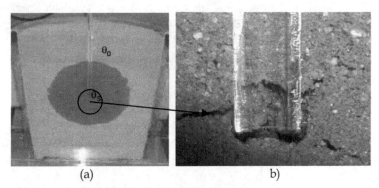

(a)   b)

Fig. 6. Water movement of SDI emitters in loamy soils: (a) wetting bulb and (b) detail of horizontal cracks in the cavity. (Note: θ_0= initial soil water content and θ_s= saturated soil).

Experiments carried out in uniform soil samples in pots show that the cavity shape tended to be spherical at small emitter discharges (Gil et al. 2010). At higher emitter discharges, horizontal cracks initially appear in the cavity, but slowly they fill with soil and, ultimately resulting in a spherical cavity (Fig.6b). The radius of the cavity linearly increases with small emitter discharges and stabilizes at higher discharges (Fig. 7).

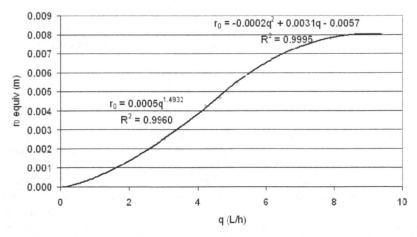

Fig. 7. Variation of the radius spherical cavity with emitter discharge in uniform soils. (Source: Gil et al. 2011).

2.1 Water distribution variability

The main causes that affect flow rate variability in buried emitters are: hydraulic variation, manufacturing variation, root intrusion, deposition of soil particles and the interaction of soil properties. If the temperature is constant and emitter clogging is negligible, water distribution in drip irrigation units follows a normal distribution. The emitter discharge variability CVq will depend on the hydraulic variation and the manufacturing variation CVm. This hypothesis is better suited when the emitter manufacturing variation is the main cause of the final variation. Normal flow distribution in the unit is characterized by two parameters: the mean and the standard deviation of flow (or its coefficient of variation). Adding these two variations to Eq. [1], results in (Bralts et al. 1981):

$$q = k \cdot h^x \cdot (1 + u \cdot CV_m) \tag{3}$$

where u is a normal random variable of mean 0 and standard deviation 1.

The coefficient of manufacturing variation CV_m is a measure of flow variability of a random emitters' sample of the same brand, model and size, as produced by the manufacturer and before any field operation or ageing has taken place (ASAE, 2003). For SDI units, Eq. [3] transforms to

$$q = k \cdot (h - h_s)^x \cdot (1 + u \cdot CV_m) \tag{4}$$

The goal of irrigation systems is to apply water with high uniformity. The system design is frequently based in a given target uniformity that is defined by uniformity indexes. One of the common indexes is the coefficient of variation of emitter flow CVq (Keller and Karmelli 1975)

$$CV_q = \frac{\sigma_q}{\bar{q}} \tag{5}$$

where σ_q is the flow standard deviation and $\bar{q}$ is the mean flow.

Other frequently used index is the Christiansen coefficient Cu (Christiansen 1942)

$$Cu = 100 \left(1 - \frac{|\bar{\sigma}_q|}{\bar{q}} \right) \tag{6}$$

where $|\bar{\sigma}_q|$ is the mean absolute flow standard deviation.

A good irrigation performance requires water distribution uniformities characterized by Cu > 0.9 or CV_q= 0.1, and emitters with CV_m < 0.10.

Observations on homogenous soils in pots, have shown that the interaction between the effect of soil on emitter discharge in low infiltration soils would work as a self-regulated mechanism (Gil et al. 2008). The soil overpressure would act as a regulator, and the emitters with a greater flow rate in surface irrigation would generate a higher overpressure in the soil, which would reduce the subsurface irrigation flow rate to a greater extent than in emitters with a lower flow rate. Consequently, the flow emitter variability would be smaller in buried emitters than in surface ones. Thus, for non-regulated emitters the uniformity of water application in SDI laterals would be greater than the uniformity of surface drip lateral irrigation in homogeneous soils as it is shown in Fig. 8.

For compensating emitters, the flow rate variability in SDI could be similar to the surface drip irrigation in loamy and sandy soils if the head pressure gradient across the emitter and the soil were above the lower limit of the emitter compensation range. Thus, the variability of soil overpressure would be offset by the elastomer regulation.

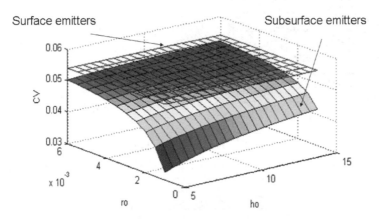

Fig. 8. Coefficient of variation of flow CV_q in laterals with 100 non regulated emitters of 2 L/h laid on a uniform loamy soil, considering hydraulic variation negligible (Note: r_0, h_0 and h_s are expressed in m). (Source: Gil et al. 2008).

3. Field performance of subsurface drip irrigation

Increments between 2.8 % and 7.0 % on emitter flow-rate have been reported in laterals when non-regulated emitters were excavated (Sadler et al. 1995). The uniformity of water application in excavated drip tapes, after three years in the field, was lower than the new tapes deployed over the soil. Likewise, field evaluations of SDI laterals resulted in Cu within the interval 75-90 (Ayars et al. 2001; Phene et al. 1996).

In loamy bare soils, regulated and non-regulated emitters have shown similar behaviour in evaluations carried out in two consecutive years (Rodríguez-Sinobas et al. 2009b). Lateral inflow decreases rapidly within the first 10 to 15 min of irrigation and it approaches a steady value as the time advances (see Fig. 9). Consequently, lateral head losses reduce and the head pressure at the lateral downstream increases. This performance highlights that overpressure h_s at the emitter outlet increases at the start of the irrigation until it stabilizes confirming the trend observed in single non-regulated emitters but contradicting the behaviour observed in single regulated emitters. Meanwhile, the variation of inlet lateral flow is larger than in non-regulated emitters and differences among laterals are noticeable. Also, the flow in these laterals decreases as the inlet head reduces.

In the one hand, the discharge of regulated emitters decreases over the operating time until it stabilizes. The elastomer material may suffer fatigue when being held under pressure and its structural characteristics may change (Rodríguez-Sinobas et al. 1999). When irrigation is shutdown pressure is cancelled and the elastomer relaxes and surmounts the deformation caused by pressure. The longer the time elapsed between successive irrigation events the longer the time for the elastomer to return to its initial condition. This behaviour would be conditioned by elastomer material and its relative size. On the other hand, soil particles suctioned when irrigation is shutdown, might deposit between the labyrinth and the elastomer and thus, cancelled out the self-regulatory effect.

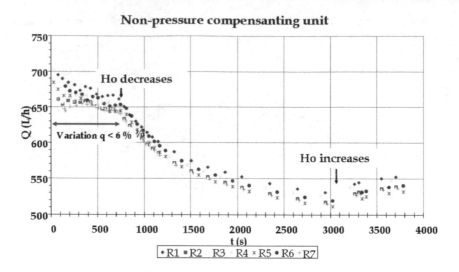

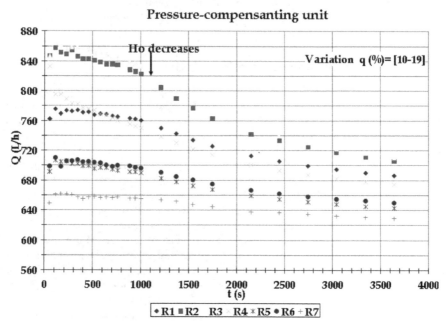

Fig. 9. Variation of lateral inlet flow of a SDI unit laid on a loamy soil during irrigation (Note: Ri= lateral).

Evaluations performed on a seven years old SDI unit of regulated emitters showed that the hydraulic variability was the major factor affecting emitter discharge, but the effect of the emitter's manufacturing and wear variation was smaller (Rodríguez-Sinobas et al. 2009b). The uniformity of water application Cu varied between 82 and 92 and the degree of clogging between 25 to 38%. The first improved when reducing pressure variability, thus the emitters behaved as a non pressure-compensating too. Their performance might be affected by: clogging (deposition of suspended particles, root intrusion) and entrapped air.

3.1 Field evaluation

Specific methodology for SDI has not been yet developed for field evaluation of SDI units (Fig. 10a). Nevertheless, in spite of the difficulty to measure the emitter discharge in field conditions, its performance may be assessed by a simulation program estimating the distribution of emitter discharges and soil pressures within the unit.

As a practical procedure, the measurement of unit inlet flow and the pressure heads at both ends in the first and last laterals of the unit could suffice to calibrate the simulation program. Both laterals could be easily located in the field and unearthed. Pressures and inlet flow can be measured by digital manometers and by Woltman or portable ultrasonic flow meter, respectively as it is shown in Fig.10b.

(a)   (b)

Fig. 10. Field evaluation of SDI units: (a) sample with unearthed emitters and (b) manometer and flow-meter for recording head pressure and inlet flow in laterals.

Field evaluations can highlight emitter clogging such as the root intrusion observed in Fig. 11.

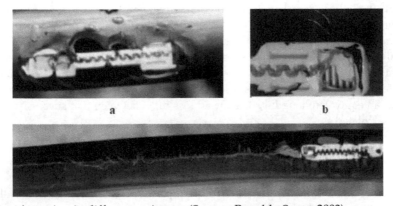

a   b

Fig. 11. Root intrusion in different emitters . (Source: Ronaldo Souza 2003)

4. Simulating subsurface drip irrigation

As it was mentioned above, simulations are a tool for estimating the performance of laterals and units. On the one hand, discharges and pressures of the emitters among the unit (generally looped units) can be computed taking into account the soil properties. On the other hand, the movement of water in the soil and wet bulb geometry can also be estimated.

4.1 Simulation of subsurface drip irrigation laterals and units

A computer program can be developed to predict water distribution along SDI laterals and units. To do so, we would need to take into account:

- Design variables: length and diameters of laterals; submain and flushing pipe; emitter discharge; emitter's manufacture and wear coefficient and local losses at the insertions of the emitters and the laterals.
- Operation variables: inlet pressure and irrigation time.
- Soil properties: at least texture and saturated hydraulic conductivity and if possible, the other soil parameters of Eq. [2]. When dealing with spatial variability, it is required the couple of a specific geo-statistical modelling software and the one predicting water distribution within the unit.

Thus for a target irrigation uniformity, different study cases can be simulated and then, the selection of proper values for the design or operation variables could be achieved. Moreover, the suitability of this method to different soil types and their best management practices can be addressed prior the unit is laid on in the field.

4.1.1 Hydraulic calculation of laterals

In this chapter, we present a simulation program developed by Rodríguez-Sinobas el al. 2009a which calculates the flow rates for surface emitters in laterals with Eq. [3] and for buried emitters with Eq. [4]. The overpressure h_s at the emitter exit was obtained with Eq. [2]. Likewise, the head pressure at each emitter insertion hi was determined by application of the energy equation as follows:

$$h_i = h_{i+1} + hf_{i-i+1} \pm I_0 \cdot s \tag{7}$$

where hf_{i-i+1} is the head loss between two consecutive emitters i-i+1; I_0 is the lateral slope and s is the emitter spacing.

Flow regime in laterals is considered smooth turbulent with Reynolds numbers within the range $3000 < R < 100\,000$ since most subsurface drip irrigation laterals are made of smooth polyethylene pipe. Thus Blasius' head loss equation provides an accurate estimation of the frictional losses inside uniform pipes. Likewise, local head losses at emitter insertions are included as an equivalent length l_e. Then head loss hf_{i-i+1} between two successive emitters i-i+1 can be calculated as

$$hf_{i-i+1} = 0.0246 \cdot v^{0.25} \frac{Q_{i-i+1}^{1.75}}{D^{4.75}} \cdot s \cdot \left(1 + \frac{l_e}{s}\right) \tag{8}$$

where Q_{i-i+1} is the flow conveying in the uniform pipe between emitters i-i+1; D is the lateral internal diameter and v the kinematic water viscosity.

In most commercial models, local head losses at the emitter insertion are within the interval of 0.2 and 0.5 m (Juana et al. 2002). Lateral head losses hf can be calculated as

$$hf = F \cdot 0.0246 \cdot v^{0.25} \frac{Q_0^{1.75}}{D^{4.75}} \cdot L \cdot \left(1 + \frac{l_e}{s}\right) \tag{9}$$

where Q_0 is the lateral inlet flow; L is the lateral length and F the reduction factor. The reduction factor F (Christiansen 1942) takes the form

$$F = \frac{1}{m+1} + \frac{1}{2 \cdot N} + \frac{\sqrt{m-1}}{6 \cdot N^2} \tag{10}$$

where N is the number of emitters and m is the flow exponent of the head loss equation. This formula applied to laterals with the first emitter located at a distant L/N from the inlet. In general, number of emitters in laterals is large, and water distribution is assumed to be continuous and uniform. Hence, F= 1/(1+m).

As said, emitter discharge in SDI can be affected by soil properties. It can be influenced by K_s and α, each of them varying throughout the field. Furthermore, the discharge can be also affected by variation in cavity radius r_0. Soil hydraulic properties from close points are expected to be more alike than those far apart, they can be correlated using variograms. Furthermore, these can aim at the estimation of soil properties distribution.

Examples

For the porpoise of illustration, the performance of a lateral in different uniform soils is represented in Fig. 12a, and the effect of soil heterogeneity is shown in Fig.12b. The comparison of the performance between SDI and surface laterals is addressed in all the examples.

The uniformity index CV_q is higher in SDI than in the surface drip irrigation in all uniform soils except for the sandy one, where uniformities are alike. For a given r_0 and h_0, the possible self-regulation due to the interaction between the emitter discharge and soil pressure is observed in low saturated hydraulic conductivity soils (loamy and clay). The larger the cavity radio the smaller the soil effect, and thus, the SDI lateral behaves as a surface lateral. The irrigation uniformity index in non-uniform soils is less than in uniform soils. It slightly reduces as r_0 and h_0 decrease.

For sandy soils the uniformity is very similar in both laterals; even the uniformity in the SDI lateral is higher for certain values of r_0 and h_0.

In loamy soils, for higher r_0 values, the uniformity reduces as emitter flow increases. However, with small r_0 this trend is not shown, and similar uniformities are observed between the smaller emitter flows. The larger the cavity the smaller the soil effect, and thus the SDI lateral would behave as a surface lateral. No difference in water application uniformity is observed between surface and buried laterals in loamy soils for values of $r_0 >$ 0.01 m. On the contrary, for small r_0 the self-regulating effect is predominant. Therefore, this is enhanced by higher emitter flows since they develop higher soil overpressures.

For heterogeneous loamy soils the tendency is similar to the homogeneous soil: irrigation uniformity decreases as emitter discharge increases (Fig. 12b). However in some scenarios, the values of CV_q are higher. The uniformity is similar in all the three discharges for $r_0 >$ 0.01; below this value, uniformity rapidly decreases.

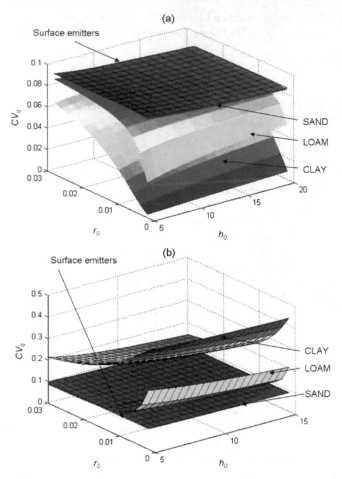

Note: Units for r_0 and h_0 are expressed in m.

Fig. 12. Irrigation uniformity index CV_q as a function of r_0 and h_0 in different soils for a SDI lateral with non-compensating emitters: (a) uniform soils; (b) non-uniform soils.

Uniformity of water application in all cases reduces as inlet pressure increases. Consequently, it would be advisable to select emitters with small discharge, and to irrigate with inlet pressure not very small (in this case above 10 m).

4.1.2 Hydraulic calculation of units

Typical SDI units are composed of a looped network (Fig. 2). Water can move either way from head to downstream lateral and reverse. The hydraulic calculation of laterals can be achieved as it was outlined in section 4.1.1. The inlet pressure h_0 at each lateral would be determined as

$$h_{0i} = h_{0i+1} + hf_{Si-i+1} \pm I_{0S} \cdot s_L \tag{11}$$

where i and $i+1$ are sub-indices for two successive laterals; hf_{Si-i+1} are the head loss in the uniform pipe between laterals i-i+1; I_{0S} is the submain slope and s_L is the lateral spacing. hf_{Si-i+1} may be determined with Eq. [8] considering the flow conveying between the laterals i and i+1.

Each lateral inflow can be calculated by adding the flow from all of its emitters. Local head losses at the connection of the lateral with the submain are smaller than those in laterals (Rodriguez-Sinobas et al. 2004), and could be left out in Eq. [8].

Two conditions are met for calculation of looped network. First, the sum of flows conveying along the submain equates the sum of the emitter's flow throughout the unit. Second, the pressure at every location within the network is the same when calculated from head towards the downstream end of the lateral than reversed. The direction of flow in a loop network is not known. A minimum energy line with minimum pressures divides the flow coming from the lateral head and the one coming from its downstream end. Thus from this line, the network calculation is accomplished as two branched networks: one whose length corresponded to the distance from the lateral inlet up to energy line; and the other whose length corresponded to the distance from the downstream end of the lateral to that line

This program predicts discharges and pressures from both directions. The iterative process stops when minimum pressures are met. If the value for minimum pressure from one way is larger than from the other, the energy line moves towards the least pressure. Likewise, the sum of the flows coming and exiting from the flushing pipe is zero.

The program ouputs are: the flows and pressures at the inlet and downstream end of the laterals; the minimum energy line; the distribution of emitter discharges and emitter pressures; and the irrigation uniformity coefficient CV_q. Simulations can be performed for irrigation units both laid on the surface and buried in the soil. In the last, emitter discharge is calculated with Eq. [4] and soil water pressure at the emitter outlet is calculated with Eq. (2).

Soil variability may be determined as detailed for laterals but in this case, the variable will be two-dimensional.

Example

A looped unit operating at several pressures in a loamy soil is selected for illustration of the performance of a typical SDI unit in Fig.13. For all cases, irrigation uniformity improves with small emitter discharges.

In non-uniform soils, the irrigation uniformity indexes are higher than in uniform soils, but their differences were small. Only the clay soil with the highest discharge displayed a significant difference.

In summary, irrigation uniformity is very good for most of the typical selected scenarios. Nevertheless, for clay soils emitter discharge reduced 14 % and this should be taken into account in the management of the irrigation system. On the other hand, if higher emitter flows would have been selected, the unit would have performed worse.

4.2 Hydrus 2D/3D

The program HYDRUS-2D/3D (Simunek et al. 1999) includes computational finite element models for simulating the two- and three-dimensional movement of water, heat, and multiple solutes in variably saturated media. Thus it can simulate the transient infiltration process by numerically solving Richards' equation.

The program is recommended as a useful tool for research and design of SDI systems. This simulation program may be used for optimal management of SDI systems (Meshkat et al.

2000; Schmitz et al. 2002; Cote et al. 2003; Li et al. 2005) or for system design purposes (Ben-Gal et al. 2004; Provenzano 2007; Gil et al 2011).

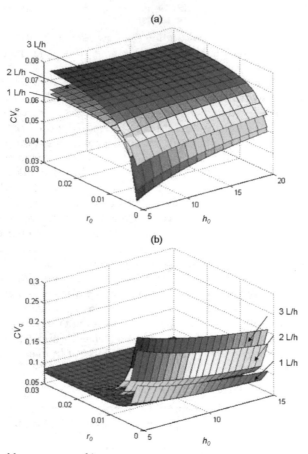

Note: Units for r_0 and h_0 are expressed in m.

Fig. 13. Irrigation uniformity coefficient CV_q as a function of r_0 and h_0 in a SDI lateral with non-compensating emitters of different discharges: (a) loamy uniform soil; (b) loamy non-uniform soil.

To simulate water movement in SDI emitters, a two-dimensional axis-symmetric water flow around a spherical surface may be simulated reproducing the hydraulic conditions. Simulations can considered a one-half cross section the soil profile. A sphere with a radius r_0 can be placed at a given depth below the soil surface, simulating the emitter and the cavity formed around a subsurface source.

A triangular mesh is automatically generated by the program. Dimensions of the finite-elements grid (triangles) may be refined around the emitter and other key points. Boundary conditions can be selected as follows: absence of flux across central axis and the opposite side; the atmospheric condition on the soil surface and free drainage along the bottom of soil profile. For compensating emitters, a constant flux boundary may be assumed around the cavity — it corresponded to the emitter discharge. For non compensating emitters, a variable flux boundary condition, function of the soil-water pressure, could be assumed calculated from the x and k coefficients of the emitter discharge equation.

Examples

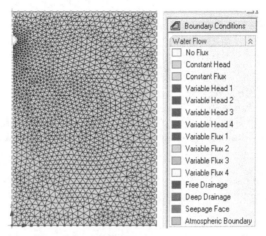

Fig. 14. Geometry, mesh and boundary conditions.

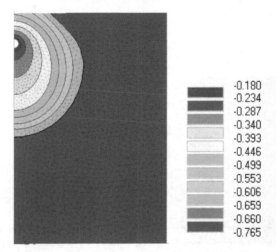

Fig. 15. Example of water distribution in a sandy soil after 4 h of irrigation (q = 10 L/h).

5. Criteria for the design and management of subsurface drip irrigation

In designing SDI systems variables such as emitter discharge, lateral depth and emitter spacing must be selected. Likewise, the management of these systems includes the selection of the operating pressure and the irrigation time. The inlet pressure in the irrigation unit determines the uniformity of water application whereas the irrigation time is a key factor to meet crop water and nutrient requirements

5.1 Determination of maximum emitter discharge

Since soil properties may reduce emitter discharge, the maximum emitter flow rate can be selected by a method called 'soil pits method' (Battam et al. 2003). A steady water flow is introduced to the soil through a narrow polyethyene tube that is located underneath the soil, during a given time. The depth at which no surfacing occurs is taken as the minimum depth for SDI laterals.

Other method proposed by Gil et al. (2011) is based on the following dimensionless emitter discharge equation:

$$q^* = k \cdot (\Delta h^*)^x \qquad (12)$$

with $q^*= q/h_0^x$ and $\Delta h^*=(h_0 - \bar{h}_s)/h_0$; where h_0 is the emitter pressure.

Values of pressure-discharge from several emitter compensating and non-compensating models, arranged according eq. [12], show the same trend. Each category follows, approximately, a single line that shows a linear relationship in flow rate variation below 50 %(see Fig.16).

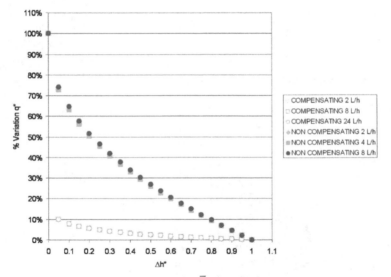

Fig. 16. Variation of q* (q/h_0^x) versus Δh^* [($h_0 - \bar{h}_s$)/h_0] for different emitter models. (Source: Gil el al. 2011).

From the above graph, a target uniformity for SDI design is selected and then, Δh^* calculated. Considering a typical variation of q*= 10%, Δh^*= 0.79 and 0.05 for non-compensating and compensating emitters, respectively. Then, the overpressure generated in the soil $\bar{h}_s$ can be calculated from Eq. [12] for different h_0 values. These would correspond to emitter pressures within the lateral to limit the reduction of emitter discharge to 10 %.

However, the performance of pressure compensating emitters differs from the non-compensating. In theory, their flow rate keeps constant within their compensating interval but for other values, these would be affected by soil pressure. Thus, considering an emitter head h_0= 10 m and Δh^* = 0.05, $h_0 - \bar{h}_s$ = 0.5 m. For anti-drain emitters, this value would not be reached since once the pressure gradient is less than 2 m, the emitter would close its discharging orifice.

Once the value $\bar{h}_s$ is known, emitter flow rate could be calculated from its relation with q as the one depicted in Fig. 17. This relation depends upon soil hydraulic properties and it could be determined by numerical models simulating soil water movement. The maximum discharge q_{max} will be obtained by adding to the assigned value the discharge emitter variation desired (10% in the given example).

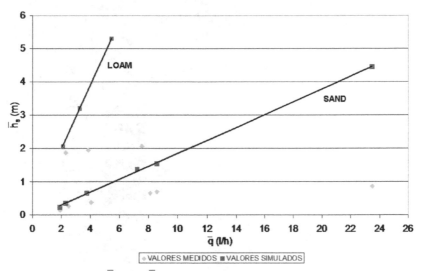

Fig. 17. Relation between $\bar{h}s$ and $\bar{q}$ for different soils considering a constant spherical radius r_0. (Source: Gil el al. 2011).

5.2 Design and management of subsurface drip irrigation

Computer programs developed for the hydraulic characterization and prediction of water distribution in SDI units have been proved as a useful tool for the design and management of SDI systems under specific scenarios. In addition, these programs might be coupled with geostatistical modelling software for the inclusion of spatial distribution of soil variability. Results show that SDI irrigation is suitable for sandy soils and is also suitable for loamy soils under specific conditions. It is advisable to select emitters with small discharge (1 L/h and 2 L/h), and to irrigate with inlet pressures not very small.

The uniformity of water application index CV_q in non-uniform soils is less than in uniform soils (Fig. 13). It is higher in SDI than in the surface drip irrigation in all uniform soils except for the sandy one, where both uniformities are alike.

On the other hand, the wetted bulb dimensions and its water distribution are two main factors determining emitter spacing and lateral depth. Both variables are selected in order to obtain an optimum water distribution within the crop root zone. The depth of wetting bulb coincides with the length of the plant roots.

As an illustration, some guidelines for the project and management of SDI units in a uniform loamy soil are introduced to highlight the applicability of finite element models for simulating water movement in SDI. The example simulates a line source that would correspond to water movement in laterals with small emitter spacing. Thus, the wetted area follows a continuous line with variable width. Two irrigation times and three possible lateral depths are compared considering an emitter pressure of 16 m and a uniform initial water content of 0.1 m³/m³.

Fig. 18 shows that for this scenario, emitter depths deeper than 10 cm are recommended to prevent soil surface wetting for irrigation times higher than 30 min. Differences observed for 0.2 and 0.3 m depths were negligible.

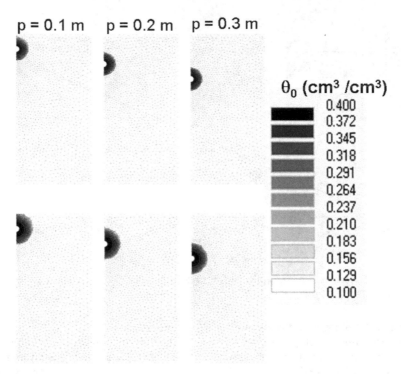

Fig. 18. Wetting bulbs of drip and a SDI in uniform loamy soil for an emitter pressure of 16 m and initial soil water content θ_0= 0.1 .

Wetting bulbs could be simulated for different scenarios by numerical methods such as Hydrus-2D. For each soil, selection of proper design variables (emitter spacing or lateral depth) and/or operation variables (inlet head and irrigation time) could be guided by figures, as those shown above, or by graphs showing wetting bulb dimensions for different conditions.

6. Perspectives for subsurface drip irrigation

Irrigation will be one key factor to sustain food supply for the world increasing population. The technical FAO study "World agriculture: towards 2015/2030" (FAO 2002) highlights that production of staple crops must follow the same increasing trend than in the last decades. Therefore, their productivity should rise and, thus irrigation will play an important role but it will have to adjust to the conditions of water scarcity and environmental and ecological sustainability.

On the other hand, many areas in the world, as their economy develops, the urban activities expand increasing water demands and develop competing uses of limited water resources. Thus, it raises uncertainty whether the volume of water used in irrigated agriculture can be sustained. Within this framework, water conservation policies are developed by policymakers reinforcing the use of new technologies and more efficient irrigation methods. Farmers are encouraged to use irrigation methods that reduce water use and increase yield in order to allow water to flow to other economic sectors.

In consequence, subsurface drip irrigation perspectives are promising. It shows higher capability for minimizing the loss of water by evaporation, runoff, and deep percolation in comparison to other methods. Thus, the irrigation water saved may become available for other uses. It may also lead to increase crop yields since it reduces fluctuations in soil water content and well aerated plan root zone. In addition, treated wastewaters could be used with risk reduction in human and animal health in arid areas with water scarcity. Moreover, it may reduce agrochemical application since agrochemicals are precisely distributed within the active roots zone.

Finally, SDI efficiency, in some scenarios, is better than surface trickle irrigation, and it requires smaller inlet heads. Thus energy savings are also enhanced. Furthermore, SDI may eliminate anaerobic decomposition of plant materials and thus substantially reduce methane gas production.

7. Acknowledgment

We would like to thank the Spanish Ministry of Science and Technology (CICYT) for its support provided through project No. AGL2008-00153/AGR.

8. References

American Society of Agricultural Engineering (ASAE). (2005). ASAE standards engineering practices data. 43rd edition: 864. Michigan. (August 2005), ISBN 1892769476

American Society of Agricultural Engineering (ASAE). (2003). Design and installation of microirrigation systems. Standard EP-405.1, (February 2003), ISSN 0032-0889

Ayars, J.E., Schoneman, R.A., Dale, F., Meson, B. & Shouse, P. (2001). Managing subsurface drip irrigation in the presence of shallow ground water. *Agricultural Water Managenemt.*, Vol. 47, No.3, (April 2001), pp. 242-264, ISSN 0378-3774

Bainbridge, A. (2001). Buried clay pot irrigation: a little known but very efficient traditional method of irrigation. *Agricultural Water Management*, Vol. 48, No.2, (June 2001), pp. 79-88, ISSN 0378-3774

Batam, M.A., Sutton, B.G. & Boughton, D.G. (2003). Soil pit as a simple design aid for subsurface drip irrigation systems. *Irrigation Science*, Vol. 22, No. 3/46, (October 2003), pp. 135-141, ISSN 0342-7188

Bralts, V.F., Wu,I.P. & Gitlin, H.M. (1981). Manufacturing variation and drip irrigation uniformity. *Transactions of the ASAE* Vol. 24, No. 1, (January 1981) pp. 113-119, ISSN 0001-2351

Ben-Gal A., Lazorovitch, N. & Shani, U. (2004). Subsurface drip irrigation in gravel filled cavities. *Vadose Zone Journal* Vol. 3, pp. 1407-1413

Christiansen , J.E. (1942). Irrigation by sprinkling. University of California Agriculture Experiment Station Bulletin, 670

Cote, C.M.; Bristow, K.L.; Charlesworth, P.B.; Cook, F.G. & Thorburn, P.J.(2003). Analysis of soil wetting and solute transport in subsurface trickle irrigation. *Irrigation Science*, Vol. 22, No.3, (May 2003), pp. 143-156, ISSN 0342-7188

FAO (May 2002). Technical FAO study "World agriculture: towards 2015/2030". Food Agriculture Organization of the United Nations, Available from http://www.fao.org/english/newsroom/news/2002/7828-en.html

Gardner, W.R. (1958). Some steady state solutions of unsaturated moisture flow equations with application to evaporation from a water table. Soil Science, Vol. 85, No.5, (May 1958), pp. 228-232, ISSN: 0038-075X

Gil, M.; Rodríguez-Sinobas, L.; Sánchez, R., Juana, L. & Losada, A. (2008). Emitter discharge variability of subsurface drip irrigation in uniform soils. Effect on water application uniformity". *Irrigation Science*, Vol. 26, No.6, (June 2008), pp. 451-458, ISSN 0342-7188

Gil, M.; Rodríguez-Sinobas, L.; Sánchez, R. & Juana, L. (2010). Evolution of the spherical cavity radius generated around a subsurface emitter. *Biogeoscience* (Open Access Journal, European Geosciences Union) Biogeosciences Discuss., Vol. 7, (January 2010), pp. 1–24, 2010. Available from www.biogeosciences-discuss.net/7/1/2010/

Gil, M.; Rodríguez-Sinobas, L.; Sánchez, R. & Juana, L. (2011). Determination of maximum emitter discharge in subsurface drip irrigation units. *Journal of Irrigation and Drainage Engineering* (ASCE), Vol. 137, No. 3 (March 2011), pp.325-33, ISSN 0733-9437

Goldberg, D. & Shmuelli, M. (1970). Drip irrigation. A method used under arid and desert conditions of high water and soil salinity. *Transactions of the ASAE*, Vol. 13, No.1 (January 1970), pp. 38-41, ISSN 0001-2351

Juana, L.; Rodríguez Sinobas, L. & Losada, A. (2002). Determining minor head losses in drip irrigation laterals II: Experimental study and validation. *Journal of Irrigation and Drainage Engineering* (ASCE), Vol. 128, No.6 (November 2011), pp.355-396, ISSN 0733-9437

Keller, J. & Karmelli, D. (1975). Trickle irrigation desing parameters. *Transactions of the ASAE*, Vol. 17, No.4 (July 1975), pp. 678-684, ISSN 0001-2351

Lanting, S. (1975). Subsurface irrigation- Engineering research. In 34th Report Hawaii Sugar Technology Annual Conference pp. 57-62. Honolulu, Hawaii

Lamm, F. R. (2002). Advantages and disadvantages of subsurface drip irrigation. *Proceedings of the Int. Meeting on Advances in Drip/Micro Irrigation*, pp. 1-13, ISBN 0-929354-62-8, Puerto de La Cruz, Tenerife, Canary Islands, Spain, Dec. 2-5, 2002

Lazarovitch N.; Simunek, J.; & Shani, U. (2005). System dependent boundary conditions for water flow from a subsurface source. *Soil Sci Soc Am J*, Vol. 69, No.1 (January 2005), pp. 46-50, ISSN 0361-5995

Lazarovitch, N.; Shani, U.; Thompson, T.L. & Warrick, A.W. (2006). Soil hydraulic properties affecting discharge uniformity of gravity-fed subsurface drip irrigation. *Journal of Irrigation and Drainage Engineering* (ASCE), Vol. 132, No. 2 (March 2006), pp. 531-536, ISSN 0733-9437

Li, J.; Zhang, J. & Rao, M.(2005). Modelling of water flow and nitrate transport under surface drip fertirrigation. *Transactions of the ASAE*, Vol. 48, No. 3 (May 2005), pp: 627-637, ISSN 0001-2351

Meshkat, M.; Warner, R.C. & Workman, S.R. (2000). Evaporation reduction potential in an undisturbed soil irrigated with surface drip and sand tube irrigation. *Transactions of the ASAE*. Vol. 43, No. 1 (January 2000), pp. 79-86, ISSN 0001-2351

Payero, J.O. (2005). Subsurface Drip Irrigation: Is it a good choice for your operation? Crop Watch news service. University of Nebraska-Lincoln, Institute of Agriculture and Natural Resources, Available from
   http://www.ianr.unl.edu/cropwatch/archives/2002/crop02-8.htm

Philip, J.R. (1992). What happens near a quasi-linear point source? *Water Resources Research* Vol. (28), No.1 (January 1992), pp. 47-52, ISSN 0043-1397

Phene, C.J.; Yue, R.; wu, I.P.; Ayars, J.E.; Shoneman, R.A. & Meso, B. M. (1992). Distribution uniformity of subsurface drip irrigation systems. *ASAE Paper* No. 92: 2569. American Society of Agricultural Engineering, St. Joseph, Michigan (May 1992), ISSN 0001-2351

Provenzano, G. (2007). Using Hydrus-2D simulation model to evaluate wetted soil volume in subsurface drip irrigation systems. *Journal of Irrigation and Drainage Engineering* (ASCE), Vol. 133, No. 4 (July 2007), pp.342-349, ISSN 0733-9437

Rodriguez-Sinobas, L., Juana, L., & Losada, A. (1999). Effects of temperature changes on emitter discharge. *Journal of Irrigation and Drainage Engineering* (ASCE), Vol. 125, No. 2 (March 1999), pp. 64-73, ISSN 0733-9437

Rodriguez-Sinobas, L.; Juana Sirgado,L.; Sánchez Calvo, R. & Losada Villasante, A. (2004). Pérdidas de carga localizadas en inserciones de ramales de goteo. *Ingeniería del Agua*, Vol. 11, No. 3 (May 2004), pp. 289-296, ISSN 1134-2196

Rodríguez-Sinobas, L.; Gil, M.; L.; Sánchez, R. & Juana, L. (2009a). Water distribution in subsurface drip irrigation Systems. I: Simulation. *Journal of Irrigation and Drainage Engineering* (ASCE), Vol. 135, No. 6 (December 2009), pp.721-728, , ISSN 0733-9437

Rodríguez-Sinobas, L.; Gil, M.; L.; Sánchez, R. & Juana, L. (2009b).Water distribution in subsurface drip irrigation Systems. II: Field evaluation. *Journal of Irrigation and Drainage Engineering* (ASCE), Vol. 135, No. 6 (December 2009), pp.729-738, ISSN 0733-9437

Sadler E.J., Camp, C.R. & Busscher, W.J. (1995). Emitter flow rate changes caused by excavating subsurface microirrigation tubing. *Proceedings of the 5th Int. Microirrigation Congress*, pp. 763-768, ISBN 0-929355-62-8, Orlando, Florida, USA, April 2-6, 1995

Shani, U.; Xue, S.; Gordin-Katz, R. & Warric, A.W. (1996). Soil-limiting from Subsurface Emitters. I: Pressure Measurements *Journal of Irrigation and Drainage Engineering* (ASCE), Vol. 122, No.2 (April 1996), pp. 291-295, ISSN 0733-9437

Simunek, J.; Sejna, M. & van Genuchten, M.Th. (1999). The Hydrus 2D software package for simulating two-dimensional movement of water, heat and multiple solutes in variably saturated media, version 2.0. Rep. IGCWMC-TPS-53, Int. Ground Water Model. Cent. Colo. Sch. of Mines, Golden, CO, p.251

Schmitz, G.H.; Schutze, N. & Petersohn, U. (2002). New strategy for optimizing water application under trickle irrigation. *Journal of Irrigation and Drainage Engineering* (ASCE), Vol. 128, No.5 (Setember 2002), pp.287-297, ISSN 0733-9437

Souza Resende, R. (2003). Intusão radicular e efeito de vácuo em gotejamiento enterrado na irrigação de cana de açúcar. PhD. Thesis. Escola Superior de Agricultura 'Luiz de Queiroz" Universidade de São Paulo, Piracicaba, November 2003, pp 143

United State Deparment of Agriculture USDA-NASS. (2009). Farm and Ranch Irrigation Survey National Agricultural Statistic Service, Available from www.nass.usda.gov

Vaziri, C.M. & Gibson, W. (1972). Subsurface and drip irrigation for Hawaiian sugarcane. In: 31st Report Hawaii Sugar Technology Annual Conference, Honolulu, 1972. Proceedings. Honolulu: Hawaiian sugar Planters Assoc., 1972, pp.18-22

Permissions

The contributors of this book come from diverse backgrounds, making this book a truly international effort. This book will bring forth new frontiers with its revolutionizing research information and detailed analysis of the nascent developments around the world.

We would like to thank Dr. Teang Shui Lee, for lending his expertise to make the book truly unique. He has played a crucial role in the development of this book. Without his invaluable contribution this book wouldn't have been possible. He has made vital efforts to compile up to date information on the varied aspects of this subject to make this book a valuable addition to the collection of many professionals and students.

This book was conceptualized with the vision of imparting up-to-date information and advanced data in this field. To ensure the same, a matchless editorial board was set up. Every individual on the board went through rigorous rounds of assessment to prove their worth. After which they invested a large part of their time researching and compiling the most relevant data for our readers. Conferences and sessions were held from time to time between the editorial board and the contributing authors to present the data in the most comprehensible form. The editorial team has worked tirelessly to provide valuable and valid information to help people across the globe.

Every chapter published in this book has been scrutinized by our experts. Their significance has been extensively debated. The topics covered herein carry significant findings which will fuel the growth of the discipline. They may even be implemented as practical applications or may be referred to as a beginning point for another development. Chapters in this book were first published by InTech; hereby published with permission under the Creative Commons Attribution License or equivalent.

The editorial board has been involved in producing this book since its inception. They have spent rigorous hours researching and exploring the diverse topics which have resulted in the successful publishing of this book. They have passed on their knowledge of decades through this book. To expedite this challenging task, the publisher supported the team at every step. A small team of assistant editors was also appointed to further simplify the editing procedure and attain best results for the readers.

Our editorial team has been hand-picked from every corner of the world. Their multi-ethnicity adds dynamic inputs to the discussions which result in innovative outcomes. These outcomes are then further discussed with the researchers and contributors who give their valuable feedback and opinion regarding the same. The feedback is then collaborated with the researches and they are edited in a comprehensive manner to aid the understanding of the subject.

Apart from the editorial board, the designing team has also invested a significant amount of their time in understanding the subject and creating the most relevant covers. They scrutinized every image to scout for the most suitable representation of the subject and create an appropriate cover for the book.

The publishing team has been involved in this book since its early stages. They were actively engaged in every process, be it collecting the data, connecting with the contributors or procuring relevant information. The team has been an ardent support to the editorial, designing and production team. Their endless efforts to recruit the best for this project, has resulted in the accomplishment of this book. They are a veteran in the field of academics and their pool of knowledge is as vast as their experience in printing. Their expertise and guidance has proved useful at every step. Their uncompromising quality standards have made this book an exceptional effort. Their encouragement from time to time has been an inspiration for everyone.

The publisher and the editorial board hope that this book will prove to be a valuable piece of knowledge for researchers, students, practitioners and scholars across the globe.

List of Contributors

Ángel Galmiche-Tejeda, José Jesús Obrador-Olán and Eustolia García-López
Colegio de Postgraduados, Campus Tabasco, Mexico

Eugenio Carrillo Ávila
Campus Campeche, México

Rafiq Ahmad
Department of Botany, University of Karachi, Pakistan

Rizwana Jabeen
Government College for Women, Shahrah-e-Liaquat, Karachi, Pakistan
Department of Botany, University of Karachi, Pakistan

Hamid Iqbal Tak and Faheem Ahmad
Department of Biological Sciences, Faculty of Agriculture, Science and Technology, North-West University, Mafikeng Campus, Mmabatho, South Africa

Yahya Bakhtiyar
Department of Zoology, Jammu University, India

Arif Inam
Department of Botany, Aligarh Muslim University, India

Roberto Wagner Lourenço and Admilson Irio Ribeiro
Department of Environmental Engineering, São Paulo States University, Sorocaba Campus, Sorocaba, SP, Brazil

André Juliano Franco
São Carlos Federal University, Sorocaba Campus, Sorocaba, SP, Brazil

Paulo Milton Barbosa Landim
Department of Applied Geology, State University of São Paulo, Rio Claro Campus, SP, Brazil

Maria Rita Donalisio and Ricardo Cordeiro
Department Social and Preventive Medicine, Campinas University, Campinas, SP, Bazil

Shahid Naseem and Erum Bashir
Department of Geology, University of Karachi, Karachi, Pakistan

Salma Hamza
Department of Geology, Federal Urdu University of Arts, Science and Technology, Karachi, Pakistan

Masoud Tabari
Tarbiat Modares University, Iran

Silvia Aparecida Martim
Instituto de Biologia, Departamento de Ciências Fisiológicas, Universidade Federal Rural do Rio de Janeiro, Brazil

Ricardo Enrique Bressan-Smith
Centro de Ciências e Tecnologia Agropecuária, Laboratório de Fisiologia Vegetal, Universidade Estadual do Norte Fluminense – Darcy Ribeiro, Brazil

Arnoldo Rocha Façanha
Centro de Biociência e Biotecnologia, Laboratório de Biologia Celular e Tecidual, Universidade Estadual do Norte Fluminense – Darcy Ribeiro, Brazil

Masoud Tabari and Mohammad Ali Shirzad
Tarbiat Modares University, Iran

Alba Leonor da Silva Martins and Aline Pacobahyba de Oliveira
Embrapa Solos – National Center of Soil Research – CNPS, Brazil

Jesús Hernan Camacho-Tamayo
Colômbia National University, Bogotá, Colombia

Emanoel Gomes de Moura
Maranhão State University, Brazil

Pirjo Mäkelä, Jouko Kleemola and Paavo Kuisma
Department of Agricultural Sciences, University of Helsinki & Potato Research Institute, Finland

Leonor Rodríguez Sinobas and María Gil Rodríguez
Research Group "Hydraulic of Irrigation" Technical University of Madrid, Spain

9 781632 391308